POWER CONVERTER CIRCUITS

ELECTRICAL AND COMPUTER ENGINEERING

A Series of Reference Books and Textbooks

FOUNDING EDITOR

Marlin O. Thurston
Department of Electrical Engineering
The Ohio State University
Columbus, Ohio

1. Rational Fault Analysis, *edited by Richard Saeks and S. R. Liberty*
2. Nonparametric Methods in Communications, *edited by P. Papantoni-Kazakos and Dimitri Kazakos*
3. Interactive Pattern Recognition, *Yi-tzuu Chien*
4. Solid-State Electronics, *Lawrence E. Murr*
5. Electronic, Magnetic, and Thermal Properties of Solid Materials, *Klaus Schröder*
6. Magnetic-Bubble Memory Technology, *Hsu Chang*
7. Transformer and Inductor Design Handbook, *Colonel Wm. T. McLyman*
8. Electromagnetics: Classical and Modern Theory and Applications, *Samuel Seely and Alexander D. Poularikas*
9. One-Dimensional Digital Signal Processing, *Chi-Tsong Chen*
10. Interconnected Dynamical Systems, *Raymond A. DeCarlo and Richard Saeks*
11. Modern Digital Control Systems, *Raymond G. Jacquot*
12. Hybrid Circuit Design and Manufacture, *Roydn D. Jones*
13. Magnetic Core Selection for Transformers and Inductors: A User's Guide to Practice and Specification, *Colonel Wm. T. McLyman*
14. Static and Rotating Electromagnetic Devices, *Richard H. Engelmann*
15. Energy-Efficient Electric Motors: Selection and Application, *John C. Andreas*
16. Electromagnetic Compossibility, *Heinz M. Schlicke*
17. Electronics: Models, Analysis, and Systems, *James G. Gottling*
18. Digital Filter Design Handbook, *Fred J. Taylor*
19. Multivariable Control: An Introduction, *P. K. Sinha*
20. Flexible Circuits: Design and Applications, *Steve Gurley, with contributions by Carl A. Edstrom, Jr., Ray D. Greenway, and William P. Kelly*
21. Circuit Interruption: Theory and Techniques, *Thomas E. Browne, Jr.*
22. Switch Mode Power Conversion: Basic Theory and Design, *K. Kit Sum*
23. Pattern Recognition: Applications to Large Data-Set Problems, *Sing-Tze Bow*

POWER CONVERTER CIRCUITS

WILLIAM SHEPHERD
Ohio University
Athens, Ohio, U.S.A.

LI ZHANG
University of Leeds
Leeds, England

CRC Press
Taylor & Francis Group
Boca Raton London New York

CRC Press is an imprint of the
Taylor & Francis Group, an **informa** business

CRC Press
Taylor & Francis Group
6000 Broken Sound Parkway NW, Suite 300
Boca Raton, FL 33487-2742

First issued in paperback 2019

© 2004 by Taylor & Francis Group, LLC
CRC Press is an imprint of Taylor & Francis Group, an Informa business

No claim to original U.S. Government works

ISBN-13: 978-0-8247-5054-1 (hbk)
ISBN-13: 978-0-367-39447-9 (pbk)

Library of Congress Cataloging-in-Publication Data
A catalog record for this book is available from the Library of Congress.

Visit the Taylor & Francis Web site at
http://www.taylorandfrancis.com

and the CRC Press Web site at
http://www.crcpress.com

Preface

This book is intended for use in junior and senior level undergraduate courses in power electronics. Classical analysis is used throughout. The coverage is concerned exclusively with power electronic circuits. No coverage is given to the physics and the fabrication of power electronic switches, nor to the detailed design of power electronic circuit protection and ancillary components.

Rectifier circuits are treated in a far more comprehensive and detailed manner than is customary in contemporary texts. Chapter 14 is devoted to matrix converters, which usually receive scant attention. Chapter 13 is devoted to envelope cycloconverters, which usually receive no attention at all. Features of the text include the large number of worked numerical examples and the very large number of end of chapter problems, with answers.

The material contained in this book has been used for classroom and examination purposes at the University of Leeds, England; the University of Bradford, England; and Ohio University, Athens, Ohio. The authors are grateful for the permissions to reproduce the relevant material.

Some short sections of this book are reproduced from earlier work by one of the authors. The writers are grateful to Cambridge University Press, England, for permission to reproduce this previously published work.

The typing of the manuscript, with its many iterations, was done by Janelle Baney, Suzanne Vazzano, Erin Dill, Juan Echeverry, and Brad Lafferty of the Schools of Electrical Engineering and Computer Science of Ohio University. We are deeply grateful for their contributions.

Much of the computer transcription of the text and diagrams was undertaken at the Instructional Media Services Unit of the Alden Library at Ohio University.

Our thanks are due to Peggy Sattler, the head of the unit, and particularly to Lara Neel, graduate assistant. Our deepest gratitude goes to them for their dedicated professionalism.

At the latter end, the final transfer of the material to computer disc, with text corrections, was performed by postgraduate students Yanling Sun and Liang-jie Zhu of the Ohio University School of Education.

William Shepherd
Li Zhang

Contents

Part 2   DC to AC Converters (Inverters)

Part 3   AC to AC Converters

Part 4   DC to DC Converters

Nomenclature

$A_1\ B_1\ C_1$	Fourier coefficients of first order
$a_n\ b_n\ c_n$	Fourier coefficients of nth order
$a_s\ b_s\ c_s$	Fourier coefficients of the supply point current
e	instantaneous emf, V
e_c	instantaneous capacitor voltage, V
e_l	instantaneous inductor voltage, V
e_L	instantaneous load voltage, V
e_r	instantaneous ripple voltage, V
$e_{ab}\ e_{bc}\ e_{ca}$	instantaneous line supply voltages in a three-phase system, V
$e_{aN}\ e_{bN}\ e_{cN}$	instantaneous phase load voltages in a three-phase system, V
$e_{AN}\ e_{BN}\ e_{CN}$	instantaneous phase supply voltages in a three-phase system, V
f	frequency, Hz
i	instantaneous current, A
$i_a\ i_b\ i_c$	instantaneous line currents in a three-phase system, A
i_c	instantaneous capacitor current, A
$i_{ca}\ i_{cb}\ i_{cc}$	instantaneous currents in the capacitor branches of a three-phase system, A
i_L	instantaneous load current, A
i_R	instantaneous resistor current, A
i_s	instantaneous supply current, A
$i_{sa}\ i_{sb}\ i_{sc}$	instantaneous supply currents in the lines of a three-phase system, A
i_{ss}	instantaneous steady-state component of current, A

i_t	instantaneous transient component of current, A
n	harmonic order
p	frequency ratio
t	time, s
I_{min}	minimum value of the rms supply current, A
I_{sc}	short circuit current, A
$\hat{I}$	peak value of the sinusoidal short-circuit current, A
L	self-inductance coefficient, H
L_s	self-inductance coefficient of supply line inductance, A
M	modulation ratio
N	number of turns
P	average power, W
$P_a\ P_b\ P_c$	average power per phase in a three-phase system, W
P_{in}	average input power, W
P_L	average load power, W
P_s	average supply point power, W
PF	power factor
PF_c	power factor in the presence of compensation
PF_{max}	maximum value of power factor
Q_L	reactive component of load voltamperes, VA
R	resistance, Ω
RF	ripple factor
S_L	apparent voltamperes of the load, VA
T	time constant, S
V	terminal voltage of battery, V
X_c	capacitive reactance, Ω
X_L	inductive reactance, Ω
X_{sc}	short-circuit (inductive) reactance, Ω
Z	impedance, Ω

GREEK SYMBOLS

α	thyristor firing angle, radian
$\alpha\ \beta$	limits of conduction in a diode battery-charger circuit (Chapter 2), radian
α'	inherent delay angle caused by supply inductance, radian
δ	dimensionless "figure of merit"
φ	instantaneous flux, Wb
ψ	displacement angle between voltage and fundamental current, radian
ψ_n	displacement angle between voltage and nth harmonic component of current, radian

ψ_s	displacement angle between voltage and supply current, radian
μ	overlap angle, radian
θ	conduction angle, radian
ω	angular supply frequency, radian/s
Δ	small value of angle, radian
Φ	phase angle to sinusoidal currents of supply frequency, radian
Φ_n	phase angle to sinusoidal currents of nth harmonic frequency, radian

1

Switching and Semiconductor Switches

1.1 POWER FLOW CONTROL BY SWITCHES

The flow of electrical energy between a fixed voltage supply and a load is often controlled by interposing a controller, as shown in Fig. 1.1. Viewed from the supply, the apparent impedance of the load plus controller must be varied if variation of the energy flow is required. Conversely, seen from the load, the apparent properties of the supply plus controller must be adjusted. From either viewpoint, control of the power flow can be realized by using a series-connected controller with the desired properties. If a current source supply is used instead of a voltage source supply, control can be realized by the parallel connection of an appropriate controller.

The series-connected controller in Fig. 1.1 can take many different forms. In alternating current (ac) distribution systems where continuous variability of power flow is a secondary requirement, electrical transformers are often the prevalent controlling elements. The insertion of reactive elements is inconvenient because variable inductors and capacitors of appropriate size are expensive and bulky. It is easy to use a series-connected variable resistance instead, but at the expense of considerable loss of energy. Loads that absorb significant electric power usually possess some form of energy "inertia." This allows any amplitude variations created by the interposed controller to be effected in an efficient manner.

1

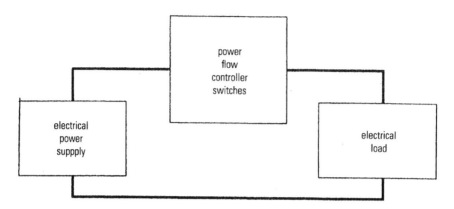

F<small>IG</small>. 1

Amplitude variations of current and power flow introduced by the controller may be realized by fractional time variation of connection and disconnection from the supply. If the frequency of such switching is so rapid that the load cannot track the switching events due to its electrical inertia then no energy is expended in an ideal controller. The higher the load electrical inertia and the switching frequency, the more the switching disturbance is reduced in significance.

1.2 ATTRIBUTES OF AN IDEAL SWITCH

The attributes of an ideal switch may be summarized as follows:

1.2.1 Primary Attributes

1. Switching times of the state transitions between "on" and "off" should be zero.
2. "On" state voltage drop across the device should be zero.
3. "Off" state current through the device should be zero.
4. Power–control ratio (i.e., the ratio of device power handling capability to the control electrode power required to effect the state transitions) should be infinite.
5. "Off" state voltage withstand capability should be infinite.
6. "On" state current handling capability should be infinite.
7. Power handling capability of the switch should be infinite.

1.2.2 Secondary Attributes

1. Complete electrical isolation between the control function and the power flow
2. Bidirectional current and voltage blocking capability

An ideal switch is usually depicted by the diagram of Fig. 1.2. This is not a universal diagram, and different authors use variations in an attempt to provide further information about the switch and its action. Figure 1.2 implies that the power flow is bidirectional and that no expenditure of energy is involved in opening or closing the switch.

1.3 ATTRIBUTES OF A PRACTICAL SWITCH

Power electronic semiconductor switches are based on the properties of very pure, monocrystalline silicon. This basic material is subjected to a complex industrial process called *doping* to form a wafer combining a p-type (positive) semiconductor with an n-type (negative) semiconductor. The dimensions of the wafer depend on the current and voltage ratings of the semiconductor switch. Wafers are usually circular with an area of about 1 mm^2/A. A 10 A device has a diameter of about 3.6 mm, whereas a 500 A device has a diameter of 25 mm (1 in.). The wafer is usually embedded in a plastic or metal casing for protection and to facilitate heat conduction away from the junction or junctions of both the p-type and n-type materials. Junction temperature is the most critical property of semiconductor operation.

Practical semiconductor switches are imperfect. They possess a very low but finite on-state resistance that results in a conduction voltage drop. The off-state resistance is very high but finite, resulting in leakage current in both the forward and reverse directions depending on the polarity of the applied voltage.

Switching-on and switching-off (i.e., commutation) actions do not occur instantaneously. Each transition introduces a finite time delay. Both switch-on and switch-off are accompanied by heat dissipation, which causes the device temperature to rise. In load control situations where the device undergoes frequent switchings, the switch-on and switch-off power losses may be added to the steady-state conduction loss to form the total incidental dissipation loss, which usually

Fig. 2

manifests itself as heat. Dissipation also occurs in devices due to the control electrode action.

Every practical switching device, from a mechanical switch to the most modern semiconductor switch, is deficient in all of the ideal features listed in Sec. 1.2

1.4 TYPES OF SEMICONDUCTOR CONVERTER

Semiconductor switching converters may be grouped into three main categories, according to their functions.

1. Transfer of power from an alternating current (ac) supply to direct current (dc) form. This type of converter is usually called a *rectifier*.
2. Transfer of power from a direct current supply to alternating current form. This type of converter is usually called an *inverter*.
3. Transfer of power from an ac supply directly into an ac load of different frequency. This type of converter is called a *cycloconverter* or a *matrix converter*.
4. Transfer of power from a direct current supply directly into a direct current load of different voltage level. This type of converter is called a *chopper converter* or a *switch-mode converter*.

1.4.1 Rectifiers

The process of electrical rectification is where current from an ac supply is converted to an unidirectional form before being supplied to a load (Fi. 1.3). The ac supply current remains bidirectional, while the load current is unidirectional. With resistive loads the load voltage polarity is fixed. With energy storage loads and alternating supply voltage the load current is unidirectional but pulsating, and the load voltage in series-connected load inductance elements may vary and alternate in polarity during the load current cycle.

In rectifier circuits there are certain circuit properties that are of interest irrespective of the circuit topology and impedance nature. These can be divided into two groups of properties, (1) on the supply side and (2) on the load side of the rectifier, respectively. When the electrical supply system has a low (ideally zero) impedance, the supply voltages are sinusodial and remain largely undistorted even when the rectifier action causes nonsinusoidal pulses of current to be drawn from the supply. For the purposes of general circuit analysis one can assume that semiconductor rectifier elements such as diodes and silicon controlled rectifiers are ideal switches. During conduction they are dissipationless and have zero voltage drop. Also, when held in extinction by reverse anode voltage, they have infinite impedance.

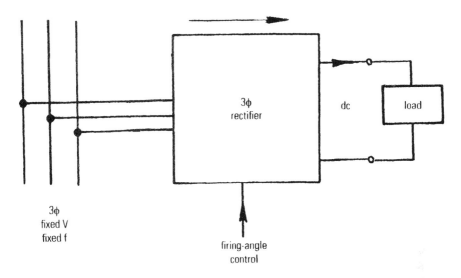

Fɪɢ. 3

In order to investigate some basic properties of certain rectifier circuits, it is convenient to consider single-phase circuits separately from three-phase circuits. Additional classifications that are helpful are to consider diode (uncontrolled rectifier) circuits separately from thyristor (controlled rectifier) circuits and to also separate resistive load circuits from reactive load circuits. These practices are followed in Chapters 2–8.

Three-phase and single-phase rectifiers are invariably commutated (i.e., switched off) by the natural cycling of the supply-side voltages. Normally there is no point in using gate turn-off devices as switches. Controlled rectifiers most usually employ silicon controlled rectifiers as switches. Only if the particular application results in a need for the supply to accept power regenerated from the load might the need arise to use gate turn-off switches.

1.4.2 Inverters

The process of transferring power from a direct current (dc) supply to an ac circuit is called a *process of inversion* (Fig. 1.4). Like rectification, the operation takes place by the controlled switching of semiconductor switching devices. Various forms of inverter circuits and relevant applications are described in Chapters 9–11.

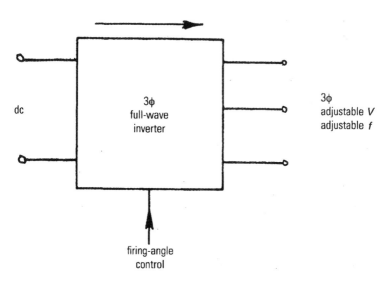

dc

3φ
full-wave
inverter

3φ
adjustable *V*
adjustable *f*

firing-angle
control

Fɪɢ. 4

1.4.3   Cycloconverters

Power can be transferred from an ac supply to an ac load, usually of lower
frequency, by the direct switching of semiconductor devices (Fig. 1.5). The commutation takes place by natural cycling of the supply-side voltages, as in rectifiers.
A detailed discussion of cycloconverter circuits and their operation is given in
Chapters 12 and 13.

1.5   TYPES OF SEMICONDUCTOR SWITCH

The main types of semiconductor switches in common use are

1. Diodes
2. Power transistors
    a. Bipolar junction transistor (BJT)
    b. Metal oxide semiconductor field effect transistor (MOSFET)
    c. Insulated gate bipolar transistor (IGBT)
    d. Static induction transistor (SIT)
3. Thyristor devices
    a. Silicon controlled rectifier (SCR)
    b. Static induction thyristor (SITH)
    c. Gate turn-off thyristor (GTO)

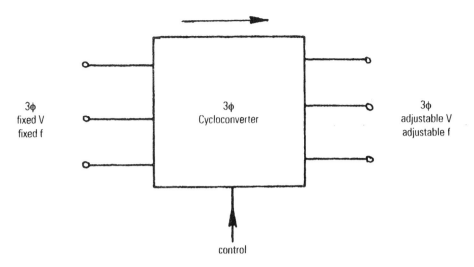

FIG. 5

d. MOS controlled thyristor (MCT)
e. Triac

Some details of certain relevant properties of these devices are summarized in Table 1.1.

1.5.1 Diodes

Diodes are voltage-activated switches. Current conduction is initiated by the application of forward voltage and is unidirectional. The diode is the basic form of rectifier circuit switch. It is regarded as an uncontrolled rectifier in the sense that it cannot be switched on or off by external signals. During conduction (Fig. 1.6), the forward current is limited only by the external circuit impedance. The forward voltage drop during conduction is of the order 1–2 V and can be ignored in many power electronics calculations.

The application of reverse voltage cuts off the forward current and results in a very small reverse leakage current, a condition known as *reverse blocking*. A very large reverse voltage would punch through the p-n junction of the wafer and destroy the device by reverse avalanching, depicted in Fig. 1.6.

1.5.2 Power Transistors

Power transistors are three-terminal rectifier devices in which the unidirectional main circuit current has to be maintained by the application of base or gate current

TABLE 1.1

Type of switch	Current	Turn-on	Turn-off	Features
Ideal switch	Bidirectional	Instantaneous	Instantaneous	Zero on-state impedance
Diode	Unidirectional	Forward voltage $(V_A > V_K)$	Reverse voltage $(V_A < V_K)$	Voltage activated Low on-state impedance Low on-state volt drop High off-state impedance
Thyristors				
Silicon controlled rectifier (SCR)	Unidirectional	Forward voltage $(V_A > V_K)$ Forward gate bias $(V_G > V_K)$	Reverse voltage $V_A < V_K$ to reduce the current	Gate turn-off is not possible
Gate turn-off devices State induction thyristor (SITH)	Unidirectional	Forward voltage $(V_A > V_K)$ turn-on is the normal state (without gate drive)	-remove forward voltage -negative gate signal $(V_G < V_K)$	Low reverse blocking voltage
Gate turn-off thyristor (GTO)	Unidirectional	Forward voltage $(V_A V_K)$ And + ve gate pulse $(I_G > 0)$	− By θ ve gate pulse $(I_G < 0)$ or by current reduction	When the reverse blocking voltage is low it is known as an *asymmetric GTO*
MOS controlled thyristor (MCT)	Unidirectional	Forward voltage $(V_A > V_K)$ − θ ve gate pulse $(V_G < V_K)$	+ve gate pulse $(V_G > V_A)$	Low reverse avalanche voltage
TRIAC	Birectional	Forward or reverse voltage $(V_A > < V_K)$ +ve or θ ve gate pulse	Current reduction by voltage reversal with zero gate signal	Symmetrical forward and reverse blocking Ideally suited to phase angle triggering
Transistors				
Bipolar junction transistor (BJT)	Unidirectional	Forward voltage $(V_C > V_E)$ +ve base drive $(V_B > V_B)$	Remove base current $(I_B=0)$	Cascading 2 or 3 devices produces a Darlington connection with high gain (low base current)
Metal-oxide-semiconductor field-effect transistor (MOSFET)	Unidirectional	Forward voltage $(V_D > V_E)$ +ve gate pulse $(V_G > V_S)$	Remove gate drive $(V_G=0)$	Very fast turn-on and turn-off
Insulated gate bipolar transistor (IGBT)	Unidirectional	Forward voltage $(V_C > V_E)$ +ve gate pulse $(V_G > V_S)$	Remove gate drive $(V_G=0)$	Low on-state losses, very fast turn-on/turn-off, low reverse blocking
Static induction transistor (SIT)	Unidirectional	Forward voltage $(V_D > V_X)$ normally on $(V_G = 0)$	+ve gate pulse $(V_G > V_S)$	Also called the *power JFET high on-state voltage drop*

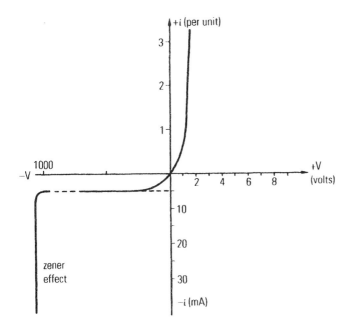

F<small>IG</small>. 6

at the control electrode. Removal of the gate or base drive results in current extinction.

The bipolar junction transistor (BJT) is a three-terminal silicon switch. If the base terminal B and collector terminal C are both positively biased with respect to the emitter terminal E (Table 1.1), switch-on occurs. Conduction continues until the base current is removed, so that the BJT is a current controlled device. It will only reverse block up to about 20 V and needs to be used with a series diode if higher reverse blocking is required.

The metal-oxide-semiconductor field-effect transistor (MOSFET) is a very fast acting, three-terminal switch. For conduction the drain voltage V_D and gate voltage V_G must both be greater than the source voltage V_S (Table 1.1). The device is voltage controlled, whereby removal of the gate voltage results in switch-off. MOSFETs can be operated in parallel for current sharing. Ratings of 500 V and 50 A are now (1999) available.

A compound device known as the *insulated gate bipolar transistor* (IGBT) combines the fast switching characteristics of the MOSFET with the power-handling capabilities of the BJT. Single device ratings in the regions 300–1600 V and 10–400 A mean that power ratings greater than 50 kW are available. The switching frequency is faster than a BJT but slower than a MOSFET. A device design that emphasizes the features of high-frequency switching or low on-state resistance has the disadvantage of low reverse breakdown voltage. This can be compensated by a reverse-connected diode.

The static induction transistor (SIT) has characteristics similar to a MOS-FET with higher power levels but lower switching frequency. It is normally on, in the absence of gate signal, and is turned off by positive gate signal. Although not in common use, ratings of 1200 V, 300 A are available. It has the main disadvantage of high (e.g., 15 V.) on-state voltage drop.

1.5.3 Thyristors

The silicon controlled rectifier (SCR) member of the thyristor family of three-terminal devices is the most widely used semiconductor switch. It is used in both ac and dc applications, and device ratings of 6000 V, 3500 A have been realized with fast switching times and low on-state resistance. An SCR is usually switched on by a pulse of positive gate voltage in the presence of positive anode voltage. Once conduction begins the gate loses control and switch-on continues until the anode–cathode current is reduced below its holding value (usually a few milliamperes).

In addition to gate turn-on (Fig 1.7), conduction can be initiated, in the absence of gate drive, by rapid rate of rise of the anode voltage, called the *dv/dt effect*, or by slowly increasing the anode voltage until forward breakover occurs. It is important to note that a conducting SCR cannot be switched off by gate control. Much design ingenuity has been shown in devising safe and reliable ways of extinguishing a conducting thyristor, a process often known as *device commutation*.

The TRIAC switch, shown in Table 1.1, is the equivalent of two SCRs connected in inverse parallel and permits the flow of current in either direction. Both SCRs are mounted within an encapsulated enclosure and there is one gate terminal. The application of positive anode voltage with positive gate pulse to an inert device causes switch-on in the forward direction. If the anode voltage is reversed, switch-off occurs when the current falls below its holding value, as for an individual SCR. Voltage blocking will then occur in both directions until the device is gated again, in either polarity, to obtain conduction in the desired direction. Compared with individual SCRs, the TRIAC combination is a low-voltage, lower power, and low-frequency switch with applications usually restricted below 400 Hz.

Certain types of thyristor have the facility of gate turn-off, and the chief of those is the gate turn-off thyristor (GTO). Ratings are now (1999) available up to 4500 V, 3000 A. with switching speeds faster than an SCR. Turn-on is realized by positive gate current in the presence of positive anode voltage. Once

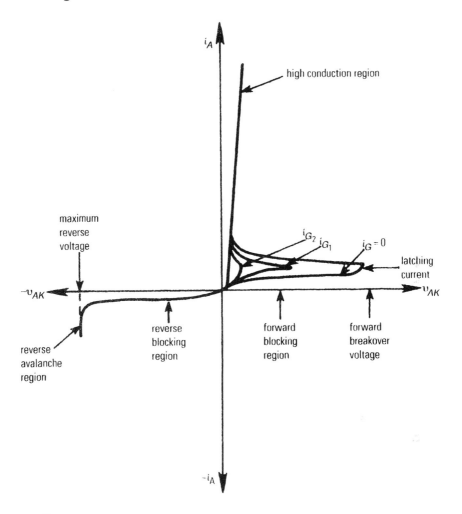

Fig. 7

ignition occurs, the anode current is retained if the gate signal is removed, as in an SCR. Turn-on by forward breakover or by *dv/dt* action should be avoided. A conducting GTO can be turned off, in the presence of forward current, by the application of a negative pulse of current to the gate. This usually involves a separate gating circuit of higher power rating than for switch-on. The facility of a high power device with gate turn-off is widely used in applications requiring forced commutation, such as dc drives.

The static induction thyristor (SITH) acts like a diode, in the absence of gate signal, conducting current from anode (A) to cathode (K) (Table 1.1). Negative gate voltage turns the switch off and must be maintained to give reverse voltage blocking. The SITH is similar to the GTO in performance with higher switching speed but lower power rating.

The MOS-controlled thyristor (MCT) can be switched on or off by negative or positive gate voltage, respectively. With high-speed switching capability, low conduction losses, low switching losses, and high current density it has great potential in high-power, high-voltage applications. The gating requirements of an MCT are easier than those of the GTO, and it seems likely that it will supplant it at higher power levels. A peak power of 1 MW can be switched off in 2 ns by a single MCT.

2

Single-Phase Uncontrolled Rectifier Circuits

A general representation of a rectifier or converter, with single-phase supply, is shown in Fig. 2.1. The contents of the rectifier "box" can take several different forms. For the purpose of analysis it is most helpful to consider first the simplest form of half-wave connection and then proceed to more complicated full-wave connections.

2.1 HALF-WAVE UNCONTROLLED RECTIFIER CIRCUITS WITH RESISTIVE LOAD

2.1.1 Single Diode Circuit

The simplest form of uncontrolled rectifier circuit consists of a single-phase sinusoidal voltage source supplying power to a load resistor R through an ideal diode D (Fig. 2.2). Conduction only occurs when the anode voltage of the diode is positive with respect to the cathode, i.e., during positive half cycles of the supply voltage. The load current and voltage therefore consist of the positive, half-sinusoidial pulses given in Fig. 2.3. By Kirchhoff's loop law the instantaneous diode voltage is the difference between the instantaneous values of the supply voltage e_s and load voltage e_L. Diode voltage e_D is therefore coincident with the negative pulses of supply voltage in Fig. 2.3.

$$e_s = E_m \sin\omega t \qquad (2.1)$$

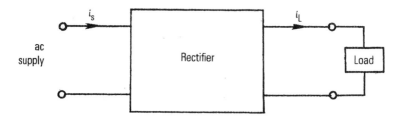

FIG. 1   General rectifier representation.

$$e_L = E_m \sin \omega t \left|_{0,2\pi}^{\pi,3\pi}\right.$$

(2.2)

$$e_D = e_s - e_L = E_m \sin \omega t \left|_{\pi,3\pi}^{2\pi,4\pi}\right.$$

(2.3)

$$i_L = i_s = \frac{e_L}{R} = \frac{E_m}{R} \sin \omega t \left|_{0,2\pi}^{\pi,3\pi} + 0\right|_{\pi,3\pi...}^{2\pi,4\pi...}$$

(2.4)

2.1.1.1 Load-Side Quantities

The average value of any function $i_L (\omega t)$ that is periodic in 2π is

$$I_{av} = \frac{1}{2\pi} \int_0^{2\pi} i_L (\omega t) \, d\omega t$$

(2.5)

Substituting Eq. (2.4) into Eq. (2.5) and integrating gives

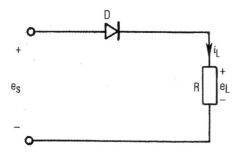

FIG. 2   Single-phase, half-wave diode rectifier.

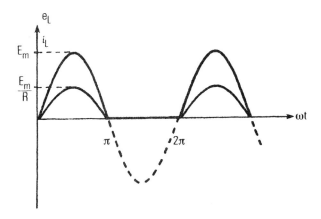

Fig. 3 Waveforms for single-phase, half-wave diode rectifier with R load.

$$I_{av} = \frac{E_m}{\pi R} = 0.318 \frac{E_m}{R} \tag{2.6}$$

Power dissipation in a series circuit can always be defined in terms of the root means square (rms) current, whatever waveshape the periodic current may be. For the load current $i_L(\omega t)$ periodic in 2π the rms current is defined by

$$I_L = \sqrt{\frac{1}{2\pi} \int_0^{2\pi} i_L^2(\omega t) \, d\,\omega t} \tag{2.7}$$

Substituting Eq. (2.4) into Eq. (2.7) and integrating gives

$$I_L = \frac{E_M}{2R} = \frac{I_m}{2} \tag{2.8}$$

The rms value $I_L = I_m/2$ for half-wave operation compares with the corresponding value $I_L = I_m/\sqrt{2}$ for sinusoidal operation.

The average power dissipation in the load resistor R is given by

$$P_L = I_L^2 R \tag{2.9}$$

Substituting Eq. (2.8) into Eq. (2.9) gives

$$P_L = \frac{E_m^2}{4R} = \frac{E_S^2}{4R} = P_{in} \tag{2.10}$$

where $E_S = E_m/\sqrt{2}$ is the rms value of the supply voltage and P_{in} is the input power. It is seen from Eq. (2.10) that the average power is one-half the value for the continuous sinusoidal operation at the same peak voltage.

The degree of distortion in a rectified current waveform can be calculated by combining the rms and average values in a "ripple factor."

$$\text{Ripple factor} = \frac{\text{rms value of ac components}}{\text{average value}}$$

$$= \frac{\sqrt{I_L^2 - I_{av}^2}}{I_{av}}$$

$$= \sqrt{\left(\frac{I_L}{I_{av}}\right)^2 - 1} \qquad (2.11)$$

For the case of half-wave rectification, substituting Eqs. (2.6) and (2.8) into Eq. (2.11) gives

$$RF = \sqrt{\left(\frac{\pi}{2}\right)^2 - 1} = 1.21 \qquad (2.12)$$

The ideal value of the ripple factor is zero, for an undistorted steady dc output. The value 1.21 for half-wave rectification is undesirably large and is unacceptable for many applications. It is found that the Fourier series for the current waveform of Fig. 2.3 is

$$i_L(\omega t) = \frac{E_m}{R}\left(\frac{1}{\pi} + \frac{1}{2}\sin \omega t - \frac{2}{3\pi}\cos 2\omega t - \frac{2}{15\pi}\cos 4\omega t - \ldots\right) \qquad (2.13)$$

The time average value $E_m/\pi R$ in Eq. (2.13) is seen to agree with Eq. (2.6). It is shown in Example 2.3 that the coefficients of the terms in Eq. (2.13) sum to the rms value of the current.

2.1.1.2 Supply-Side Quantities

In the series circuit Fig. 2.2 the supply current is also the load current. Equations (2.6) and (2.8) therefore also define the supply current. Because the diode is presumed ideal, the input power from the supply is equal to the power dissipated in the load resistor.

From the supply side, the circuit of Fig. 2.2 is nonlinear; that is, the impedance of the diode plus resistor cannot be represented by a straight line in the voltage-current plane. When a sinusoidal voltage supplies a nonlinear impedance the resulting current is periodic but nonsinusoidal. Any function that is periodic can be represented by a Fourier series, defined in the Appendix, and reproduced below, which enables one to calculate values for the harmonic components of the function.

$$i(\omega t) = \frac{a_0}{2} + \sum_{n=1}^{\infty} \left(a_n \cos n\omega t + b_n \sin n\omega t \right)$$

$$= \frac{a_0}{2} + \sum_{n=1}^{\infty} c_n \sin(\omega t + \psi_n) \tag{2.14}$$

where

$$c_n = \sqrt{a_n^2 + b_n^2} \tag{2.15}$$

$$\psi_n = \tan^{-1}\left(\frac{a_n}{b_n} \right) \tag{2.16}$$

When function $i(\omega t)$ is periodic in 2π the coefficients of (2.14) are given by

$$\frac{a_0}{2} = \frac{1}{2\pi} \int_0^{2\pi} i(\omega t)\, d\omega t \tag{2.17}$$

$$a_n = \frac{1}{\pi} \int_0^{2\pi} i(\omega t) \cos n\omega t\, d\omega t \tag{2.18}$$

$$b_n = \frac{1}{\pi} \int_0^{2\pi} i(\omega t) \sin n\omega t\, d\omega t \tag{2.19}$$

Comparing Eq. (2.17) with Eq. (2.5) shows that the Fourier coefficient $a_0/2$ defines the time average or dc value of the periodic function. The fundamental (supply frequency) components are obtained when $n = 1$ in Eqs. (2.16) and (2.17). For the current of Fig. 2.3, defined in Eq. (2.4), it is found that

$$a_1 = \frac{1}{\pi} \int_0^{2\pi} i(\omega t) \cos \omega t\, d\omega t$$

$$= \frac{E_m}{R} \int_0^{\pi} \sin \omega t \cos \omega t\, d\omega t$$

$$= 0 \tag{2.20}$$

Similarly,

$$b_1 = \frac{1}{\pi} \int_0^{2\pi} i(\omega t) \sin \omega t\, d\omega t$$

$$= \frac{E_m}{\pi R} \int_0^{2\pi} \sin^2 \omega t\, d\omega t$$

$$= \frac{E_m}{2R} \tag{2.21}$$

The peak amplitude c_1 of the fundamental frequency sine wave component of Fig. 2.3 is therefore

$$c_1 = \sqrt{a_1^2 + b_1^2} = \frac{E_m}{2R} \tag{2.22}$$

Angle ψ_1 defines the displacement angle between the fundamental component of $i(\omega t)$ and the supply current origin. In this case

$$\psi_1 = \tan^{-1}\left(\frac{a_1}{b_1}\right) = 0 \tag{2.23}$$

Equation (2.23) shows that the fundamental component of current is in time phase with the supply voltage.

2.1.1.3 Power Factor

The power factor of any circuit is the factor by which the apparent voltamperes must be multiplied to obtain the real or time average power. For the supply side of Fig. 2.1, the power factor is given by

$$PF = \frac{P}{EI_s} \tag{2.24}$$

where P is the average power and E and I_s are rms values of the supply voltage and current. The universal definition of Eq. (2.24) is independent of frequency and of waveform.

In most nonlinear circuits supplied by a sinusoidal voltage the supply current contains a supply frequency component (or fundamental harmonic component) of rms value I_{s_1} that is phase displaced from the supply voltage by angle ψ_1. The average input power P can then be written

$$P = EI_{s_1} \cos \psi_1 \tag{2.25}$$

Combining Eqs. (2.14) and (2.15) gives

$$PF = \frac{I_{s_1}}{I_s} \cos \psi_1 \tag{2.26}$$

The ratio I_{s_1}/I_s is the *current distortion factor* and arises chiefly, but not entirely, because of the nonlinear load (i.e., rectifier) impedance. The term $\cos \psi_1$ in (2.26) is called the *current displacement factor* and may be partly or wholly attributable to reactive components of the load impedance. It should be noted, however, that in some rectifier circuits the current displacement angle ψ_1 is nonzero even with resistive loads.

Although the product terms of Eq. (2.26) are entirely analytical, they are useful because the relative values sometimes suggest the best approach to the problem of power factor correction. The two terms of Eq. (2.26) are valid only if the supply voltage is sinusoidal whereas Eq. (2.24) is universally true in any passive circuit, irrespective of supply voltage waveform.

For the half-wave rectifier circuit of Fig. 2.2 the Fourier coefficients of the fundamental current wave showed that the displacement angle ψ_1 is zero [Eq. (2.23)]. This means that the displacement factor, $\cos \psi_1$, is unity. The power factor, in this case, is therefore equal to the distortion factor and is due entirely to the nonlinear rectifier impedance. Substituting values from Eq. (2.1) and (2.8) into Eq. (2.24), noting that $E_s = E_m \sqrt{2}$ gives the result

$$PF = \frac{1}{\sqrt{2}} \tag{2.27}$$

Since $\psi_1 = 0$ the power factor is neither leading nor lagging. Because of this one would expect that the power factor could not be improved by the connection of energy storage devices across the supply terminals. In fact, the connection of a capacitance across the supply terminals in Fig. 2.2 is found to make the overall power factor less (i.e., worse).

2.1.2 Single-Diode Circuit with Load-Side Capacitor

In a half-wave diode controlled resistor circuit the load current conduction period can be extended and its average value increased by the use of a parallel-connected capacitor (Fig. 2.4). The capacitor stores energy while the diode is conducting and releases it, through the load resistor, while the diode is in extinction. With appropriate values of R and C, the load current can be made continuous even though the supply current remains of pulse waveform.

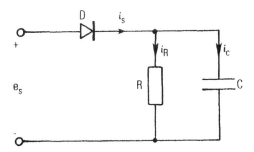

FIG. 4 Single-phase, half-wave diode rectifier with load-side capacitor.

In Fig. 2.4 the instantaneous currents in the load (resistor), capacitor and supply branches are I_R, I_c, and I_s, respectively.

The circuit operates in two modes, according as to whether the rectifier is conducting. While current flows into the circuit from the supply, the following equations are true

$$i_R = \frac{E_m}{R} \sin \omega t \qquad (2.28)$$

$$i_c = C \frac{de}{dt} = \omega C E_m \cos \omega t \qquad (2.29)$$

$$i_s = i_R + i_c = \frac{E_m}{|Z|} \sin(\omega t + \psi_s) \qquad (2.30)$$

where

$$|Z| = \sqrt{\frac{R^2}{1 + \omega^2 C^2 R^2}} \qquad (2.31)$$

$$\psi_s = \tan^{-1}(-\omega CR) \qquad (2.32)$$

The two analytical components of the supply current are shown in Fig. 2.5 for the case when $R = X_c$. Extinction of the supply current i_s occurs when the two components i_R and i_c are equal and opposite, at angle x radians from the origin of the supply voltage wave. The diode conduction mode thus terminates at $\omega t = x$. At this angle it is seen that

$$\frac{E_m}{R} \sin x + \omega C E_m \cos x = 0 \qquad (2.33)$$

from which

$$x = \tan^{-1}(-\omega CR) \qquad (2.34)$$

Comparison of Eqs. (2.32) and (2.34) shows that

$$x = \pi - \psi_s = \pi - \tan^{-1}(-\omega CR) \qquad (2.35)$$

At $\omega t = x$ diode conduction ceases. The instantaneous load voltage is then

$$e_L = e_c = E_m \sin x \qquad (2.36)$$

The capacitor, charged to voltage $E_m \sin x$, begins to discharge through the series R-C circuit of the load branches. Since the R and C elements are linear the load current decays exponentially with a time constant $\tau = RC$

Therefore, in Fig. 2.5, for $\omega t > x$

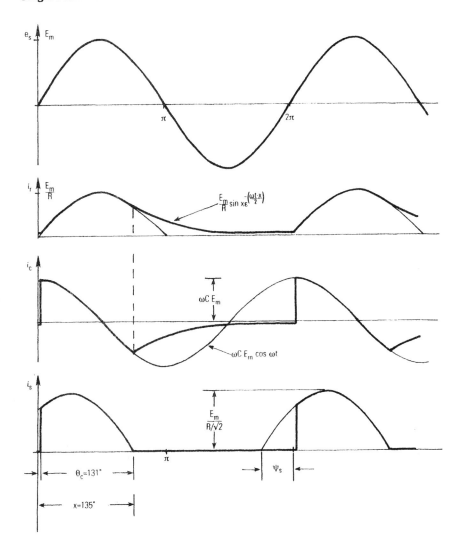

Component currents in circuit of Fig. 2.4. When R = x_c, x = 135°, θ_c = 131°

Fig. 5   Component currents in the circuit of Fig. 2.4: $R = X_c$, $x = 135°$, $\theta_c = 131°$.

$$i_R = \frac{E_m}{R} \sin x \varepsilon^{-\frac{(\omega t - x)}{\omega CR}} \tag{2.37}$$

Since $\sin x = \sin(\pi - \psi_s) = \sin \psi_s$, then

$$i_L = \frac{E_m}{R} = \sin \psi_s \varepsilon^{-\cot \psi_s (\omega t - \pi + \psi_s)} \tag{2.38}$$

The load voltage, during supply current extinction, is therefore

$$e_L = i_L R = E_m \sin x \varepsilon^{-\cot \psi_s (\omega t - x)} \tag{2.39}$$

Ignition of the diode recommences when the supply voltage becomes instantaneously greater than the load voltage. In Fig. 2.5 this occurs at the instants $\omega t = x - \theta_c$ and $2\pi + x - \theta_c$, where θ_c is the conduction angle of the diode

$$\omega t \big|_{ig} = 2\pi + x - \theta_c \tag{2.40}$$

In terms of ψ_s, the ignition angle is

$$\omega t \big|_{ig} = 3\pi - \psi_s - \theta_c \tag{2.41}$$

The conduction angle may be found by equating the load voltage Ri_R from Eq. (2.37) to the supply voltage from Eq. (2.28), putting $\omega t = 2\pi + x - \theta_c$,

$$E_m \sin x \varepsilon^{-[(2\pi - \theta_c)/\omega CR]} = E_m \sin(2\pi + x - \theta_c)$$

or

$$\sin x \varepsilon^{-[(+2\pi - \theta_c)/\omega CR]} = \sin(x - \theta_c) \tag{2.42}$$

Equation (2.42) is transcendental and must be solved by iteration. For values of ωCR less than unity (i.e., $\psi_s < 45°$) the ignition angle $x - \theta_c$ becomes small for half-wave rectifier operation. Increase of the capacitance tends to shorten the conduction angle.

The three instantaneous currents in the circuit of Fig. 2.4 are described, for the period $0 < \omega t < 2\pi$, by

$$i_R = \frac{E_m}{R} \sin \omega t \Big|_{\pi - \psi_s - \theta_c}^{\pi - \psi_s} + \frac{E_m}{R} \sin \psi_s \varepsilon^{-\cot \psi_s (\omega t - \pi + \psi_s)} \Big|_{0, \pi - \psi_s}^{\pi - \psi_s - \theta_c, 2\pi} \tag{2.43}$$

$$i_c = \omega CE_m \cos \omega t \Big|_{\pi - \psi_s - \theta_c}^{\pi - \psi_s} - \frac{E_m}{R} \sin \psi_s \varepsilon^{-\cot \psi_s (\omega t - \pi + \psi_s)} \Big|_{0, \pi - \psi_s}^{\pi - \psi_s - \theta_c, 2\pi} \tag{2.44}$$

$$i_s = \frac{E_m}{|Z|} \sin\left(\omega t + \psi_s\right) \Big|_{\pi - \psi_s - \theta_c}^{\pi - \psi_s} + 0 \Big|_{0, \pi - \psi_s}^{\pi - \psi_s - \theta_c, 2\pi} \qquad (2.45)$$

The average value of the load current is found to be

$$I_{av} = \frac{1}{2\pi} \int_0^{2\pi} i_L\left(\omega t\right) d\omega t = \frac{E_m}{|Z|} \frac{1 - \cos\theta_c}{2\pi} \qquad (2.46)$$

A numerical example for operation of this circuit is given in Example 2.4.

2.1.3 Single-Diode circuit for Battery Charging

A simple diode circuit containing a current limiting resistance R can be used to charge a battery of emf V from a single-phase supply (Fig. 2.6). The battery opposes the unidirectional flow of current so that the net driving voltage is $e - V$. Neglecting any voltage drop on the diode (which is likely to be of the order 1–2 V) the current is therefore

$$i_L = \frac{e - V}{R} = \frac{1}{R}\left(E_m \sin\omega t - V\right) \Big|_\alpha^\beta \qquad (2.47)$$

where α and β define the current pulse in Fig. 2.7. Current flows only in the positive voltage direction when $e - V > 0$. Angles α and β are defined by

$$e - V = E_m \sin \alpha - V = 0 \qquad (2.48)$$

Therefore,

$$\alpha = \sin^{-1} \frac{V}{E_m} \qquad (2.49)$$

By symmetry

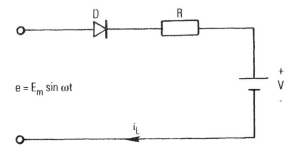

$e = E_m \sin \omega t$

FIG. 6 Single-phase, half-wave diode battery charging circuit.

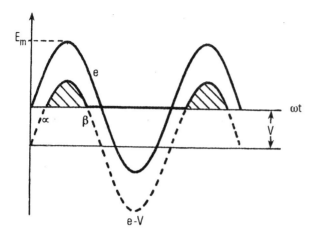

Fig. 7 Voltage and current waveforms for the battery-charger circuit of Fig. 2.6: $V \cong 0.6E_m$.

$$\beta = \pi - \alpha \tag{2.50}$$

The average value I_{av} of the battery charging current is defined by

$$I_{av} = \frac{1}{2\pi} \int_0^{2\pi} i_L(\omega t) \ d\omega t$$

Substituting Eq. (2.47) into the above defining integral expression gives

$$I_{av} = \frac{1}{2\pi R} \int_\alpha^\beta \left(E_m \sin \omega t - V \right) \ d\omega t$$

$$= \frac{1}{2\pi R} \left[E_m \left(\cos\alpha - \cos\beta \right) + V\left(\alpha - \beta \right) \right] \tag{2.51}$$

Eliminating β between Eqs. (2.50) and (2.51) gives

$$I_{av} = \frac{1}{2\pi R} \left[2E_m \cos\alpha + V\left(2\alpha - \pi \right) \right] \tag{2.52}$$

A relevant numerical calculation is given in Example 2.5.

2.1.4   Worked Examples

Example 2.1   An ideal single-phase source, 240 V, 60 Hz, supplies power to a load resistor $R = 100 \ \Omega$ via a single ideal diode. Calculate the average and rms values of the current and the power dissipation. What must be the rating of the diode?

The circuit is shown in Fig. 2.2. The specified voltage of 240 V can be presumed to be the rms value. The average circuit current, from Eq. (2.6), is

$$I_{av} = \frac{E_m}{\pi R} = \frac{240\sqrt{2}}{\pi \times 10} = 1.08 \text{ A}$$

The rms value of the current is given by Eq. (2.8)

$$I_L = \frac{E_m}{2\pi} = \frac{240\sqrt{2}}{2 \times 100} = 1.7 \text{ A}$$

In the circuit of Fig. 2.2 power dissipation takes place only in the load resistor and is given by Eq. (2.10)

$$P_L = I_L^2 (1.7)^2 \ 100 = 298 \text{ W}$$

The diode must be rated in terms of a peak reverse voltage and a mean forward current.

From Eq. (2.3) it can be seen that the peak diode voltage is equal to the peak supply voltage.

$$\text{Diode } PRV = E_m = 240\sqrt{2} = 339.4 \text{ V}$$

A convenient commercial rating would be to choose a diode rated at 400 V. Either the rms or the mean (average) current could be used as a basis of current rating. Since $I_L = 1.7$ A a convenient commercial rating would be 2 A.

Example 2.2   The rms value I of a periodic waveform that consists of an average or dc value lay plus a sum of harmonic components with rms values I_1, I_2, $\cdots$ is given by

$$I^2 = I_{av}^2 + I_1^2 + I_2^2 + \cdots$$

Show that the rms value of the current in the half-wave rectifier circuit, Fig. 2.2, can be obtained in terms of its harmonic components and that the value obtained agrees with the integration method.

In terms of the rms values of the harmonic components,

$$I^2 = I_{av}^2 + I_1^2 + I_2^2 + I_3^2 + \cdots$$

In terms of the peak values of $\hat{I}_1$, $\hat{I}_2$, $\cdots$, $\hat{I}_n$ of the sinusoidal harmonics, and since the peak, average, and rms values of the dc term I_{av} are identical, then

$$I^2 = I_{av}^2 + \frac{1}{2}\left(\hat{I}_1^2 + \hat{I}_2^2 + \ldots + \hat{I}_n^2\right)$$

where $\hat{I}_1 = I_1/\sqrt{2}$, etc. The Fourier series for this waveform is given by Eq. (2.13).

The rms value I_L of the function $i_L(\omega t)$, Eq. (2.13), for half-wave rectifiction is seen to be

$$I_L^2 = \left(\frac{E_m}{R}\right)^2 \left\{ \left(\frac{1}{\pi}\right)^2 + \frac{1}{2}\left[\left(\frac{1}{2}\right)^2 + \left(\frac{2}{3\pi}\right)^2 + \left(\frac{2}{15\pi}\right)^2 + \cdots \right] \right\}$$

from which

$$I_L = \frac{E_m}{2R}$$

This result is seen to agree with that of Eq. (2.8).

Example 2.3   The single-phase diode resistor circuit of Fig. 2.2 is supplied with power from an ideal voltage source of rating 240 V, 50 Hz. Calculate the circuit power factor. If an ideal capacitor C is now connected across the supply where $X_c = R = 100\ \Omega$, calculate the new value of power factor.

From Eq. (2.27), the power factor is seen to be

$$PF = \frac{1}{\sqrt{2}} = 0.707$$

Alternatively, substituting the values of I_L and P_L from Example 2.1 into Eq. (2.24) gives

$$PF = \frac{P_L}{E_L I_L}$$

$$= \frac{289}{240 \times 1.7} = 0.707$$

If a capacitor of reactance X_c is connected across the supply, as in Fig. 2.13, the capacitor instantaneous current is given by

$$i_c = \frac{E_m}{X_c} \sin(\omega t + 90)$$

$$= \frac{E_m}{X_c} \cos \omega t$$

The resultant instantaneous supply current is therefore

$$i_s = i_c + i_L$$

$$= \frac{E_m}{X_c} \cos \omega t + \frac{E_m}{R} \sin \omega t \begin{vmatrix} \pi, 3\pi, \ldots \\ 0, 2\pi, \ldots \end{vmatrix}$$

The rms value of the overall supply current is given, in general, by

$$I_s = \sqrt{\frac{1}{2\pi} \int_0^{2\pi} i_s^2 \left(\omega t\right) d\omega t}$$

In the present case, since $X_c = R$, eliminating X_c gives

$$
\begin{aligned}
I_s^2 &= \frac{E_m^2}{2\pi R^2} \left[\int_0^\pi \left(\cos\omega t + \sin\omega t\right)^2 + \int_\pi^{2\pi} \cos^2\omega t\, d\omega t \right] \\
&= \frac{E_m^2}{2\pi R^2} \left[\int_0^{2\pi} \cos^2\omega t\, d\omega t + \int_0^\pi \left(\sin^2\omega t + \sin 2\omega t\right) \right] \\
&= \frac{E_m^2}{2\pi R^2} \left(\pi + \frac{\pi}{2} + 0 \right)
\end{aligned}
$$

$$I_s = \frac{\sqrt{3}}{2} \frac{E_m}{R} = 0.866 \frac{240\sqrt{2}}{100} = 2.94 \text{ A}$$

Which compares with a load current of 1.7 A.

The load current and load power are not affected by the terminal capacitance. The new power factor is therefore

$$PF = \frac{P_L}{E_s I_s} = \frac{289}{240 \times 2.94} = 0.41$$

The introduction of the capacitor has caused a considerable reduction of the power factor.

Example 2.4   A single-phase supply of voltage $e_s = 380 \sin 100\,\pi t$ supplies power to a load resistor $R = 40\ \Omega$ via on ideal diode valve. Calculate the average current in the load. Repeat the calculation if an ideal capacitor $C = 138$ μF is connected across the resistor

In the absence of the capacitor, the average load current in the circuit of Fig. 2.2 is given by Eq. (2.6)

$$I_{av} = \frac{E_m}{\pi R} = 3.024 \text{A}$$

With a capacitor of 138 μF, at 50 Hz, in Fig. 2.4, the use of Eq. (2.32) gives a value for the magnitude of the input phase angle ψ_s.

$$\psi_s = \tan^{-1} \left(2\pi \times 50 \times 138 \times 10^{-6} \times 40\right) = \tan^{-1} (1.734) = 60°$$

Therefore,

$$\sin\psi_s = 0.866 \quad \text{and} \quad \cot\psi_s = 0.577$$

With the component values given, $E_m/R = 9.5$ A and $\omega CE_m = 16.47$ A. From Eq. (2.35)

$$x = \pi - \psi_s = 120°$$

Iterative solution of the transcendental equation, Eq. (2.42), gives a value for the conduction angle θ_c of the supply current

$$\theta_c = 115.75°$$

The circuit waveforms are shown in Fig. 2.5.
    For the specified values of R and C the input impedance $|Z|$, Eq. (2.31), is

$$|Z| = \sqrt{\frac{40^2}{1 + (1.734)^2}} = 20 \,\Omega$$

The average load current, Eq. (2.46), is therefore

$$I_{av} = \frac{380}{20} \left(\frac{1 - \cos 115.75°}{2} \right) = 4.336 \text{ A}$$

Introduction of the capacitor filter has therefore resulted in about 43% increase of the original average load current of 3.024 A.

    Example 2.5   In the battery charger circuit, Fig. 2.6, the supply voltage is given by $e = 300 \sin \omega t$ and resistor $R = 10 \,\Omega$. Calculate the average charging current if $V = 150$ V. What is the operating power factor?
    From Eqs. (2.49) and (2.50),

$$\alpha = \sin^{-1}\left(\frac{150}{300}\right) = 30°$$

$$\beta = \pi - \alpha = 150°$$

Substituting values $E_m = 300$, $V = 150$, $R = 10$, $\alpha = 30° = \pi/6$ into Eq. (2.52) gives

$$I_{av} = 3.27 \text{ A}$$

The rms current is given by integrating Eq. (2.47)

$$I_L^2 = \frac{1}{2\pi} \int_0^{2\pi} i_L^2 (\omega t) \, d\omega t$$

$$= \frac{1}{2\pi R^2} \int_\alpha^\beta (E_m \sin \omega t - V)^2 \, d\omega t$$

from which it is found that, with $\beta = \pi - \alpha$, then

$$I_L^2 = \frac{1}{2\pi R^2}\left[\left(\frac{E_m^2}{2}+V^2\right)(\pi-2\alpha)+\frac{E_m^2}{2}\sin 2\alpha - 4E_m V \cos\alpha\right]$$

Substituting numerical values gives

$$I_L = 4.79 \text{ A}$$

The total power delivered to the circuit from the ac supply is

$$P = I_L^2 R + I_{av}V$$

In this application the power factor is therefore

$$PF = \frac{P}{EI_L} = \frac{I_L^2 R + I_{av}V}{\left(E_m/\sqrt{2}\right)I_s}$$

Now

$$P_L = I_L^2 R = 229.4 \text{ W}$$

The total power delivered by the supply is

$$P = 229.4 + (3.27)(150) = 720 \text{ W}$$

The total power factor seen from the supply is given by Eq. (2.24)

$$PF = \frac{720}{\left(300/\sqrt{2}\right)(4.79)} = 0.71$$

2.2 FULL-WAVE DIODE CIRCUIT WITH RESISTIVE LOAD

By the use of four diodes, Fig. 2.8a, rectifier circuit performance can be greatly improved. All of the supply voltage wave is utilized to impress current through the load, Fig. 2.8b, and the circuit, which is very widely used, is called a *single-phase, full-wave diode rectifier*.

The various properties of the full-wave circuit can be evaluated in precisely the same way as those for the half-wave circuit.

2.2.1 Load-Side Quantities

The load current waveform, Fig. 2.8b is represented, for the first supply voltage cycle, by

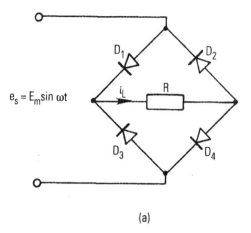

(a)

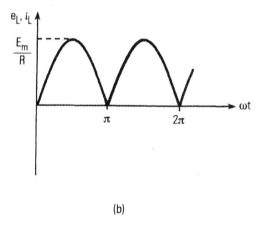

(b)

FIG. 8   Single-phase, full-wave diode rectifier: (a) circuit diagram and (b) load voltage and current waveforms for R load.

$$i_L(\omega t) = \frac{E_m}{R} \sin \omega t \Big|_0^\pi + \frac{E_m}{R} \sin(\omega t - \pi) \Big|_\pi^{2\pi} \tag{2.53}$$

The average value of function $i_L(\omega t)$ is

$$
\begin{aligned}
I_{av} &= \frac{1}{2\pi} \int_0^{2\pi} i_L(\omega t)\, d\omega t \\
&= \frac{E_m}{2\pi R} \left[\int_0^\pi \sin \omega t\, d\omega t + \int_0^{2\pi} \sin(\omega t - \pi)\, d\omega t \right] \\
&= \frac{E_m}{2\pi R} \left\{ (-\cos \omega t) \Big|_0^\pi + \left[-\cos(\omega t - \pi) \right] \Big|_0^\pi \right\} \\
&= \frac{2E_m}{\pi R} = 0.636 E_m / R
\end{aligned}
\tag{2.54}
$$

The value of Eq. (2.54) is seen to be twice the corresponding value for the half-wave circuit given in Eq. (2.6).

The rms value of function $i_L(\omega t)$ is given by

$$
\begin{aligned}
I_L^2 &= \frac{1}{2\pi} \int_0^{2\pi} i_L^2(\omega t)\, d\omega t \\
&= \frac{E_m^2}{2\pi R^2} \left[\int_0^\pi \sin^2 \omega t\, d\omega t + \int_\pi^{2\pi} \sin^2(\omega t - \pi)\, d\omega t \right] \\
&= \frac{E_m^2}{2\pi R^2} \left[\frac{1}{2} \int_0^\pi (1 - \cos 2\omega t)\, d\omega t + \frac{1}{2} \int_0^{2\pi} 1 - \cos 2(\omega t - \pi)\, d\omega t \right] \\
&= \frac{E_m^2}{4\pi R^2} \left\{ \left(\omega t - \frac{\sin 2\omega t}{2} \right) \Big|_0^\pi + \left[\omega t - \frac{\sin 2(\omega t - \pi)}{2} \right] \Big|_\pi^{2\pi} \right\} \\
&= \frac{E_m^2}{2R^2}
\end{aligned}
\tag{2.55}
$$

Therefore,

$$I_L = \frac{E_m}{\sqrt{2} R} \tag{2.56}$$

Value I_L in Eq. (2.56) is seen to be $\sqrt{2}$ times the corresponding value for half-wave operation, given by Eq. (2.8). It should also be noted that I_L has the same rms value as a sinusoidal current of the same peak value; an rms value is not affected by the polarity of the waveform.

The load current ripple factor is

$$RF = \sqrt{\left(\frac{I_L^2}{I_{av}}\right)^2 - 1} = \sqrt{\left(\frac{5.66}{5.1}\right)^2 - 1} = 0.48 \tag{2.57}$$

which compares with a value 1.21 for half-wave operation.

The load power is given once again by

$$P_L = I_L^2 R \tag{2.58}$$

Combining Eqs. (2.56) and (2.58) gives

$$P_L = \frac{1}{2}\frac{E_m^2}{R} = \frac{E^2}{R} \tag{2.59}$$

The load power dissipation, Eq. (2.59), is twice the value obtained with half-wave operation, Eq. (2.10), and is equal to the dissipation obtained with sinusoidal load current of the same peak value.

2.2.2　Supply-Side Quantities

The application of sinusoidal voltage to the resistive circuit of Fig. 2.8a causes a sinusoidal supply current in time phase with the voltage. The time average value of the supply current I_s is therefore zero over any complete number of cycles and its rms value equals that of the load current.

$$I_{s(av)} = \frac{1}{2}\int_0^{2\pi} \frac{E_m^2}{R^2}\sin^2 \omega t \; d\omega t = 0 \tag{2.60}$$

$$I_s = \sqrt{\frac{1}{2\pi}\int_0^{2\pi} \frac{E_m^2}{R^2}\sin^2 \omega t \; d\omega t} = \frac{E_m}{\sqrt{2R}} \tag{2.61}$$

If the bridge diodes are ideal, the input power must be equal to the load power

$$P_{in} = P_L = \frac{E_m^2}{2R} \tag{2.62}$$

For a circuit where the input voltage and current are both sinusoidal and in time phase, the power factor is unity

$$PF = 1.0 \tag{2.63}$$

The properties of the full-wave diode bridge with resistive load are summarized in Table 2.1

TABLE 2.1   Single-Phase Diode Rectifier Circuits with Resistive Load

Property	Half-wave bridge	Full-wave bridge
Average load current	1.35	$\dfrac{2}{\pi}\dfrac{E_m}{R}$
RMS load current	$\dfrac{E_m}{2R}$	$\dfrac{E_m}{\sqrt{2}R}$
Power	$\dfrac{E_m^2}{4R}$	$\dfrac{E_m^2}{2R}$
RMS supply current	$\dfrac{E_m}{2R}$	$\dfrac{E_m}{\sqrt{2}R}$
Power factor	$\dfrac{1}{\sqrt{2}}$	1.0
Ripple factor of load current	1.21	0.47

2.3   HALF-WAVE DIODE CIRCUITS WITH SERIES R-L LOAD

2.3.1   Single-Diode Circuit

The action of the series R-L circuit with a single-diode rectifier valve illustrates a lot of the important features of rectifier circuit operation. In Fig. 2.9 the current is unidirectional and the polarity of the voltage drop e_R on the resistor R is always as indicated. The polarity of the instantaneous emf e_L of the series inductor L varies cyclically as does the total load voltage e_L. Where the application of sinusoidal voltage e_s results in an instantaneous current i_L, this is given by solution of the first-order linear differential equation [Eq. (2.65)].

$$e = E_m \sin \omega t \tag{2.64}$$

$$e_L = i_L R + L\frac{di_L}{dt} \tag{2.65}$$

Neglecting the voltage drop on the diode, $e = i_L$ during conduction. If no current is initially present conduction begins when the anode voltage of the diode becomes positive, that is, at the beginning of the supply voltage cycle in Fig. 2.10. An expression for the current i_L may be thought of, at any instant, as the sum of hypothetical steady-state and transient components i_{ss} and i_t, respectively.

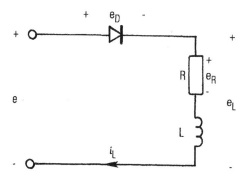

Fig. 9 Single-phase, half-wave diode rectifier with series R-L load.

$$i_L(\omega t) = i_{ss}(\omega t) + i_t(\omega t)$$
$$= I_m \sin(\omega t - \Phi) + i_t(\omega t) \qquad (2.66)$$

where Φ is the phase angle for sinusoidal currents of supply frequency

$$\Phi = \tan^{-1}\frac{\omega L}{R} \qquad (2.67)$$

At instant $\omega t = 0$ in Fig. 2.10 the total current $i_L(0) = 0$ but the steady-state component $i_{ss}(0) = I_m \sin(0 - \Phi) = -I_m \sin\Phi$, so that the instantaneous transient current is

$$i_t(0) = I_m \sin\Phi \qquad (2.68)$$

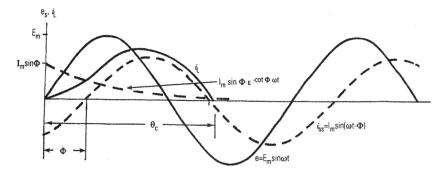

Fig. 10 Current waveform for the single-phase, half-wave circuit of Fig. 2.9: $\Phi = 60°$.

For $\omega t = 0$ the transient current decays exponentially through the series R-L load impedance

$$I_t\left(\omega t\right) = i_m \sin\Phi\varepsilon^{-(\omega t/\omega T)} = I_m \sin\Phi\varepsilon^{\cot\Phi\cdot\omega t} \tag{2.69}$$

where

$$\frac{1}{\omega t} = \frac{R}{\omega L} = \cot\Phi \tag{2.70}$$

and

$$I_m = \frac{E_m}{\sqrt{R^2 + \omega^2 L^2}} = \frac{E_m}{|Z|} \tag{2.71}$$

The instantaneous current at any time interval during conduction is therefore

$$i_L\left(\omega t\right) = I_m \sin(\omega t - \Phi) + I_m \sin\Phi\varepsilon^{-\cot\Phi\cdot\omega t} \tag{2.72}$$

It is seen in Fig. 2.10 that the total current i_L consists of unidirectional, nonsinusoidal pulses lying outside the envelope of the steady-state sinusoid. The conduction angle θ_c when $i_L\left(\omega t\right) = 0$ in Eq. (2.56) satisfies

$$\sin(\theta_c - \Phi) + \sin\Phi\varepsilon^{-\cot\Phi\theta_c} = 0 \tag{2.73}$$

Equation (2.73) is transcendental and must be solved iteratively for θ_c. For values of phase angle Φ up to about $60°$ one can estimate θ_c fairly accurately by the relation

$$\theta_c = \pi + \Phi + \Delta \tag{2.74}$$

where Δ is of the order a few degrees. With highly inductive loads the conduction angle increases until, at $\Phi = \pi/2$, the conduction takes the form of a continuous, unidirectional sinusoidal oscillation of mean value $I_m = E_m/\omega L$.

The instantaneous voltage drop e_R on the load resistor has the value $i_L R$ and must have the same waveform as $i_L\left(\omega t\right) = 0$. Now by Kirchhoff's law, in Fig. 2.9,

$$e_L\left(\omega t\right) = e_R\left(\omega t\right) + e_L\left(\omega t\right)$$
$$= E_m \sin\omega t \quad \left(\text{during conduction}\right) \tag{2.75}$$

At the value ωt_1, where $e_L\left(\omega t\right) = e_R\left(\omega t\right)$ in Fig. 2.11a, $e_L\left(\omega t\right) = L\, di/dt = 0$. The derivative di/dt is zero at a current maximum, and therefore the crossover of the $e_R\left(\omega t\right)$ curve with the $e(\omega t)$ curve in Fig. 2.11b occurs when $e_R\left(\omega t\right)$ has its maximum value. Time variations of the circuit component voltages are shown in Fig. 2.11 for a typical cycle. The polarity of the inductor voltage $e_L\left(\omega t\right)$ is such as, by Lenz's law, to oppose the change of inductor current. While the current is increasing, $0 < \omega t < \omega t_1$, the induced emf in the inductor has its positive point nearest the cathode. When the inductor current is decreasing, $\omega t_1 < \omega t < \omega t_3$, the induced emf in the inductor tries to sustain the falling current by presenting its positive pole furthest from the cathode in Fig. 2.9.

The average value of the rectified current is the mean value of the $i_L(\omega t)$ curve (Fig. 2.10) over 2π radians and is given by

$$
I_{av} = \frac{1}{2\pi} \int_0^{\theta_c} i_L(\omega t)\, d\omega t
$$

$$
= \frac{E_m}{2\pi|Z|} \left[\cos\Phi(1-\cos\theta_c) - \sin\Phi\sin\theta_c - \frac{\sin\Phi}{\cot\Phi}\left(\varepsilon^{-\cot\Phi\theta_c} - 1\right) \right]
\tag{2.76}
$$

Eliminating the exponential component between Eq. (2.73) and Eq. (2.76) gives, after some manipulation,

$$
I_{av} = \frac{E_m}{2\pi R}(1-\cos\theta_c)
$$

$$
= \frac{E_m}{2\pi|Z|} \frac{1-\cos\theta_c}{\cos\Phi}
\tag{2.77}
$$

Since the average value of the inductor voltage is zero,

$$
E_{av} = I_{av} R
$$

$$
= \frac{E_m}{2\pi}(1-\cos\theta_c)
\tag{2.78}
$$

The rms value of the load current is found using Eqs. (2.72) and (2.73):

$$
I_L^2 = \frac{1}{2\pi} \int_0^{\theta_c} i_L^2(\omega t)\, d\omega t
$$

$$
= \frac{E_m^2}{4\pi|Z|^2} \{ \theta_c - \sin\theta_c \cos(\theta_c - 2\Phi)
$$

$$
+ \tan\Phi \left[\sin^2\Phi - \sin^2(\theta_c - \Phi) \right] + 4\sin\Phi\sin\theta_c \sin(\theta_c - \Phi) \}
\tag{2.79}
$$

The corresponding ripple factor can be obtained by the substitution of Eqs. (2.78) and (2.79) into Eq. (2.11). Both the average and rms values of the current increase for increased values of phase angle.

2.3.2 Worked Examples

Example 2.6 A single-phase, full-wave diode bridge (Fig. 2.8a) is used to supply power to a resistive load of value $R = 50\ \Omega$. If the supply voltage

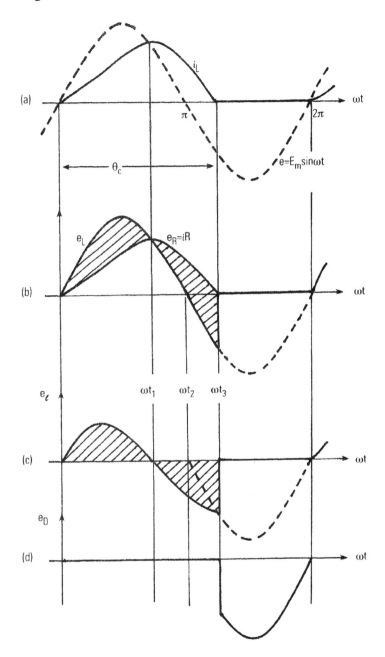

Fig. 11 Component voltage waveforms for the single-phase, half-wave circuit of Fig. 2.9: $\Phi = 60°$.

remains sinusoidal and has a peak value 400 V, calculate the average and rms values of the load current. Calculate the ripple factor for the load current and compare this with half-wave operation.

The average value of the load current (Fig. 2.8b) is given by Eq. (2.54),

The corresponding rms value I_L is the value of a sinusoidal of the same peak height, as given in Eq. (2.56).

$$I_L = \frac{1}{\sqrt{2}} \frac{E_m}{R} = 5.66 \text{A}$$

The ripple factor is

$$RF = \sqrt{\left(\frac{I_L^2}{I_{av}}\right)^2 - 1} = \sqrt{\left(\frac{5.66}{5.09}\right)^2 - 1} = 0.48$$

which compares with a value 1.21 for half-wave operation.

Example 2.7   For the single-phase bridge of Example 2.6 (i.e., Fig. 2.8) evaluate the voltage and current ratings required of the bridge diodes

The diodes in Fig. 2.8a conduct current only in their respective forward directions. When the top terminal of Fig. 2.8a is positive, current flows through D_1, R, and D_4. The current of diodes D_1 and D_4 is therefore given by

$$I_{D_1} = \frac{E_m}{R} \sin \omega t \left|\begin{matrix} \pi, 3\pi... \\ 0, 2\pi... \end{matrix}\right.$$

The diode rms current is $1/\sqrt{2}$ or 0.707 of the rms supply current. A diode current waveform consists of only the positive pulses of current and is therefore similar to the waveform of Fig. 2.3. Because a diode current waveform contains only one half the area of the load current its mean current rating is one half the mean load current

$$I_{av} (\text{diode}) = \frac{1}{2} I_{av} (\text{load})$$

$$= \frac{1}{\pi} \frac{E_m}{R}$$

$$= 2.55 \text{ A}$$

While current is passing through D_1, R and D_4 in Fig. 2.8a, the diodes D_2 and D_3 are held in extinction due to reverse anode voltage. The peak value PRV of this reverse voltage is the peak value of the supply voltage.

$$PRV = E_m = 400 \text{ V}$$

Example 2.8   A series R-L load with $R = 10 \, \Omega$ and $X_L = 20 \, \Omega$ at $f = 50$ Hz and supplied from an ideal single-phase supply $e = 300 \sin 2\pi ft$ through an ideal diode. Determine the peak value of the current pulses, the conduction angle and the average load current

$$\Phi = \tan^{-1} \frac{X_L}{R} = \tan^{-1} 2 = 63.4°$$

$\cot \Phi = 0.5$, $\cos \Phi = 0.447$, $\sin \Phi = 0.894$

Conduction angle θ_c is determined by iterative solution of the transcendental equation, Eq. (2.73). We start by estimating a value of θ_c using Eq. (2.74)

$$\theta_c = 180° + 63.4° + \Delta \cong 243.4° + \Delta$$

The algebraic sum of the two parts of Eq. (2.73) is given in the *RH* column. From the results obtained for the estimated values $\theta_c = 250°$ and $249°$ one can make a linear interpolation that the actual value of θ_c is $249.25°$. This is shown in the final row to be almost correct.

$$\theta_c = 249.25°$$

The maximum value of the current pulses occurs when $di_L/d\omega t = 0$. From Eq. (2.72)

$$\frac{di_L}{dt} = I_m \cos\left(\omega t - \Phi\right) - I_m \sin \Phi \cot \Phi \varepsilon^{-\cot \Phi \omega t} = 0$$

from which it is seen that

$$\cos(\omega t - \Phi) = \cot \Phi \, \varepsilon^{-\cot \Phi \, \omega t}$$

Iterative solution of this, along the lines above, yields

$$\omega t \Big|_{\frac{di_L}{dt}=0} = 146.2°$$

Estimated value of θ_c					$\sin(\theta_c - \Phi) +$
deg.	rad.	$\sin(\theta_c - \Phi)$	$\varepsilon^{-\cot \Phi \, \theta_c}$	$\sin \Phi \varepsilon^{-\cot \Phi \, \theta_c}$	$\sin \Phi \varepsilon^{-\cot \Phi \, \theta_c}$
246°	4.294	−0.045	0.117	0.105	0.06
248°	4.33	−0.08	0.115	0.103	0.023
250°	4.363	−0.115	0.113	0.101	−0.014
249°	4.316	−0.0976	0.114	0.102	0.0044
249.25°	4.35	−0.1019	0.1136	0.1016	−0.003

Substituting $\omega t = 146.2°$ into Eq. (2.72) gives

$$I_{L\,max} = 1.2417\,I_m = 1.2417\frac{300}{\sqrt{10^2 + 20^2}} = 16.7\text{ A}$$

Now $\cos\theta_c = \cos249.25° = -0.354$. In Eq. (2.77) therefore,

$$I_{av} = \frac{300}{2\pi\times10}(1+0.354) = 6.465\text{ A}$$

This compares with the value I_m/π or 9.55 A that would be obtained with only the 10-Ω resistor as load.

If the load consisted of a resistor of the same value as the present load impedance the average current would be $300/(\pi \times 22.36) = 4.27$ A.

2.4 FULL-WAVE DIODE CIRCUITS WITH SERIES R-L LOAD

The average load current is increased and the ripple factor reduced by the introduction of a freewheel diode across the load. This provides a relaxation path for the load current during the intervals of negative supply voltage and assists toward continuous load current conduction even though the supply current is discontinuous with high harmonic content.

Consider the circuit of Fig. 2.12 in which the load now consists of inductance L in series with resistance R. This circuit is similar in action to two half-wave circuits in series.

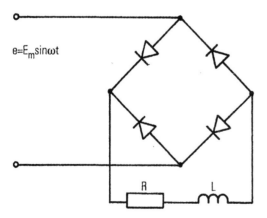

FIG. 12   Single-phase, full-wave diode rectifier circuit with series R-L load.

During conduction one of the opposite pairs of diodes connects the supply voltage across the load so that the load voltage e_L (ωt) retains the form shown in Fig. 2.8b for resistive load. In typical steady-state operation the load current becomes continuous, Fig. 2.13b, with identical successive portions each half cycle as the current transfers from diodes D_1 and D_4 to diodes D_2 and D_3. The supply current i_s (ωt) assumes the bidirectional (alternating) form of Fig. 2.13c. Although the supply current is subject to abrupt transitions the load current (passing through the inductor) remains quite smooth and large induced emfs are avoided. When the load inductance is very large such that almost perfect smoothing is realized, the load current assumes a constant value and the supply current becomes a rectangular wave. These aspects of bridge operation are discussed more fully in the following chapters.

Example 2.9 In the single-phase, full-wave bridge circuit of Fig. 2.14 the load current is 50 A, and the maximum peak to peak ripple voltage across the load is to be 10 V. If $E_m = 240\sqrt{2}$ V at 50 Hz, estimate suitable values for L and C. What will be the current rating of the capacitor?

In a single-phase, full-wave rectified sinusoid (Fig. 2.8b)

$$E_{av} = \frac{2}{\pi} E_m$$

and

$$E = \frac{E_m}{\sqrt{2}}$$

so that

$$E_{av} = \frac{2\sqrt{2}}{\pi} E = 0.9E$$

In this case

$$E_{av} = 0.9 \times 240 = 216 \text{ V}$$

The principal higher harmonic of a full-wave rectified waveform is the second harmonic (since there is no fundamental component), in this case of frequency 100 Hz.

Peak 100Hz component of the ac ripple $\approx E_m - E_{av} \approx 240\sqrt{2} - 216$ = 123.4 V.
Peak-to-peak ripple ≈ 246.8 V.
Ripple reduction factor required = 246.8:10.

Now the excess ripple voltage has to be filtered by the series choke. Therefore, $X_L = 24.68X_c$.

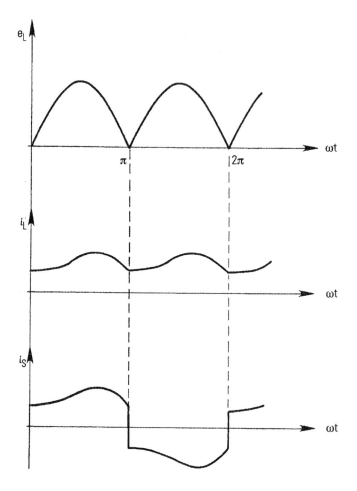

FIG. 13   Waveforms for the full-wave circuit of Fig. 2.12.

The load current is 50 A. It is good practice to allow a safety factor of 2 and permit the inductor (and capacitor) current ripple to have a peak value of 25 A. The inductor ripple voltage has already been estimated as 123.4 V.

$$X_L = \frac{123.4}{25} = 4.84 \ \Omega$$

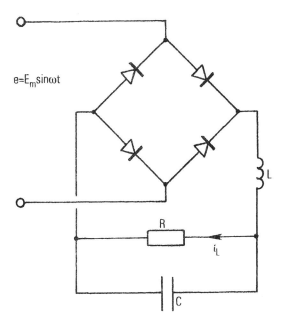

FIG. 14 Circuit for Example 2.9.

At the second harmonic frequency,

$$L = \frac{4.84}{200\pi} = 7.7 \text{ mH}$$

$$X_L = \frac{4.84}{25.68} = 188 \times 10^{-3} \ \Omega$$

from which

$$C = 8460 \ \mu\text{F}$$

Since the peak capacitor current ripple has been estimated as 25 A, the rms capacitor current is

$$I_c = \frac{25}{\sqrt{2}} = 17.7 \text{ A}$$

PROBLEMS

Single-Phase Diode Rectifier Circuits with Resistive Load

2.1 Sketch the load current waveforms of the diode rectifier circuits shown in Fig. 2.15, with sinusoidal supply voltage and resistive load.

2.2 Obtain expressions for the average and rms currents in the circuits of Fig. 2.2 if $e = E_m \sin \omega t$. Show that the load power is one half the value for sinusoidal operation and calculate the ripple factor.

2.3 Derive expressions for the average and rms currents in a single-phase, full-wave diode bridge circuit with resistive load. Show that the ripple factor is less than one-half the value for half-wave rectification. What is the ideal value of ripple factor?

2.4 Calculate the power factor of operation for the half-wave rectifier circuit of Fig. 2.2.

2.5 For a single-phase, half-wave diode rectifier circuit with resistive load and supply voltage $e = E_m \sin \omega t$, show that the load voltage $e_L (\omega t)$ can be represented by the Fourier series

$$e_L \left(\omega t \right) = E_m \left(\frac{1}{\pi} + \frac{1}{2} \sin \omega t - \frac{2}{3\pi} \cos 2\omega t + \ldots \right)$$

2.6 A single-phase supply $e = E_m \sin \omega t$ supplies power to a resistor R through an ideal diode. Calculate expressions for the Fourier coefficients a_1, b_1, and c_1 of the fundamental component of the circuit current and hence show that the displacement factor is unity.

2.7 A resistive load $R = 10 \ \Omega$ is supplied with rectified current from a single-phase ac supply through a full-wave diode bridge. The supply voltage is given by $e = E_m \sin \omega t$. Sketch the circuit arrangement and load voltage waveform. Calculate the average and rms values of the load voltage if $E_m = 330$ V. Define a ripple factor for this load voltage and calculate its value.

2.8 A single-phase, full-wave diode bridge supplies power to a resistive load from a sinusoidal voltage source $e = E_m \sin \omega t$. Show that the load current waveform created by the natural commutation of the diodes is such that its Fourier series does not contain a term of supply frequency.

2.9 Sketch the current waveform of the rectifier circuit (Fig. 2.16) if $e = E_m \sin \omega t$ and $V = E_m/2$. If $V = E_m$, what is the value of the average current compared with the case $V = 0$?

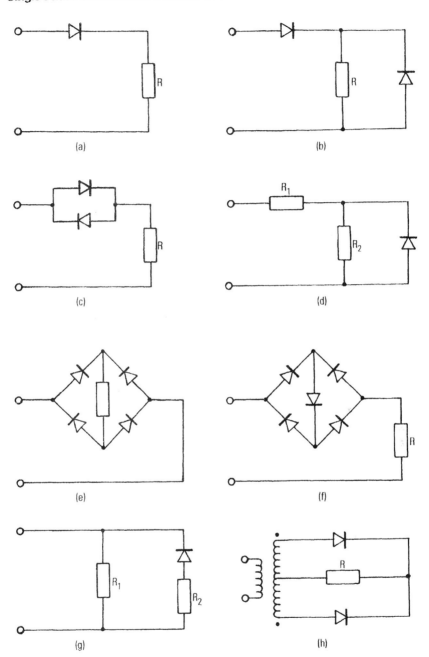

FIG. 15 Circuits for Problem 2.1.

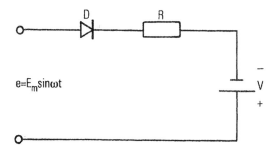

FIG. 16 Circuit for Problem 2.9.

2.10 Sketch the current waveform for the battery charger operation shown in Fig. 2.6 if $E_m > V$. Derive an expression for the instantaneous current I, and calculate the average value of charging current if $E_m = 2$ V, where $e = E_m \sin \omega t$. What is the value of the average current if $E_m = 100$ V, $R = 100$ Ω?

2.11 In Problem 2.10 above, calculate the value of the rms current and hence the power factor.

2.12 An ideal single-phase supply voltage $e = E_m \sin \omega t$ supplies energy to a battery of terminal voltage V through the full-wave rectifier circuit of Fig. 2.17. Sketch in proportion the waveforms of e, V, and load current I_L if $V = E_m/2$. Calculate the average value of the supply current and the load current if $E_m = 100$ V and $R = 25$ Ω. What effect would there be on the circuit function and average load current if an open circuit failure occurred on (a) diode D_1 and (b) diode D_3?

2.13 Sketch typical steady-state current waveforms for the diode circuit with capacitor smoothing shown in Fig. 2.4.

2.14 Deduce and sketch the waveforms of the voltages across the supply, the diode, and the load for the rectifier circuit of Fig. 2.4 if $R = X_c$.

2.15 An ideal supply voltage $e_s = E_m \sin \omega t$ supplies power to a resistor R through an ideal diode D. If an ideal capacitor C is connected across the resistor, show that the average resistor current is given by $(E_m/2\pi[Z]) (1 - \cos\theta_c)$, where Z is the input impedance during conduction and θ_c is the conduction angle of the supply current.

2.16 In the circuit of Fig. 2.4 if the resistor R is very large, the load consists essentially of a pure capacitance. Deduce and sketch the waveforms of the current, the capacitor voltage, and the diode voltage.

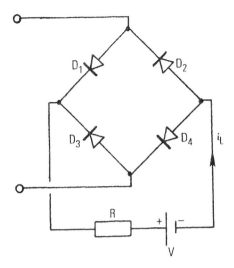

Fɪɢ. **17**   Circuit for Problem 2.12.

2.17   A half-wave diode circuit with load resistor R is supplied from an ideal voltage source $e_s = E_m \sin \omega t$. A capacitor C is connected across the supply terminals (Fig. 2.18). What is the effect on the overall circuit power factor if $R = X_c$?

Single-Phase Diode Rectifier Circuits Series R-L Load

2.18   Show, from first principles, that the current in a series R-L circuit containing a diode with sinusoidal supply is given by

$$i = I_m \sin(\omega t - \Phi) + I_m \sin \Phi \varepsilon^{-1/\tau}$$

where

$$I_m = \frac{E_m}{\sqrt{R^2 + \omega^2 L^2}}$$

$$\Phi = \tan^{-1} \frac{\omega L}{R}$$

$$\tau = \frac{L}{R}$$

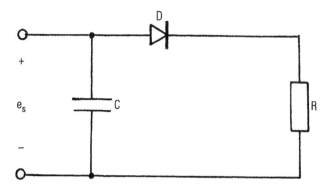

FIG. **18** Circuit for Problem 2.17.

2.19 Derive the following transcendental equation for conduction angle θ_c in a series R-L diode circuit with sinusoidal phase angle Φ and a supply

$$e = E_m \sin \omega t$$
$$\sin(\theta_c - \Phi) + \sin \Phi \varepsilon^{-\cot\Phi\theta_c} = 0$$

If $\Phi = 45°$, solve the equation by iteration to obtain θ_c.

2.20 A circuit consists of a resistor R, inductor L, and diode D in series, supplied from an ideal sinusoidal of instantaneous value $e = E_m \sin \omega t$.

a. Derive an expression for the time variation $i(\omega t)$ of the instantaneous current in terms of E_m, $|Z|$ $(= \sqrt{R^2 + \omega^2 L^2})$, and Φ $(= \tan^{-1} \omega L/R$.

b. Sketch, roughly to scale, consistent time variations of e, i, and the inductor-induced emf e_L if $\Phi = 60°$.

c. Show that with a load time constant $\tau = L/R$, the instant of the cycle when the inductor emf is zero is given by the transcendental equation $\cos(\omega t - \Phi) = \cos \Phi \, \varepsilon^{-t/\tau}$.

2.21 Explain the basis of the equal-area criterion for a series R-L circuit with diode (Fig. 2.19). Sketch the variation of e_R and e_L with time over a typical cycle, assuming switch-on occurs at a positive going zero of the supply. Also sketch the time variation of the diode voltage.

2.22 A series R-L circuit of phase angle $\Phi = \tan^{-1} \omega L/R$ is supplied with current from an ideal voltage source $e = E_m \sin \omega t$ through an ideal diode D. Derive or state expressions for the instantaneous current $i(\omega t)$

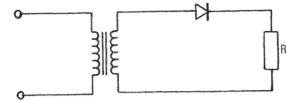

FIG. **19**   Circuit for Problem 2.29.

and the conduction angle θ_c. Sketch waveforms of e and i if $\Phi = 30°$; estimate the value of θ_c. Show that the current has a maximum value I_m at $\omega t = A$ given by

$$I_m = \frac{E_m}{|Z|}\left[\sin(A-\Phi)+\tan\Phi\cos(A-\Phi)\right]$$

where $|Z| = \sqrt{R^2 + \omega^2 L^2}$.

2.23   In the diode-controlled single-phase circuit of Fig. 2.9 the load phase angle $\Phi = \tan^{-1}(\omega L/R)$ and the supply voltage $e = E_m \sin \omega t$. Show that the average current is given by

$$I_{av} = \frac{E_m}{2\pi R}(1-\cos\theta_c)$$

Calculate I_{av} if $E_m = 230\sqrt{2}$ V, $\Phi = 30°$, and $R = 20\ \Omega$.

2.24   A series R-L circuit of phase angle $\Phi = \tan^{-1}(\omega L/R) = 45°$ is supplied with current from an ideal supply $e = E_m \sin \omega t$ through an ideal diode D. Derive expressions for the instantaneous current $i(\omega t)$ and the conduction angle θ_c Give an estimate for the value of θ_c in this case.

Sketch in correct proportion waveforms for e, i, and the instantaneous voltage $e_L(\omega t)$ across the load inductor component. Derive an expression for e_L in terms of i.

2.25   In the diode-controlled single-phase circuit. of Fig. 2.9 the load phase angle $\Phi = \tan^{-1}(\omega L/R)$ and the supply voltage $e = E_m \sin \omega t$. If the conduction angle of the load current is θ_c, show that its rms value is given by Eq. (2.79). Calculate I_L if $E_m = 200$ V, $R = 10\ \Omega$ and $X_L = 20\ \Omega$ at 50 Hz.

2.26   A series resistance–inductance circuit of phase angle $\Phi = \tan^{-1}(\omega L/R)$ is supplied with current from an ideal single-phase voltage supply

$e = E_m \sin \omega t$ through an ideal diode D. Derive or state expressions for the instantaneous current $i(\omega t)$ if $\Phi = 45°$, and estimate (do not calculate) the value of θ_c. Derive an expression for the instantaneous emf $e_L(\omega t)$ induced across inductor L, and sketch the waveform of this consistent with your waveforms of e and i. Explain how you would calculate the maximum positive value of $e_L(\omega t)$ and show that, in this case, it has a value (nearly) equal to $0.5 E_m$.

2.27   Sketch the current waveform for a series R-L circuit controlled by a series diode. If the load phase-angle Φ to fundamental currents is $30°$, what is the extinction angle?

2.28   A supply voltage $e = 220\sqrt{2} \sin \omega t$ is applied to the series R-L load of Fig. 2.9. Calculate the average load current and the rms values of a few harmonic current components using the Fourier series for the load voltage if $R = \omega L = 110 \ \Omega$. Hence, calculate the load current and the current ripple factor.

2.29   A resistive load is supplied through a diode connected in the secondary circuit of a single-phase transformer with low leakage reactance (Fig. 2.19). Sketch the forms of the load current, supply current, and diode voltage over a supply voltage period.

2.30   Deduce and sketch current waveforms for the half-wave rectifier circuit of Fig. 2.9 if L is large.

2.31   Show that the Fourier series for the full-wave rectified sinusoid of Fig. 2.8b is given by

$$i = \frac{E_m}{R}\left(\frac{2}{\pi} - \frac{4}{3\pi}\cos 2\omega t - \frac{4}{15\pi}\cos 4\omega t - \ldots\right)$$

2.32   Sketch current and load voltage waveforms for the full-wave bridge circuit of Fig. 2.12 if inductor L is very large.

2.33   Sketch the current and voltage waveforms for a nominated diode from the full-wave bridge circuit of Fig. 2.20 if $R = \omega L$.

2.34   In the full-wave diode bridge circuit of Fig. 2.20 the supply voltage is given by $e = 400 \sin 100\pi t$. The load impedance consists of $R = \omega L = 25 \ \Omega$. Use the Fourier series method (taking the first three terms of the series) to calculate the rms load current, the rms supply current, and the power dissipation. Hence, calculate the bridge power factor, seen from the supply point.

(Hint: Use the Fourier series of Problem 2.31 to calculate the coefficients of the corresponding current terms with inductive load.)

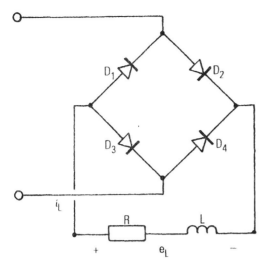

FIG. 20   Circuit for Problems 2.33 and 2.34

3

Single-Phase Controlled Rectifier Circuits

By the use of delayed triggering of a controlled rectifier the average and rms load voltages in rectifier circuits can be smoothly adjusted.

3.1 SINGLE-PHASE CONTROLLED CIRCUITS WITH RESISTIVE LOAD

3.1.1 Voltage and Current Relations for a Half-Wave Controlled Rectifier

If the circuit of Fig. 3.1 is used with an arbitrary firing angle α for the SCR switch, the waveforms obtained are shown in Fig. 3.2 for half-wave operation. The load voltage and current consist of pieces of sinusoid defined by the relationship

$$e_L\left(\omega t\right) = E_m \sin \omega t \Big|_{\alpha, 2\pi + \alpha, \ldots}^{\pi, 3\pi, \ldots} \tag{3.1}$$

having an average value

$$E_{av} = \frac{1}{2\pi} \int_\alpha^\pi e_L\left(\omega t\right) d\omega t$$

$$= \frac{E_m}{2\pi}\left(1 + \cos\alpha\right) \tag{3.2}$$

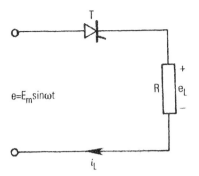

FIG. 1 Single-phase, half-wave controlled rectifier circuit.

When $\alpha = 0$, Eq. (3.2) reduces to Eq. (2.5b).

The rms value E_L of the load voltage in the circuit of Fig. 3.1 is obtained by the use of $e_L(\omega t)$ as follows

$$E_L^2 = \frac{1}{2\pi} \int_\alpha^\pi e_L^2(\omega t)\, d\omega t \tag{3.3}$$

Therefore,

$$I_L = \frac{E_L}{R} = \frac{E_m}{2R}\sqrt{\frac{1}{2\pi}\left[2(\pi-\alpha)+\sin 2\alpha\right]} \tag{3.4}$$

Combining Eqs. (3.2) and (3.4) gives a value for the ripple factor *RF*

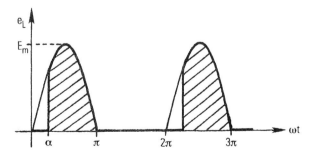

FIG. 2 Load voltage waveform for single-phase, half-wave controlled rectifier circuit with R load and $\alpha = 60°$.

$$RF = \sqrt{\left(\frac{E_L}{E_{av}}\right)^2 - 1}$$

$$= \sqrt{\frac{\pi\left[(\pi-\alpha)+(1/2)\sin 2\alpha\right]}{(\cos\alpha+1)^2} - 1} \tag{3.5}$$

At $\alpha = 0$, the ripple factor reduces to Eq. (2.8), having a value 1.21. When $\alpha = \pi/2$, the ripple factor increases to 1.98. Retardation of the switching angle as a means of controlling the average value of the load voltage therefore also results in the undesirable effect of increasing the rms value of the ac components.

The time average power dissipation in the load is found to be

$$P_L = I_L^2 R$$

$$= \frac{E_m^2}{4R}\frac{1}{2\pi}\left[2(\pi-\alpha)+\sin 2\alpha\right] \tag{3.6}$$

The power factor PF of the single-phase, half-wave controlled rectifier circuit can be obtained by combining Eqs. (3.1)(3.4), and (3.6):

$$PF = \frac{P_L}{E_s I_L} = \frac{P_L}{S_L}$$

$$= \frac{1}{\sqrt{2}}\sqrt{\frac{2(\pi-\alpha)+\sin 2\alpha}{2\pi}} \tag{3.7}$$

When $\alpha = 0$, Eq. (3.7) reduces to $1/\sqrt{2}$, which is the value for an uncontrolled rectifier in Eq. (2.27).

Fourier coefficients $a_1 b_1$ for the fundamental or supply frequency component of the load voltage are found to be

$$a_I = \frac{1}{\pi}\int_0^{2\pi} e_L(\omega t)\cos\omega t \, d\omega t$$

$$= \frac{1}{\pi}\int_\alpha^\pi E_m \sin\omega t\cos\omega t \, d\omega t$$

$$= \frac{E_m}{2}\left[\frac{\cos 2\alpha - 1}{2\pi}\right]$$

$$b_I = \frac{1}{\pi}\int_0^{2\pi} e_L(\omega t)\sin\omega t \, d\omega t \tag{3.8}$$

$$= \frac{1}{\pi} \int_\alpha^\pi E_m \sin^2 \omega t \, d\omega t$$

$$= \frac{E_m}{2} \frac{2(\pi - \alpha) + \sin 2\alpha}{2\pi} \tag{3.9}$$

The peak value c_1 of the fundamental load voltage is therefore

$$c_1 = \sqrt{a_1^2 + b_1^2}$$

$$= \frac{E_m}{4\pi} \sqrt{(\cos 2\alpha - 1)^2 + [2(\pi - \alpha) + \sin 2\alpha]^2} \tag{3.10}$$

Correspondingly, the time phase angle ψ_{L_1} between the sinusoidal supply voltage and the fundamental component of the load voltage (and current) is

$$\psi_{L_1} = \tan^{-1} \frac{a_1}{b_1}$$

$$= \tan^{-1} \frac{\cos 2\alpha - 1}{2(\pi - \alpha) + \sin 2\alpha} \tag{3.11}$$

For $0 < \alpha < 180°$ the phase angle is negative. In a linear, sinusoidal circuit a negative phase angle is associated with energy storage in a magnetic field. But in the circuit of Fig. 3.1 no energy storage is possible. The delayed switching causes a phase lag of the fundamental current component which represents a power factor problem. The power factor reduction, however, is not attributable to an energy storage phenomenon and the instantaneous voltamperes remains positive at all times, as in any resistive circuit. Nevertheless, it is found that some improvement of power factor can be obtained by the connection of capacitance across the circuit terminals.

3.1.2 Power and Power Factor in Half-Wave Rectifier Circuits

The load power can be written in the form of Eq. (2.25), using the subscript L for load

$$P_L = E_s I_{L_1} \cos \psi_{L_1} \tag{3.12}$$

Similarly, one can write an expression for the reactive voltamperes Q_L of the controlled load

$$Q_L = E_s I_{L_1} \sin \psi_{L_1} \tag{3.13}$$

where

$$I_{L_1} = \frac{c_1}{\sqrt{2}} \tag{3.14}$$

$$E_s = \frac{E_m}{\sqrt{2}} \tag{3.15}$$

The apparent voltamperes S_L at the circuit terminals is given by

$$S_L = E_s I_L \tag{3.16}$$

But from Eqs. (3.12) and (3.13),

$$P_L^2 + Q_L^2 = E_s^2 I_{L_1}^2 \tag{3.17}$$

It is found that (3.17) accounts for only part of the apparent voltamperes S_L, given in Eq. (3.16).

The analytical difference between S_L^2 and $P_L^2 + Q_L^2$ is sometimes expressed in terms of the distortion or harmonic voltamperes D_L.

$$S_L^2 = P_L^2 + Q_L^2 + D_L^2 \tag{3.18}$$

So that

$$D_L = E_s \sqrt{I_L^2 - I_{L_1}^2} \tag{3.19}$$

The power factor of this circuit can also be expressed in terms of the current distortion factor I_{L_1}/I_L and current displacement factor $\cos \psi_{L_1}$. In this case, however, the expressions for distortion factor and displacement factor, in terms of switching-angle, do not offer any advantage over Eq. (3.7).

3.1.3 Capacitance Compensation of Rectifier Power Factor

Consider operation of the circuit with a capacitor C across the supply terminals, Fig. 3.3. The instantaneous capacitor current is given by the continuous function

$$i_c(\omega t) = \omega C E_m \sin\left(\omega t + \frac{\pi}{2}\right) \tag{3.20}$$

The instantaneous load branch current is

$$i_L(\omega t) = \frac{E_m}{R} \sin \omega t \begin{vmatrix} \pi, 3\pi, \ldots \\ \alpha, 2\pi + \alpha, \ldots \end{vmatrix} \tag{3.21}$$

By Kirchhoff's law the instantaneous supply current is given by

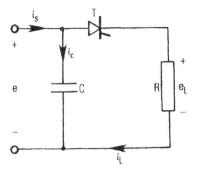

FIG. 3 Single-phase, half-wave controlled rectifier circuit with supply side capacitance.

$$i_s(\omega t) = i_L(\omega t) + i_c(\omega t) \tag{3.22}$$

Because the load current and supply current are nonsinusoidal, the three branch currents cannot be described by phasor relationships. For the capacitor compensated circuit the real power P_L, terminal voltage $e(\omega t)$, and the load branch current $i_L(\omega t)$ are unchanged, but the power factor PF is now a function of the rms value of the supply current I_s (not the load current I_L).

$$PF = \frac{P_L}{E_s I_s} \tag{3.23}$$

In Fig. 3.3 the intention is that the power factor PF seen from the supply point be improved (i.e. increased) with respect to the uncompensated value PF_L.

The rms supply current I_s is given by

$$I_s^2 = \frac{1}{2\pi} \int_0^{2\pi} i_s^2(\omega t)\, d\omega t \tag{3.24}$$

Substituting (3.20–3.22) into (3.23) gives

$$I_s = \frac{E_m}{2R}\left[\frac{2R^2}{X_c^2} + \frac{2R}{X_c}\frac{\cos 2\alpha - 1}{2\pi} + \frac{2(\pi - \alpha) + \sin 2\alpha}{2\pi}\right]^{1/2} \tag{3.25}$$

The corresponding expression for power factor is then

$$PF = \frac{\left(1/\sqrt{2}\right)\left[2(\pi - \alpha) + \sin 2\alpha\right]/2\pi}{\sqrt{\left(2R^2/X_c^2\right) + \left(2R/X_c\right)(\cos 2\alpha - 1)/2\pi + \left\{\left[2(\pi - \alpha) + \sin 2\alpha\right]/2\pi\right\}^2}} \tag{3.26}$$

Even if the value of C is adjusted continuously to give the best power factor for any fixed value of α, the degree of power factor improvement realisable is only of the order of a few percent.

Maximization of the power factor can be achieved by minimization of the rms supply current since P_L is not affected by terminal capacitance. Differentiating Eq. (3.25) with respect to C gives

$$\frac{dI_s}{dC} = \frac{E_m}{2R}\left(4\omega^2 R^2 C + 2R\frac{\cos 2\alpha - 1}{2\pi}\right)\left[\sqrt{\frac{2R^2}{X_c^2} + \frac{2R}{X_c}\frac{\cos 2\alpha - 1}{2\pi} + \left(\frac{2(\pi - \alpha) + \sin 2\alpha}{2\pi}\right)^2}\right]^{-1}$$

(3.27)

For I_s minimum, $dI_s/dC = 0$, which leads to the condition

$$\frac{1}{X_c} = \omega C = \frac{1}{2R}\frac{1 - \cos 2\alpha}{2\pi}$$

(3.28)

With optimum capacitance the minimum value $I_{s_{min}}$ of I_s that will supply the specified load power is obtained by substituting the above expression for X_c into Eq. (3.25)

$$I_{s_{min}} = \frac{E_m}{2R}\sqrt{-\frac{2R^2}{X_c^2} + \left(\frac{2(\pi - \alpha) + \sin 2\alpha}{2\pi}\right)}$$

(3.29)

The optimum power factor PF_{max} is then given by

$$PF_{max} = \frac{P_L}{E_s I_{s_{min}}}$$

(3.30)

3.1.4 Single-Phase Full-Wave Bridge Rectifier Circuit

Full-wave, controlled rectification of the load voltage and current can be obtained by use of the alternative configurations of Fig. 3.4 in which the controlled switches are shown as thyristors. With a fixed value of switching angle and sinusoidal supply voltage the load voltage waveform (Fig. 3.5) is defined by

$$e_L(\omega t) = E_m \sin \omega t \Big|_\alpha^\pi + E_m \sin(\omega t - \pi)\Big|_{\pi + \alpha}^{2\pi}$$

(3.31)

which has an average value twice as large as Eq. (3.2) for half-wave operation

$$E_{av} = \frac{E_m}{\pi}\left(1 + \cos\alpha\right) \tag{3.32}$$

When $\alpha = 0$, Eq. (3.32) reduces to the value $2E_m/\pi$ for a half sine wave. The rms value of the load current in the circuits of Fig. 3.4 is

$$I_L = \frac{E_m}{R} = \sqrt{\frac{1}{2\pi R}\int_0^{2\pi} e_L^2\left(\omega t\right)\,d\omega t}$$

$$= \frac{E_m}{\sqrt{2}R}\sqrt{\frac{1}{2\pi}\left[2\left(\pi - \alpha\right) + \sin 2\alpha\right]} \tag{3.33}$$

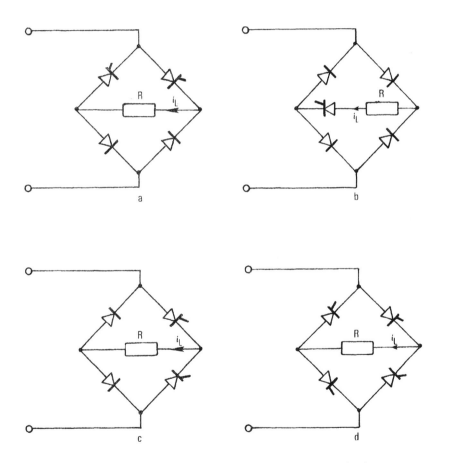

FIG. 4 Single-phase, full-wave controlled rectifier circuits with R load.

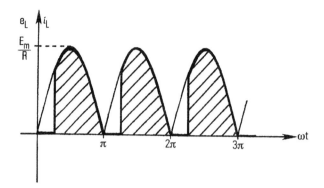

FIG. 5   Load voltage (and current) waveforms for single-phase, full-wave controlled rectifier with R load, $\alpha \approx 50°$.

The load current waveform (Fig. 3.5) has a ripple factor obtained by substituting Eqs. (3.32) and (3.33) into eq. (2.7)

$$\text{Rippel factor} = \sqrt{\left(\frac{I_L}{I_{av}}\right)^2 - 1}$$

$$= \sqrt{\frac{\pi}{4}} \sqrt{\frac{2(\pi - \alpha) + \sin 2\alpha}{1 + \cos \alpha^2} - 1} \qquad (3.34)$$

When $\alpha = 0$, Eq. (3.34) reduces to the value $RF = 0.48$, which was given previously in Eq. (2.24) for uncontrolled operation.

Comparison of (3.33) with (3.4) shows that the rms value of the load current with full-wave rectification is $\sqrt{2}$ times the value for half-wave rectification, at any fixed firing angle. The average power dissipation in the load is given by

$$P_s = P_L = I_L^2 R$$

$$= \frac{E_m^2}{2R} \frac{1}{2\pi} \left[2(\pi - \alpha) + \sin 2\alpha \right] \qquad (3.35)$$

which is twice the value for corresponding half-wave operation.

Operation of the rectifier circuits of Fig. 3.4 results in the supply current waveform of Fig. 3.6, which is identical to the waveform for a single-phase ac chopper circuit, discussed in Chapter 14. This waveform has zero average value over any number of complete cycles and the supply current contains no dc or even harmonic terms. The rms value I_s of the supply current is

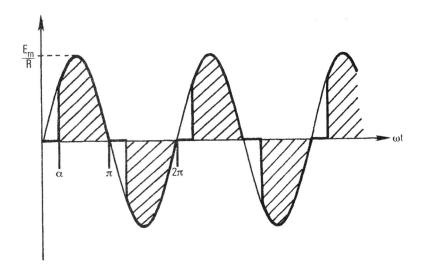

Fɪɢ. 6   Supply current waveforms for the single-phase, full-wave controlled rectifier with R load, $\alpha \approx 50°$.

$$I_s = \sqrt{\frac{1}{2\pi} \int_0^{2\pi} i_s^2(\omega t)\, d\omega t}$$

where

$$i_s(\omega t) = \frac{E_m}{R} \sin \omega t \left|\begin{matrix} \pi, 2\pi, \ldots \\ \alpha, \pi + \alpha, \ldots \end{matrix}\right. \tag{3.36}$$

The rms value is given by

$$I_s = \frac{E_m}{\sqrt{2}R} \sqrt{\frac{1}{2\pi}\left[2(\pi - \alpha) + \sin 2\alpha\right]} \tag{3.37}$$

The rms value of the supply current therefore has the same magnitude as the rms value of the load current. This is to be expected since, neglecting rectifier switch losses, the supply point power and the load power are identical.

The power factor of the full-wave, controlled rectifier circuit can be obtained by substituting Eqs. (3.1)(3.35), and (3.37) into Eq. (3.16).

$$PF = \sqrt{\frac{1}{2\pi}\left[2(\pi - \alpha) + \sin 2\alpha\right]} \tag{3.38}$$

At $\alpha = 0$, the supply point power factor is unity since the supply current is then sinusoidal and in time phase with the supply voltage.

Comparison of Eqs. (3.35) to (3.38) shows that

$$PF = \sqrt{P_s(pu)} = I_s(pu) \tag{3.39}$$

Expression (3.39) is true for any single-phase resistive circuit, irrespective of waveform, and therefore also applies to half-wave rectifier circuits, controlled or uncontrolled.

Because the supply voltage remains sinusoidal, the power factor of the full-wave controlled rectifier circuit may be interpreted as a product of displacement factor and distortion factor. This involves calculation of the Fourier components $a_1 b_1$, and c_1 of the fundamental component i_{s_1} of the supply current.

$$a_1 = \frac{1}{\pi}\int_0^{2\pi} i_s(\omega t)\cos \omega t \, d\omega t$$

$$= \frac{E_m}{2\pi R}(\cos 2\alpha - 1)$$

$$b_1 = \frac{1}{\pi}\int_0^{2\pi} i_s(\omega t)\sin d\omega t \tag{3.40}$$

$$= \frac{E_m}{2\pi R}\left[2(\pi - \alpha) + \sin 2\alpha\right]$$

$$c_1 = \sqrt{a_1^2 + b_1^2} \tag{3.41}$$

$$= \frac{E_m}{2\pi R}\sqrt{(\cos 2\alpha - 1)^2 + \left[2(\pi - \alpha) + \sin 2\alpha\right]^2} \tag{3.42}$$

Coefficient c_1, in (3.42), represents the peak value of the fundamental component of the supply current. RMS supply current I_{s_1} therefore has the value

$$I_{s_1} = \frac{c_1}{\sqrt{2}} \tag{3.43}$$

Combining Eqs. (3.18)(3.37)(3.42), and (3.43) gives a value for the current distortion factor

$$\text{Current distortion factor} = \frac{I_{s_1}}{I_s}$$

$$= \sqrt{\frac{1}{2\pi}\frac{(\cos 2\alpha - 1)^2 + \left[2(\pi - \alpha) + \sin 2\alpha\right]^2}{\left[2(\pi - \alpha) + \sin 2\alpha\right]}} \tag{3.44}$$

The displacement factor is given by

Current displacement factor $= \cos \psi_{s_1}$

$$= \cos \left(\tan^{-1} \frac{a_1}{b_1} \right)$$

$$= \frac{b_1}{c_1}$$

$$= \frac{2(\pi - \alpha) + \sin 2\alpha}{\sqrt{(\cos 2\alpha - 1)^2 + \left[2(\pi - \alpha) + \sin 2\alpha \right]^2}} \quad (3.45)$$

At $\alpha = 0$, both the current displacement factor and the current distortion factor are unity resulting in unity power factor, which compares with the corresponding value $1/\sqrt{2}$ for uncontrolled half-wave operation in Eq. (2.27). Displacement angle ψ_{s_1} between the supply voltage and the fundamental component of the supply current becomes progressively more lagging as the switching angle is further retarded. The consequent displacement factor contribution to the progressively decreasing power factor can be compensated by the connection of shunt capacitance at the circuit terminals. With resistive load the current distortion factor and current displacement factor contribute roughly equal amounts to the circuit power factor.

An alternative to the bridge converter for producing full-wave rectification is the push–pull converter which uses a transformer with a centre-tapped secondary winding, Fig. 3.7. This transformer has the same voltampere rating as the load. If the leakage inductance of the transformer is small, the load current and voltage, with resistive load, have the form shown in Fig. 3.5. The equations of Sec. 3.1.3 also apply to this circuit if transformer losses are negligible. Where an input transformer is required for isolation purposes the circuit of Fig. 3.7 is obviously applicable. If the requirement is to effectively increase the pulse or phase number from unity (as for single-phase, half-wave rectifiers) to two, then a bridge circuit of the type in Fig. 3.4 is likely to be cheaper.

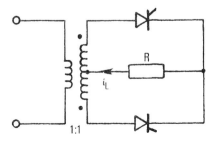

FIG. 7 Single-phase, full-wave rectifier circuit using a center-tapped transformer.

3.1.5 Worked Examples

Example 3.1   A voltage supply $e_s = 283\sin \omega t$ is applied to a circuit consisting of an ideal SCR, gated at $\alpha = 90°$, and a 20-Ω resistor. Calculate the magnitude and phase angle of the fundamental current and sketch this on the same scale as the load current.

From Eq. (3.10) the peak value of the fundamental current is

$$I_{L_1} = \frac{E_m}{4\pi R} \sqrt{(\cos 2\alpha - 1)^2 + [2(\pi - \alpha) + \sin 2\alpha]^2}$$

$$\frac{283}{4\pi 20} = \sqrt{4 + \pi^2} = 4.19 \text{ A}$$

This compares with the value 14.15 A peak current for sinusoidal operation. From Eq. (3.11) the phase angle is found to be

$$\psi_{L_1} = \tan^{-1}\left(-\frac{2}{\pi}\right) = -32.5°$$

The pulses of load current with the superposed fundamental harmonic component are given in Fig. 3.8. It should be noted that the fundamental current component has no physical existence. It is merely an analytical concept that is found to be useful in some aspects of circuit analysis, particularly in the consideration of power factor. In mathematical terms the fundamental current component in Fig. 3.8 may be written

$$i_{L_1}(\omega t) = I_{L_1} \sin(\omega t + \psi_{L_1})$$

$$= 4.194 \sin(\omega t - 32.5°)$$

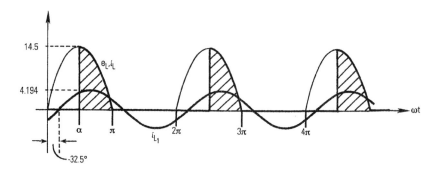

FIG. 8   Analytical component $i_L(\omega t)$ in Example 3.1; $\alpha = 90°$.

Example 3.2   A single-phase, half-wave controlled rectifier circuit has a resistive load R and an ideal sinusoidal supply $e_s = E_m \sin \omega t$. A capacitor C of adjustable value is connected across the supply terminals. With a fixed value of thyristor firing angle what is the value of C that will result in maximum power factor operation? If $\alpha = 90°$, $E_m = 283$ V, and $R = 10\ \Omega$, what value of C will give maximum power factor at $f = 50$ Hz? What degree of power factor improvement is realisable by the use of capacitance compensation?

The optimum value of C is given in Eq. (3.28). At 50 Hz, $R = 10\ \Omega$, $\alpha = \pi/2$,

$$C = \frac{2}{2000\pi \times 2\pi} = 50.65\mu F$$

With this value of capacitance, $\omega C = 0.0159$ mho and $X_c = 62.9\ \Omega$. Minimum supply current $I_{s_{min}}$, from Eq. (3.29), is found to be

$$I_{s_{min}} = \frac{283}{20}\left(-\frac{200}{3956} + \frac{1}{2}\right) = 6.36\ A$$

In the absence of the capacitor the rms load current is obtained from Eq. (3.4) or from Eq. (3.29), with $1/X_c = 0$:

$$I_L = \frac{283}{20} 1/2 = 7.075\ A$$

The average load power is therefore

$$P_L = I_L^2 R = (7.075)^2 \times 10 = 500\ W$$

The uncompensated power factor is

$$PF_L = \frac{P_L}{E_s I_L} = \frac{500}{(283/\sqrt{2}) \times 7.075} = 0.353\ \text{lagging}$$

In the presence of optimal capacitance the power factor becomes

$$PF_{max} = \frac{P_L}{E_s I_{s_{min}}} = 0.393\ \text{lagging}$$

The power factor has therefore been improved by $0.393/0.353$, or about 11%, due to the optimal capacitance compensation.

Example 3.3   A rectifier circuit containing two ideal diodes D and an ideal semiconductor switch (shown as an SCR) is given in Fig. 3.9. Sketch, on squared paper, a consistent set of waveforms for the supply voltage, the three resistor cur-

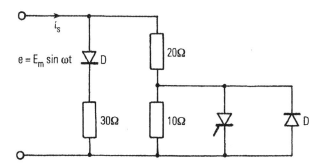

FIG. 9   Single-phase rectifier circuit of Example 3.3.

rents, the switch current i_T, and the supply current i_s if the SCR firing angle $\alpha =$ 60°. Write a mathematical expression $i_s(\omega t)$ to define the waveform of the supply current and calculate its average value over a supply cycle in terms of E_m.

The waveforms of the various branch currents are shown in Fig. 3.10.

$$i_s(\omega t) = \frac{E_m}{15}\sin\omega t\bigg|_0^\alpha + \frac{E_m}{12}\sin\omega t\bigg|_\alpha^\pi + \frac{E_m}{15}\sin\omega t\bigg|_\pi^{2\pi}$$

$$I_{s_{av}} = \frac{1}{T}\int_0^T i_s(\omega t)\,d\omega t$$

$$= \frac{E_m}{2\pi}\left(\int_0^\alpha \frac{\sin\omega t}{15}\,d\omega t + \int_\alpha^\pi \frac{\sin\omega t}{12}\,d\omega t + \int_\pi^{2\pi} \frac{\sin\omega t}{20}\,d\omega t\right)$$

$$= \frac{E_m}{2\pi}\left[\frac{1}{15}(-\cos\omega t)\bigg|_0^\alpha + \frac{1}{12}(-\cos\omega t)\bigg|_\alpha^\pi + \frac{1}{20}(-\cos\omega t)\bigg|_\pi^{2\pi}\right]$$

$$= \frac{E_m}{2\pi}\left[\frac{1}{15}(1-\cos\alpha) + \frac{1}{12}(1+\cos\alpha) + \frac{1}{20}(-1-1)\right]$$

$$I_{s_{av}} = \frac{E_m}{2\pi}\left[\frac{1}{15}\frac{1}{2} + \frac{1}{12}\frac{3}{2} + \frac{1}{20}(-2)\right]\frac{E_m}{2\pi}(0.0583) = 0.0093E_m$$

Example 3.4   A purely resistive load is supplied with power from an ideal single-phase supply of fixed voltage and frequency, denoted by $e = E_m \sin\omega t$, through an arrangement of four ideal diodes and an ideal SCR switch (Fig. 3.11a). The switch is gated during every half cycle of its positive anode voltage state. Sketch (1) the supply current waveform and (2) the load current waveform for a switching angle α of 60°. Derive an expression for the rms value of the load current in terms of E_m and α. What is the per-unit value of this (compared with $\alpha = 0$) when α is 60°?

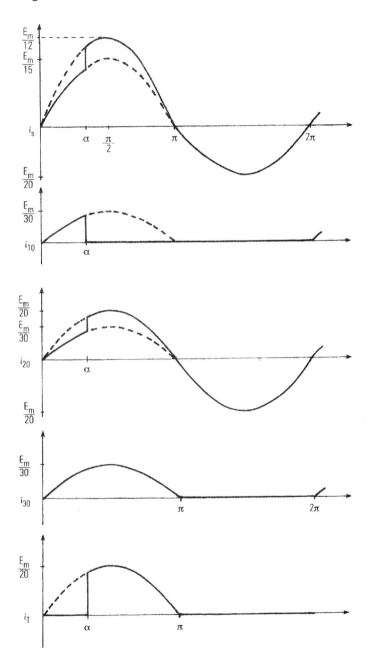

FIG. 10 Branch current waveforms for the single-phase rectifier circuit of Fig. 3.9.

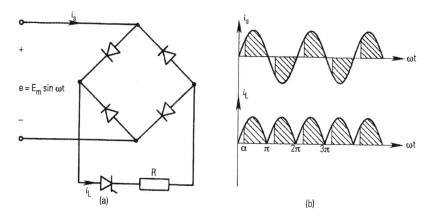

FIG. 11   Single-phase, full-wave controlled rectifier of Example 3.4; $\alpha \approx 50°$.

The waveforms are shown in Fig. 3.11(b). Assuming ideal rectifiers

$$i_L = \frac{E_m}{R}\sin\omega t \Big|_{\alpha}^{\pi} + \frac{E_m}{R}\sin(\omega t - 180°)\Big|_{\pi+\alpha}^{2\pi}$$

Now $\sin(\omega t - 180°) = -\sin\omega t$. Therefore,

$$I_L = \sqrt{\frac{1}{2\pi}\int_0^{2\pi} i_L^2\,d\omega t} \ \ = \left(\frac{E_m^2}{2\pi R^2}\left\{\int_\alpha^\pi \sin^2\omega t + \int_{\pi+\alpha}^{2\pi}\sin^2\omega t\right\}\right)^{1/2}$$

But

$$\int \sin^2\omega t = \int \frac{1-\cos 2\omega t}{2}\,d\omega t = \frac{\omega t}{2} - \frac{\sin 2\omega t}{4}$$

Therefore,

$$I_L^2 = \frac{E_m^2}{2\pi R^2}\left(\frac{\omega t}{2} - \frac{\sin 2\omega t}{4}\right)\Big|_\alpha^\pi + \frac{E_m^2}{2\pi R^2}\left(\frac{\omega t}{2} - \frac{\sin 2\omega t}{4}\right)\Big|_{\pi+\alpha}^{2\pi}$$

$$= \frac{E_m^2}{2\pi R^2}\left\{\left(\frac{\pi-\alpha}{2} + \frac{\sin 2\alpha}{4}\right) + \left[\frac{2\pi - \pi - \alpha}{2} + \frac{\sin 2(\pi+\alpha)}{4}\right]\right\}$$

$$= \frac{E_m^2}{2\pi R^2}\left(\frac{\pi-\alpha}{2} + \frac{\sin 2\alpha}{4} + \frac{\pi-\alpha}{2} + \frac{\sin 2\alpha}{4}\right)$$

Therefore,

$$I_L = \frac{E_m}{\sqrt{2}R}\sqrt{\frac{1}{2\pi}\left[2(\pi-\alpha)+\sin 2\alpha\right]}$$

When $\alpha = 0$,

$$I_{L_{0^\circ}} = \frac{E_m}{\sqrt{2}R}$$

When $\alpha = 60°$,

$$I_L = \frac{E_m}{\sqrt{2}R}\sqrt{\frac{1}{2\pi}(4.188+0.866)}$$

$$= \frac{E_m}{\sqrt{2}R}\sqrt{\frac{5.054}{2\pi}} = \frac{E_m}{\sqrt{2}R}\sqrt{0.804}$$

Therefore,

$$I_{L_{60^\circ}} = \frac{E_m}{\sqrt{2}R}(0.897)$$

and

$$\text{and } I_{L_{pu}} = I_{L_{60^\circ}} / I_{L_{0^\circ}} = 0.897$$

3.2   SINGLE-PHASE CONTROLLED RECTIFIER CIRCUITS WITH SERIES *R-L* LOAD

3.2.1   Half-Wave Controlled Rectifier Circuit

An ideal sinusoidal supply voltage $e = E_m \sin \omega t$ is applied to a series R-L circuit, Fig. 3.12, in which the current level is adjusted by the controlled switching of (for example) a thyristor. For switching angle α and phase angle Φ the instantaneous current $i_L(\omega t)$ is given by

$$i_L(\omega t) = \frac{E_m}{|Z|}\left[\sin(\omega t - \Phi) - \sin(\alpha - \Phi)\varepsilon^{+\cot\Phi(\alpha-\omega t)}\right] \tag{3.46}$$

When $\alpha = 0$, Eq. (3.46) reduces to the expression Eq. (2.72) previously deduced for diode operation.

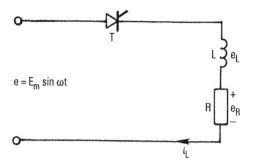

FIG. 12   Single-phase, half-wave controlled rectifier with series R-L load.

The extinction angle x of the load current is defined by a transcendental equation [Eq. (3.47)], obtained by putting $i_L(x) = 0$ in Eq. (3.46)

$$\sin(x - \Phi) - \sin(\alpha - \Phi)\varepsilon^{+\cot \Phi (\alpha - x)} \tag{3.47}$$

The conduction angle θ_c per cycle is given by

$$\theta_c = x - \alpha \tag{3.48}$$

If x is eliminated between Eqs. (3.47) and (3.48), it is seen that

$$\sin(\theta_c + \alpha - \Phi) = \sin(\alpha - \Phi)\varepsilon^{-\theta_c \cot \Phi} \tag{3.49}$$

Waveforms of the load voltage and current are given in Fig. 3.13 for a case when $\alpha > \Phi$, and it may be seen that the nonsinudoidal, unidirectional current pulses lie within the envelope of the steady-state sinusoidal component of current. A first approximation value of θ_c may be obtained from the simple relation

$$\theta_c = \pi + \Phi - \alpha - \Delta \tag{3.50}$$

where Δ is of the order a few degrees for loads with $\Phi < 70°$.

For $0 < \alpha < \Phi$ the current pulses lie outside the sinusoidal current envelope in a similar manner to that illustrated in Fig. 2.10. In this latter case the conduction angle is given by the simple, approximate relation

$$\theta_c\big|_{\alpha<\Phi} = \pi + \Phi - \alpha + \Delta \tag{3.51}$$

Note that Eq. (3.51) is not the same as Eq. (3.50).

When $\alpha = \Phi$, the rectified current pulses become half sinusoids displaced from the supply voltage by angle Φ. Calculated characteristics of the conduction angle versus switching angle for a full range of load phase angles are given in Fig. 3.14. The isosceles triangle at the top left-hand corner, bounded by the $\alpha = \Phi$ line and the $\Phi = 90°$ line, represents conditions when $\alpha < \Phi$ and conduction angles greater than π radians are realized. The scalene triangle, bounded by the $\Phi = 0$, $\Phi = 90°$, and $\alpha = \Phi$ lines represents the mode when $\alpha > \Phi$ and

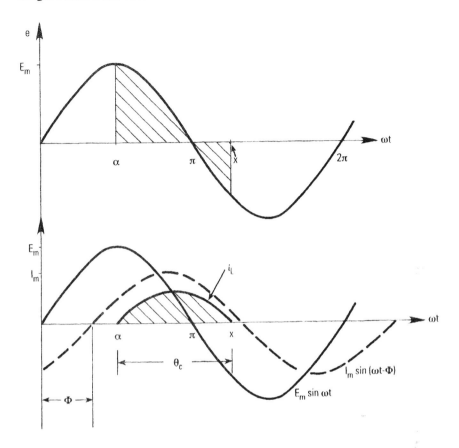

Fig. 13   Waveforms for single-phase, half-wave controlled rectifier with series R-L load; $\alpha = 90°$.

conduction angles smaller than π radians are realized. For both modes of operation the average value I_{av} of the rectified load current is defined in the standard manner

$$I_{av} = \frac{1}{2\pi} \int_{\alpha}^{\theta_c + \alpha} i_L(\omega t)\, d\omega t$$

$$= \frac{E_m}{2\pi|Z|}\Big[(1 - \cos\theta_c)\cos(\alpha - \Phi) + (\sin\theta_c - \tan\Phi)\sin(\alpha - \Phi)$$

$$+ \sin(\alpha - \Phi)\tan\Phi\varepsilon^{-\cot\Phi\theta_c}\Big] \qquad (3.52)$$

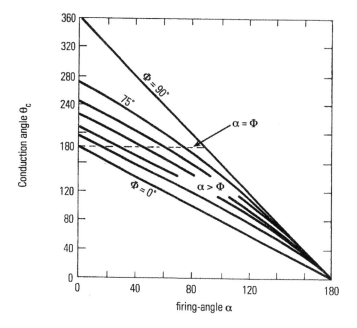

FIG. 14 Conduction angle θ_c thyristor switching-angle for the single-phase, half-wave controlled rectifier circuit.

An alternative form for I_{av} may be obtained by eliminating the exponential component between Eqs. (3.52) and (3.49), but this does not appear to render any computational advantage.

The rms value I_L of the supply current is defined by

$$I_L^2 = \frac{1}{2\pi} \int_\alpha^{\theta_c+\alpha} i_L^2(\omega t)\, d\omega t \tag{3.53}$$

The substitution of Eq. (3.46) into Eq. (3.53) yields an expression containing exponential terms which cannot be eliminated by the use of the relationships Eq. (3.47) or (3.49)

$$\begin{aligned}
I_L^2 = \frac{E_m^2}{4\pi|Z|^2}\Big[&\theta_c - \sin\theta_c\,\cos(2\alpha - 2\Phi + \theta_c) - \sin^2(\theta_c - \Phi)\tan\Phi\varepsilon^{-2\cot\Phi\alpha} \\
&+ \sin^2(\alpha - \Phi)\tan\Phi \\
&+ 4\sin(\theta_c - \Phi)\sin\Phi\sin(\theta_c + \alpha)\varepsilon^{-\cot\Phi\alpha} - 4\sin\alpha\sin\Phi\sin(\alpha - \Phi)\Big]
\end{aligned} \tag{3.54}$$

When $\alpha = 0$, Eq. (3.54) reduces to Eq. (2.63).

Conduction angle θ_c can be extended, and thus the average and rms load currents increased by the use of a freewheel diode FWD across the load (Fig. 3.15). If the load is sufficiently inductive the load current will become continuous. Instant commutation (i.e., switch-off) of the thyristor T will then occur at the instants where the supply voltage goes negative.

In general the load current can be thought to consist of two modal elements due to two modes of circuit operation. While the freewheel diode conducts, the load voltage is zero and the current decays exponentially with a time constant $\tau = L/R$. Therefore, for $\pi < \omega t < 2\pi + \alpha$, in Fig. 3.16,

$$i_L(\omega t) = i_L(\pi)\varepsilon^{\frac{R}{\omega L}(\omega t - \pi)} = i_L(\pi)\varepsilon^{-\cot\Phi(\omega t - \pi)} \tag{3.55}$$

While the switch conducts in Fig. 3.15, the diode blocks and the operation is similar to that of Fig. 3.12. Therefore, for $\alpha \leq \omega t \leq \pi$, in Fig. 3.16, the current is defined by the expression Eq. (3.46) plus a transient that has a value $i_L(\alpha)$ at $\omega t = \alpha$ and decays exponentially thereafter.

$$i_L(\omega t) = \frac{E_m}{|Z|}\left[\sin(\omega t - \Phi) - \sin(\alpha - \Phi)\varepsilon^{-\cot\Phi(\omega t - \alpha)}\right]$$

$$+ i_L(\alpha)\varepsilon^{-\cot\Phi(\omega t - \alpha)} \tag{3.56}$$

When $\omega t = \pi$, the two modal current equations, Eqs. (3.55) and (3.56), are equal. Equation (3.56) then becomes

$$i_L(\pi) = \frac{E_m}{|Z|}\left\{\sin(\pi - \Phi) - \varepsilon^{-\cot\Phi(\pi - \alpha)}\left[i_L\alpha - \sin(\alpha - \Phi)\right]\right\} \tag{3.57}$$

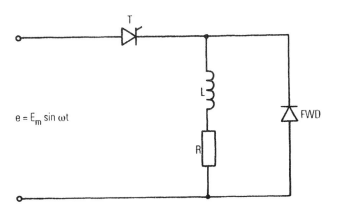

FIG. 15   Single-phase, full-wave controlled rectifier circuit with freewheel diode.

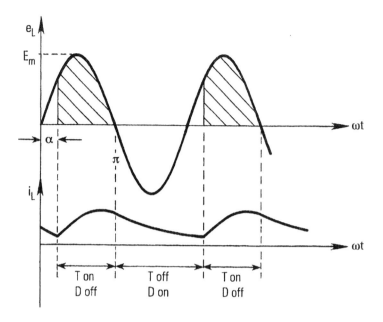

Fig. 16 Steady-state load current and voltage waveforms for the freewheel diode circuit of Fig. 3.15; $\alpha \approx 50°$.

For cyclic operation $i_L(\alpha) = i_L(2\pi + \alpha)$, so that in Eq. (3.55),

$$i_L(2\pi + \alpha) = i_L(\alpha) = i_L(\pi)\varepsilon^{-\cot \Phi (\pi + \alpha)} \tag{3.58}$$

Equations (3.57) and (3.58) are simultaneous and can be solved to give values for $i_L(\pi)$ and $i_L(\alpha)$. For a complete supply voltage cycle $\alpha \leq \omega t \leq 2\pi + \alpha$, the load current is given by adding, Eqs. (3.55) and (3.56):

$$i_L(\omega t) = \frac{E_m}{|Z|} \left\{ \sin(\omega t - \Phi) - \left[\sin(\alpha - \Phi) - \frac{i_L(\alpha)|Z|}{E_m} \right] \varepsilon^{-\cot \Phi(\omega t - \alpha)} \right\}_{\alpha}^{\pi}$$
$$+ i_L(\pi)\varepsilon^{-\cot \Phi(\omega t - \pi)} \Big|_{\pi}^{2\pi + \alpha} \tag{3.59}$$

The average and rms values of $i_L(\omega t)$ can be obtained from the respective defining integrals in Eqs. (3.52) and (3.53).

3.2.2 Single-Phase, Full-Wave Controlled Circuits

Each of the circuit configurations of Fig. 3.4 can be used for full-wave rectification of a series R-L load, as in Fig. 3.17. The cost of a diode switch is much less than that of a controlled switch of the same rating, and it also does not require an

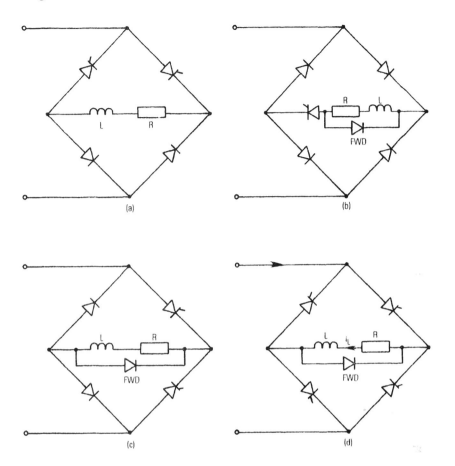

FIG. 17 Single-phase, full-wave controlled rectifier circuits with series R-L load.

associated firing circuit. From the cost viewpoint therefore it is desirable to use a circuit with few controlled switches. The single-switch circuit of Fig. 3.17b is likely to be the most economic. For comprehensive control, however, including the facility to act as an inverter, it is necessary to use the fully controlled bridge circuit of Fig. 3.17d.

In the arrangement of Fig. 3.17 the two diodes freewheel the load current when one or both of the switches is in extinction. Continuity of the load current is enhanced, with R-L loads, by the use of a freewheel diode across the load impedance, as shown in Fig. 3.17b–d. In many applications the load-side inductance is made deliberately large to ensure a continuous flow of load current. It is also usually desirable that the load current be largely dc (i.e., the load current should have the largest realizable average value) with low ripple content. The ideal load current ripple factor, Eq. (2.11) is zero, which occurs when the load current is pure dc.

The four-switch circuit of Fig. 3.17d has the load voltage and current waveforms shown in Fig. 3.18 for $\alpha = 30°$, with large L. Commutation of the two conducting switches occurs naturally at the end of a supply voltage half cycle. The load voltage is seen to be the full-wave equivalent of the corresponding half-wave voltage waveform of Fig. 3.16. From Fig. 3.18 it is seen that the average load voltage is given by

$$E_{av} = \frac{1}{\pi} \int_{\alpha}^{\pi} E_m \sin \omega t \, d\omega t = \frac{E_m}{\pi} \left(1 + \cos \alpha \right) \tag{3.60}$$

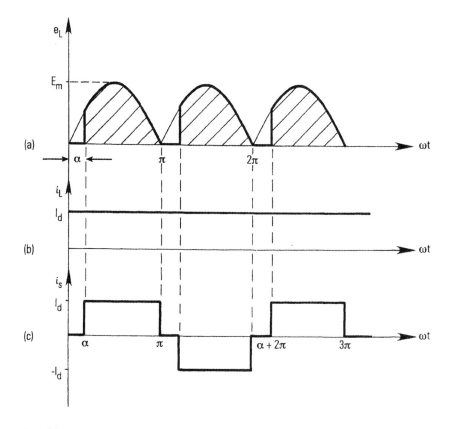

Fig. 18 Voltage and current waveforms for the single-phase, full-wave circuit of Fig. 3.17d, large L, $\alpha = 30°$.

Also,

$$I_{av} = I_d \qquad (3.61)$$

The output power is dissipated in resistor R. With ideal switches and a lossless inductor L the output power also equals the input power

$$P_{out} = P_{in} = I_d^2 R \qquad (3.62)$$

The input current is defined by the expression

$$i_s(\omega t) = I_d \Big|_\alpha^\pi - I_d \Big|_{\pi+\alpha}^{2\pi} \qquad (3.63)$$

The rms value of the input current is given by

$$\begin{aligned} I_s &= \sqrt{\frac{1}{2\pi} \int_0^{2\pi} i_s^2(\omega t)\, d\omega t} \\ &= \sqrt{\frac{1}{\pi} \int_\alpha^\pi I_d^2\, d\omega t} \\ &= I_d \sqrt{\frac{\pi - \alpha}{\pi}} \end{aligned} \qquad (3.64)$$

When $\alpha = 0$, $I_s = I_d$. The fundamental component of the supply current is found to have the rms magnitude

$$I_{s_1} = \frac{2\sqrt{2}}{\pi} I_d \cos\frac{\alpha}{2} \qquad (3.65)$$

By inspection of the supply current waveform in Fig. 3.18, it is seen that the displacement angle is ψ_1 is $\alpha/2$. The input power may therefore be alternatively defined as

$$\begin{aligned} P &= E I_{s_1} \cos\psi_1 \\ &= E \frac{2\sqrt{2}}{\pi} I_d \cos^2\frac{\alpha}{2} \end{aligned} \qquad (3.66)$$

Note that real or average power P is associated only with the combinations of voltage and current components of the same frequency. Since the supply voltage is sinusoidal and therefore of single frequency, it combines only with the fundamental (supply frequency) component of the input current. The combination of the fundamental voltage component with higher harmonic components of the current produces time-variable voltamperes but zero average power.

The power factor of the bridge circuit is obtained using Eqs. (3.64) and (3.66)

$$PF = \frac{P}{EI_s}$$

$$= \frac{E \dfrac{2\sqrt{2}}{\pi} I_d \cos^2 \alpha/2}{EI_d \sqrt{\dfrac{\pi-\alpha}{\pi}}}$$

$$= 2\sqrt{\frac{2}{\pi}} \frac{\cos^2 \alpha/2}{\sqrt{\pi-\alpha}} \qquad (3.67)$$

When $\alpha = 0$, it is seen that $PF = 2\sqrt{2}/\pi = 0.9$. The displacement factor and distortion factor components of the power factor can be obtained via the Fourier components $a_1 b_1$ of i_s, (ωt), Sec. 3.13.

When a freewheel diode is not used the symmetrical triggering of opposite pairs of switches in Fig. 3.17d (or Fig. 3.4d) results in the waveforms of Fig. 3.19, for a highly inductive load. Compared with Fig. 3.16 the load current is unchanged and so therefore is the power dissipation. But note that the rms value of the supply current is increased so that the power factor is reduced. It is seen from Fig. 3.19 that the average load voltage is given by

$$E_{av} = \frac{1}{\pi} \int_{\alpha}^{\pi+\alpha} E_m \sin \omega t \, d\omega t = \frac{2E_m}{\pi} \cos \alpha \qquad (3.68)$$

Also, the average voltage of the load inductor is zero so that

$$I_{av} = I_d = \frac{E_{av}}{R} \qquad (3.69)$$

The rms value E_L of the load voltage in Fig. 3.19 is given by

$$E_L^2 = \frac{1}{\pi} \int_{\alpha}^{\pi+\alpha} E_m^2 \sin^2 \omega t \, d\omega t$$

Or

$$E_L = \frac{E_m}{\sqrt{2}} \qquad (3.70)$$

The harmonic nature of the load current in Fig. 3.17 could be obtained by calculating the Fourier series for the periodic function i_L (ωt), for an arbitrary R-L load,

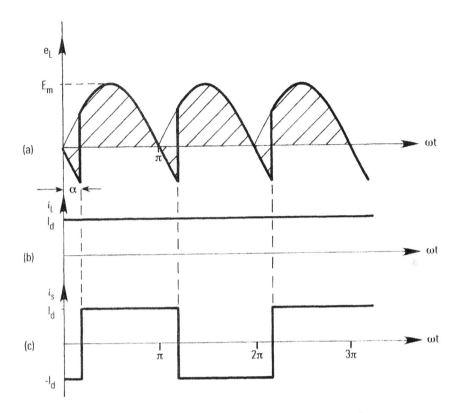

Fig. 19 Voltage and current waveforms for the single-phase, full-wave circuit of Fig. 3.17d without freewheel diode FWD, large L, $\alpha = 30°$.

with or without the diode. Alternatively, the various harmonic terms of the Fourier series for the periodic voltage e_L (ωt) (Fig. 3.18 or 3.19) can be applied since they are valid for any load impedance. The Fourier series of the load voltage may be written

$$e_L(\omega t) = E_{av} + \sum_1^n \hat{E}_n \cos(n\,\omega t - \psi_n)$$
$$= E_{av} + \hat{E}_2 \cos(2\omega t - \psi_2) + \hat{E}_4 \cos(4\omega t - \psi_4)$$
$$+ \hat{E}_6 \cos(6\omega t - \psi_6) + \cdots \tag{3.71}$$

In the case of waveform e_L in Fig. 3.19a,

$$\hat{E}_2 = \frac{E_m}{2\pi}\sqrt{1 + \frac{1}{9} - \frac{2}{3}\cos 2\alpha}$$

$$\hat{E}_4 = \frac{E_m}{2\pi}\sqrt{\frac{1}{9} + \frac{1}{25} - \frac{2}{15}\cos 2\alpha}$$

$$\hat{E}_6 = \frac{E_m}{2\pi}\sqrt{\frac{1}{25} + \frac{1}{49} - \frac{2}{35}\cos 2\alpha}$$

...

$$\hat{E}_n = \frac{E_m}{2\pi}\sqrt{\frac{1}{(n+1)^2} + \frac{1}{(n-1)^2} - \frac{2\cos 2\alpha}{n^2 - 1}} \tag{3.72}$$

and

$$\psi_2 = \tan^{-1}\frac{\sin\alpha - (\sin 3\alpha)/3}{\cos\alpha - (\cos 3\alpha)/3}$$

$$\psi_4 = \tan^{-1}\frac{(\sin 3\alpha)/3 - (\sin 5\alpha)/5}{(\cos 3\alpha)/3 - (\cos 5\alpha)/5}$$

...

$$\psi_n = \tan^{-1}\frac{[\sin(n-1)\alpha]/(n-1) - [\sin(n+1)\alpha]/(n+1)}{[\cos(n-1)\alpha]/(n-1) - [\cos(n+1)\alpha]/(n+1)} \tag{3.73}$$

When the harmonic voltage terms of Eq. (3.71) are applied to the series R-L load of Fig. 3.17, the following series is obtained for the load current

$$i_L(\omega t) = I_{av} + \sum_1^n \hat{I}_n \cos(n\omega t - \psi_L - \Phi_L)$$

$$= I_{av} + \hat{I}_2 \cos(2\omega t - \psi_2 - \Phi_2) + \hat{I}_4 \cos(4\omega t - \psi_4 - \Phi_4)$$

$$+ \hat{I}_6 \cos(6\omega t - \psi_6 - \Phi_6) + \cdots \tag{3.74}$$

where I_{av} is given by Eq. (3.69)
and

$$\hat{I}_2 = \frac{\hat{E}_2}{\sqrt{R^2 + (2\omega L)^2}}$$

$$\hat{I}_4 = \frac{\hat{E}_4}{\sqrt{R^2 + (4\omega L)^2}}$$

...

$$\hat{I}_n = \frac{\hat{E}_n}{\sqrt{R^2 + (n\omega L)^2}}$$ (3.75)

Also,

$$\Phi_2 = \tan^{-1}\frac{2\omega L}{R}$$

$$\Phi_4 = \tan^{-1}\frac{4\omega L}{R}$$

...

$$\Phi_n = \tan^{-1}\frac{n\omega L}{R}$$ (3.76)

A close approximation to the rms current I_L for any R-L load can be obtained by using the square law relationship of Eq. (2.12). If the load impedance is highly inductive, the harmonic terms in Eqs. (3.71)–(3.76) are negligibly small. The load current is then constant at the value I_L or I_d, which also becomes its rms value I_L.

A further application of this bridge circuit in the form of a dual converter is described in Sec. 12.2.1.

3.2.3 Worked Examples

Example 3.5   A series semiconductor switching circuit, Fig. 3.12, has a load in which the resistance is negligibly small. Deduce and sketch the waveforms of the load voltage and current for a firing angle α smaller than the load phase angle Φ. What are the average values of the load current and voltage?

When $\Phi = 90°$, $\cot\Phi = 0$ and Eq. (3.46) reduces to

$$i_L(\omega t) = \frac{E_m}{|Z|}(-\cos\omega t + \cos\alpha)$$ (3.77)

The variation of $i_L(\omega t)$ is shown in Fig. 3.20 for the case when $\alpha = 60°$. It is seen that the current waveform is symmetrical about π so that extinction angle x is given by

$$x = 2\pi - \alpha$$

From Eq. (3.48), therefore,

$$\theta_c = 2(\pi - \alpha)$$ (3.78)

Equation (3.52) is indeterminate for $\Phi = 90°$, but a solution for the time average current can be obtained by integrating Eq. (3.77).

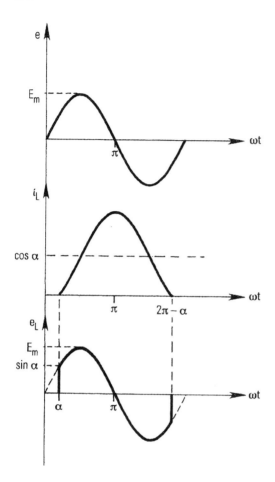

FIG. 20 Voltage and current waveforms for the single-phase, full-wave controlled rectifier circuit with highly inductive load; $\alpha = 30°$.

$$i_{av} = \frac{1}{2\pi} \int_\alpha^{\alpha+\theta_c} \frac{E_m}{|Z|} (\cos\alpha - \cos\omega t)\, d\omega t$$

$$= \frac{E_m}{2\pi|Z|} \int_\alpha^{2\pi-\alpha} \frac{E_m}{|Z|} (\cos\alpha - \cos\omega t)\, d\omega t$$

$$= \frac{E_m}{\pi|Z|} \big[(\pi - \alpha)\cos\alpha + \sin\alpha\big]$$

When $\alpha = 0$, $I_{av} = E_m/\omega L$. The output average current is then constant at its maximum realizable value, which represents ideal rectifier operation. The voltage across the load inductor may be obtained by differentiating the current expression Eq. (3.77), noting that $|Z| = \omega L$:

$$e_L(\omega t) = L\frac{di_L}{dt} = \omega L\frac{di_L}{d(\omega t)}$$

$$= E_m \sin \omega t \qquad \text{for } \alpha < \omega t < 2\pi - \alpha$$

The average value E_{av} of $e_L(\omega t)$ may be obtained by the usual integration method and is found to be zero, as can be seen by inspection in Fig. 3.20.

Example 3.6   In the series R-L circuit of Fig. 3.12$R = 25\ \Omega$, $L = 150$ mH. The supply voltage is given by $e = E_m \sin \omega t$, where $E_m = 400$, V at a frequency of 50 Hz. Calculate the average load current for the SCR firing angles (1) $30°$ and (2) $120°$.

At 50 Hz,

$$\omega L = 100\pi \frac{150}{1000} = 47.12\ \Omega$$

$$\Phi = \tan^{-1}\frac{\omega L}{R} = \tan^{-1}\frac{47.12}{25} = 62°$$

$$|Z| = \sqrt{R^2 + \omega^2 L^2} = 53.34\ \Omega$$

$$\cot \Phi = \cot 62° = 0.532$$

$$\tan \Phi = \tan 62° = 1.88$$

$$\sin(\alpha - \Phi) = \sin(30° - 62°) = -0.53$$
$$\sin(\alpha - \Phi) = \sin(120° - 62°) = +0.85$$
$$\cos(\alpha - \Phi) = \cos(-32°) = +0.85$$

or

$$\cos(\alpha - \Phi) = \cos(58°) = +0.53$$

1. $\alpha = 30°$: In this case $\alpha < \Phi$ and an estimation of the conduction angle can be made from Eq. (3.51)

$$\theta_c = \pi + \Phi - \alpha + \Delta = 180° + 62° - 30° + 7° \text{ (say)} = 219°$$

The value $\theta_c = 219°$ is used as a first guess in Eq. (3.49). By iteration it is found that $\theta_c = 216.5°$. The accuracy with which θ_c can be read from Fig. 3.18 is sufficient for most purposes.

$$\cos\theta_c = -0.804$$

$$\sin\theta_c = -0.595$$

$$I_{av} = \frac{400}{2\pi \times 53.34}[1.804(0.85)-(-0.595-1.88)(-0.53)+0.85(1.88)(0.134)]$$

$$= 1.194 \times 3.059$$

$$= 3.65 \text{ A}$$

2. $\alpha = 150°$: In this case $\alpha > \Phi$ and the use of Eq. (3.50) gives

$$\theta_c = \pi + \Phi - \alpha - \Delta$$

$$= 180° + 62° - 150° - 15° \text{ (say)} = 77°$$

The characteristics of Fig. 3.14 suggest that this figure is high and should be about 60°. Iteration from Eq. (3.49) gives a value $\theta_c = 64°$. In Eq. (3.52),

$$\cos\theta_c = 0.44$$

$$\sin\theta_c = 0.9$$

$$I_{av} = \frac{400}{2\pi \times 53.34}[0.56(0.53)+(0.9-1.88)(0.85)+0.85(1.88)(0.552)]$$

$$= 1.194 \times 0.346 = 0.413 \text{ A}$$

Example 3.7 A half-wave, controlled rectifier circuit has a series R-L load in which $\Phi = \tan^{-1}(\omega L/R) \approx 80°$. The ideal single-phase supply voltage is given by $e = E_m \sin \omega t$. Explain the action for a typical steady-state cycle when $\alpha = 60°$ and the circuit includes a freewheel diode (Fig. 3.15). What effect does the diode have on circuit power factor?

The load voltage and current are shown in Fig. 3.16. The supply current i_s (ωt) is given by the portions of the current curve between $\alpha \rightarrow \pi$, $2\pi + \alpha \rightarrow 3\pi$, etc., noting that $i_s(\alpha) = i_s(2\pi + \alpha) = 0$. The nonzero value of the load current at $\omega t = \alpha$ is due to residual current decaying through the diode during the extinction of the thyristor switch. In the presence of the diode, the energy stored in the magnetic field of the inductor is dissipated in resistor R rather than being returned to the supply. Current and power flow from the supply to the load only occur during the conduction intervals of the thyristor. Because of the significantly increased rms load current (compare the load currents in Figs. 3.14 and 3.16) however, the load power dissipation is significantly increased. All of this power must come from the supply, although not at the instants of time in which it is dissipated. The supply voltage remains sinusoidal at all times.

The power factor, seen from the supply point, is

$$PF = \frac{P}{EI}$$

The rms supply voltage $E = E_m/\sqrt{2}$ is constant. The rms supply current I probably increases by (say) 20%. But the power dissipation can be assessed in terms of the rms value of the load current. Comparing $i_L (\omega t)$ in Figs. 3.13 and 3.16 suggests that rms value I_L is at least doubled and the power increases at least four times. The presence of the freewheel diode therefore causes the power factor to increase.

Example 3.8   A single-phase, full-wave bridge circuit, Fig. 3.21, has four ideal thyristor switches and a highly inductive load. The electrical supply is ideal and is represented by $e = E_m \sin \omega t$. Sketch waveforms of the load current, supply current and load voltage for $\alpha = 60°$ Calculate the rms value of the supply current and the power factor of operation, in terms of α.

The waveforms of operation with a highly inductive load are given in Fig. 3.19. The supply current is represented by the relation

$$i_s(\omega t) = I_d \left.\right|_\alpha^{\pi+\alpha} - I_d \left.\right|_{0,\pi+\alpha}^{\alpha,2\pi}$$

This has an rms value defined by

$$I_s = \sqrt{\frac{1}{2\pi} \int_0^{2\pi} i_s^2(\omega t)\, d\omega t}$$

Therefore,

$$I_s^2 = \frac{1}{2\pi}\left[\int_\alpha^{\pi+\alpha} I_d^2\, d\omega t + \int_{0,\pi+\alpha}^{\alpha}(-I_d)^2\, d\omega t\right]$$

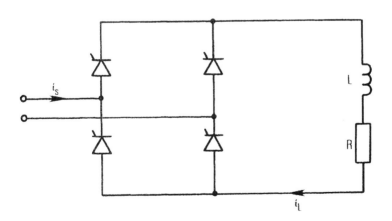

FIG. 21   Single-phase, full-wave controlled rectifier circuit with highly inductive load.

The rms value of the negative parts of the wave is equal to the rms value of the positive parts so that

$$I_s^2 = \frac{1}{\pi}\int_\alpha^{\pi+\alpha} I_d^2\, d\omega t = \frac{1}{\pi}I_d^2(\pi+\alpha-\alpha)$$

Therefore,

$$I_s = I_d$$

As α varies, the waveform of $i_s\,(\omega t)$ is unchanged and so is its rms value, but the switchover from positive to negative value occurs at the firing points.

Power dissipation is determined by the rms value of the load current

$$P = I_L^2 R = I_d^2 R$$

The power factor is found to be

$$PF = \frac{P}{EI} = \frac{I_d^2 R}{EI_d} = \frac{I_d R}{E}$$

where $E = E_m\sqrt{2}$.

But from Eq. (3.68),

$$I_d = \frac{E_{av}}{R} = \frac{2E_m}{\pi R}\cos\alpha$$

The power factor is therefore

$$PF = \frac{2\sqrt{2}}{\pi}\cos\alpha$$

When $\alpha = 0$, the $PF = 2\sqrt{2}/\pi$, which agrees with the value for a full-wave diode bridge.

Example 3.9   In the single-phase, full-wave rectifier of Fig. 3.21 the load consists of $R = 10\ \Omega$ and $L = 50$ mH. The ideal sinusoidal supply voltage is defined as $e_s = 240\sqrt{2}\sin\omega t$ at 50 Hz. Calculate values for the average and rms load currents, the power dissipation and the power factor at the supply terminals if the thyristor firing angle $\alpha = 45°$.

The circuit of Fig. 3.21 is seen to be a topological rearrangement of Fig. 3.17d, without the freewheel diode. The load voltage has the segmented sinusoidal form of Fig. 3.19a for any R-L load if the load current is continuous. The load current will be of some waveform intermediate between that of Fig. 3.19b (which is only valid for highly inductive loads) and the corresponding segmented sinusoidal waveform (not given) obtained with purely resistive loads.

At supply frequency the phase-angle of the load impedance is given by

$$\Phi = \tan^{-1}\frac{\omega L}{R}$$

$$= \tan^{-1}\frac{100 \times \pi \times 50}{1000 \times 10} = \tan^{-1} 1.57 = 57.5°$$

Since $\alpha < \Phi$, the mode of operation is that of continuous load current and the equations of the present Sec. 3.2.2 are valid.

The average load current is given by Eqs. (3.68) and (3.69):

$$I_{av} = \frac{2E_m}{\pi R}\cos\alpha = \frac{2 \times 240\sqrt{2}}{\pi \times 10} \times \cos 57.5° = 11.61\ A$$

In order to calculate the rms load current, it is necessary to calculate its Fourier harmonics. From Eq. (3.72), $\cos 2\alpha = \cos 90° = 0$, and the peak values of the lower order load voltage harmonics are

$$\hat{E}_2 = 56.92\ V \qquad \hat{E}_4 = 20.99\ V \qquad \hat{E}_6 = 13.27\ V$$

Also, in Eq. (3.62) the following impedances are offered by the load at the specified harmonic frequencies.

$$Z_2 = \sqrt{R^2 + (2\omega L)^2} = 32.97\Omega$$

$$Z_4 = \sqrt{R^2 + (4\omega L)^2} = 63.62\Omega$$

$$Z_6 = \sqrt{R^2 + (6\omega L)^2} = 94.78\Omega$$

The peak values of the current harmonics are therefore

$$\hat{I}_2 = \frac{56.92}{32.97} = 1.726\ A$$

$$\hat{I}_4 = \frac{20.99}{63.62} = 0.33A$$

$$\hat{I}_6 = \frac{13.27}{94.78} = 0.14A$$

From Eq. (2.12) the rms value of the load current is given by

$$I_L^2 = I_{av}^2 + \frac{1}{2}\left(\hat{I}_2^2 + \hat{I}_4^2 + \hat{I}_6^2 + ...\right)$$

and

$$I_L = \sqrt{(11.61)^2 + \frac{1}{2}\left[(1.726)^2 + (0.33)^2 + (0.14)^2\right]}$$
$$= \sqrt{134.79 + 1.554} = 11.68 \ A$$

It is seen that the effects of the fourth and sixth harmonic current components on the total rms value are negligible. The load power is seen, from Eq. (3.64), to be

$$P_L = (11.68)^2 \times 10 = 1364.22 \ W$$

and this is also the power entering the circuit terminals, neglecting rectifier element losses. The rms value of the supply current is equal to the rms value of the load current. The power factor at the supply point is, therefore,

$$PF_s = \frac{P_L}{E_s I_s} = \frac{1364.22}{240 \times 11.68} = 0.487$$

This compares with the value cos Φ = cos $57.5°$ = 0.537 for the load impedance alone.

PROBLEMS

Single-Phase Controlled Rectifier Circuits: *R* Load

3.1 A single controlled rectifier controls the current to a resistive load in a single-phase sinusoidal circuit (Fig. 3.1). Derive expressions for the average value and rms value of the load current at an arbitrary firing angle α if $e = E_m \sin \omega t$. What is the lowest value of ripple factor for this system? Sketch the variations of I_{av} and I_{rms} versus firing angle.

3.2 Sketch the currents for the circuit of Fig. 3.22 if $e = E_m \sin \omega t$; $R_1 = R_2$ and $\alpha = 60°$. What is the average value of the supply current if $E_m = 100$ V and $R_1 = R_2 = 10 \ \Omega$?

3.3 In the circuit of Fig. 3.22; $R_1 = R_2 = 10 \ \Omega$ and the supply voltage is $e = E_m \sin \omega t$. Sketch, roughly in proportion, corresponding waveforms of the supply voltage and the three branch currents for a typical supply voltage cycle if $\alpha = 90°$. Calculate the power dissipation in the circuit for firing angles of $0°$ and $180°$ and thereby estimate the dissipation when $\alpha = 90°$. What harmonic frequency components would you expect to find in the supply current?

3.4 The circuit of Fig. 3.23 consists of identical resistors $R_1 = R_2 = 10 \ \Omega$, an ideal diode *D*, and an ideal SCR designated T supplied from an ideal

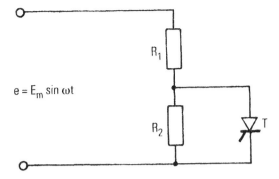

FIG. 22 Circuit for Problem 3.3.

power supply $e = E_m \sin \omega t$. Sketch, roughly to scale, corresponding waveforms of the supply voltage and the currents in R_1, R_2, D, and T if the thyristor firing angle is 90°. What is the average value of the current in R_1 for a supply cycle?

3.5 Derive expressions for the magnitude and phase angle of the fundamental (i.e., supply frequency) component of the voltage for the circuit of Fig. 3.1.

3.6 For the half-wave rectifier circuit of Fig. 3.1 derive an expression for the nth harmonic of the current in terms of α. If $\alpha = 90°$, what are the values of the first, second, and third harmonics?

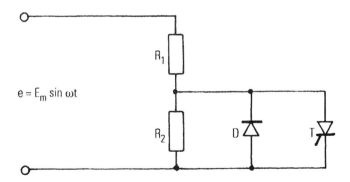

FIG. 23 Circuit for Problem 3.4.

3.7 Derive an expression for the average power dissipation in the half-wave resistor circuit of Fig. 3.1. At $\alpha = 90°$, what proportion of the total power dissipation is associated with the fundamental component of current?

3.8 An ideal voltage source $e = E_m \sin \omega t$ supplies power to a single-phase, half-wave controlled rectifier circuit with resistive load (Fig. 3.1). Obtain an expression for the circuit power factor in terms of E_m and thyristor switch firing angle α.

3.9 What is the effect on the supply current of connecting a capacitor across the supply terminals of a thyristor controlled resistive circuit (Fig. 3.3)? If $X_c = R$, is there a value of firing angle α that results in unity power factor operation?

3.10 A 100 Ω resistor is supplied with power from an ideal voltage source $e = 330 \sin 314t$ via a thyristor. At what value of firing angle α is the power dissipation one-third of the value for sinusoidal operation?

3.11 Show that the power factor of the single-phase, half-wave rectifier circuit of Fig. 3.1 is given by

$$PF = \frac{1}{\sqrt{2}} \sqrt{\frac{2(\pi - \alpha) + \sin 2\alpha}{2\pi}}$$

3.12 For the half-wave rectifier with resistive load (Fig. 3.1), show that the phase-angle of the fundamental component of the current, with respect to the supply voltage, is given by

$$\psi_1 = \tan^{-1} \frac{\cos 2\alpha - 1}{2(\pi - \alpha) + \sin 2\alpha}$$

Sketch the variation of ψ_1 with firing angle α for $0 < \alpha < \pi$.

3.13 The single-phase half-wave rectifier circuit of Fig. 3.1 is compensated by a terminal capacitance as in Fig. 3.3, where $X_c = R$. Sketch waveforms of the load current, capacitor current, and supply current for $\alpha = 90°$. Is it possible to assess the effect on the overall power factor from a consideration of the three waveforms?

3.14 Derive expressions for the current distortion factor and current displacement factor of the half-wave rectifier circuit of Fig. 3.1. Sketch the variations of these for $0 < \alpha < \pi$. If a power factor correction capacitor is used with $X_c = R$, derive new expressions for the distortion factor and displacement factor of the input current. How has the connection of the capacitor affected (a) the distortion factor and (b) the displacement factor?

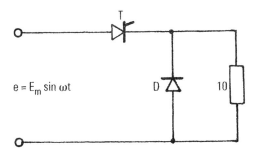

FIG. 24   Circuit for Problem 3.15.

3.15   A thyristor switch T controls the load current in the resistor circuit of Fig. 3.24. What is the effect on the load current of connecting diode D in the circuit?

3.16   A resistive load is supplied through a thyristor connected in the secondary circuit of a single-phase transformer with low leakage reactance (Fig. 3.25). Sketch the forms of the load current, supply current, and thyristor voltage over a supply voltage period.

3.17   A full-wave controlled rectifier (Fig. 3.11a) operates from a sinusoidal supply $e = E_m \sin \omega t$. Derive an expression for the ripple factor (RF) of the load voltage with resistive load. Calculate values of this at the firing angles 0, 30°, 60°, 90°, 120°, and 180°, and sketch the variation of RF versus α.

3.18   Show that the power factor of a single-phase, full-wave controlled bridge rectifier circuit with resistive load is given by

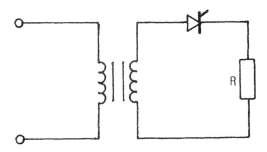

FIG. 25   Circuit for Problem 3.16.

$$PF = \sqrt{\frac{1}{2\pi}\left[2(\pi-\alpha)+\sin 2\alpha\right]}$$

Sketch the variation of PF with firing-angle α over the range $0 < \alpha < \pi$.

3.19 Show that the distortion factor and the displacement factor of the supply current to the full-wave bridge circuits of Fig. 3.4 are given by Eqs. (3.44) and (3.35), respectively. What are the values of the distortion factor, displacement factor, and power factor at $\alpha = 90°$? What value of shunt connected terminal capacitance would give the maximum realistic improvement of power factor at $\alpha = 90°$ if $E_m = 400$ V, $R = 100$ Ω, and $f = 50$ Hz?

3.20 The single-phase bridge rectifier circuit of Fig. 3.4b has the load current waveform of Fig. 3.5. Calculate the average and rms values of this waveform and hence calculate the rms value of the ac ripple components.

3.21 In the circuit of Fig. 3.4(d) the current in load resistor R is controlled by the switching of two ideal thyristors. The ideal power supply has an instantaneous voltage $e = E_m \sin \omega t$.

a. Sketch waveforms of the load current and supply current when each thyristor is gated at $\alpha = 90°$ of its respective anode voltage.

b. If one thyristor fails by open circuit, sketch the resulting supply current waveform, and calculate its average value at $\alpha = 90°$ if $R = 10$ Ω and $E_m = 100$ V.

3.22 The single-phase bridge circuits of Fig. 3.4 are alternative configurations for achieving the same circuit performance. What are the advantages and disadvantages of each circuit?

3.23 The single-phase, full-wave bridge circuit of Fig. 3.4(d) has a resistive load. Opposite pairs of thyristors in the bridge arms are triggered symmetrically at a firing-angle $\alpha = 45°$. Sketch waveforms of the load current, supply current, and load voltage.

3.24 For the single-phase, full-wave bridge of Problem 3.23, obtain expressions for the average and rms values of the load current and hence the load current ripple factor. Calculate the power dissipation if $e = E_m \sin \omega t$, where $E_m = 400$ V, $R = 100$ Ω, and $\alpha = 45°$.

3.25 Calculate the Fourier coefficients a and b for the fundamental component of the line current in Problem 3.23. Hence calculate the fundamental supply current and the distortion factor in terms of E_m, R and α.

3.26 Obtain an expression for the power factor of the resistively loaded, full-wave, single-phase bridge of Problem 3.23. Sketch the variation of power factor with firing angle for $0 < \alpha < \pi$.

Single-Phase Controlled Rectifier Circuits: *R-L* Load

3.27   Deduce and sketch waveforms of the current, inductor voltage, and rectifier voltage when a thyristor is gated at angle α in the series R-L circuit of Fig. 3.12 if $\alpha > \Phi$, where $\Phi = \tan^{-1}(\omega L/R)$

3.28   Derive expressions for the average and rms values of the rectified current pulses in the circuit of Fig. 3.12 for an arbitrary value of firing angle α $(> \Phi)$.

3.29   For the single-phase series R-L circuit (Fig. 3.12), show that if $\alpha > \Phi$, the current extinction occurs at an angle $\omega t = x$ after the supply voltage is zero, where $\sin(x - \Phi) = \sin(\alpha - \Phi)\varepsilon^{-\cot \Phi(x - \Phi)}$. Calculate the value of x if $\Phi = 30°$ and α is (a) $60°$, (b) $90°$, and (c) $120°$.

3.30   Sketch the current waveform for a series R-L circuit controlled by a series thyristor. If the load phase angle ϕ to fundamental currents is $60°$, what is the approximate extinction angle if $\alpha = 30°$? What difference is made to the current waveforms by the connection of freewheel diode as in Fig. 3.15?

3.31   In the circuit of Fig. 3.12, $R = 10\ \Omega$ and $L = 50$ mH. If $e_s = 240\sqrt{2}$ $\sin \omega t$ at 50 Hz, calculate (a) average and rms values of the current, (b) ripple factor, (c) power dissipation, (d) power factor for a thyristor firing angle $\alpha = 60°$.

3.32   In Problem 3.31 a flywheel diode is now connected across the load impedance, as in Fig. 3.15. Calculate the average value of the load current.

3.33   For the circuit of Fig. 3.15, sketch waveforms of the supply voltage, thyristor voltage, load voltage, and the three branch currents for $\alpha = 60°$, assuming continuous load current.

3.34   Repeat Problem 3.33 for a case when the combination of thyristor firing angle and load impedance phase angle are such that the load current is discontinuous.

3.35   In the circuit of Fig. 3.15, $R = 10\ \Omega$, $L = 50$ mH and $e = E_m \sin \omega t$ at 50 Hz. If a capacitor is connected across the supply terminals, is this likely to give any power factor correction?

3.36   A single-phase, full-wave bridge (Fig. 3.17d) has a highly inductive load. Sketch waveforms of the load voltage and current (a) with the freewheel diode and (b) without the freewheel diode. In each case derive an expression for the average load voltage in terms of R, α, and E_m if $e = E_m \sin \omega t$.

3.37 For the single-phase bridge circuit of Fig. 3.17d, with highly inductive load, obtain an expression for the rms value of the supply current and hence show that the power factor is given by Eq. 3.67.

3.38 For the single-phase, full-wave bridge circuit of Fig. 3.17d with highly inductive load, calculate the Fourier coefficients a_1 and b_1 of the fundamental component of the supply current. Hence calculate the distortion factor, displacement factor, and power factor.

3.39 For the single-phase full-wave bridge circuit of Fig. 3.17d, a capacitor C is connected across the terminals of the ideal supply. If the capacitor is of such a value that $I_c = I_d$, sketch waveforms of the supply current i_s (ωt) with and without the capacitor. Do the waveforms indicate if any power factor correction is likely to be obtained?

3.40 A single-phase, full-wave bridge circuit (Fig. 3.17d) has the waveforms of Fig. 3.18 when the load is highly inductive. Obtain expressions for the rms value I_{s_1} and displacement angle ψ_1 of the fundamental component i_{s_1} (ωt) of the supply current. Sketch fundamental component i_{s_1} (ωt) onto the waveform i_s (ωt) in Fig. 3.18c.

3.41 Use the values of I_{s_1} and ψ_1 from Problem 3.40 to obtain an expression for the reactive voltamperes Q entering the circuit where $Q = EI_{s_1} \sin \psi_1$.

3.42 The total apparent voltamperes S entering the circuit of Fig. 3.17 from the sinusoidal supply is given by

$$S^2 = E^2 I_s^2 = P^2 + Q^2 + D^2$$

where P is the average power, Q is the reactive voltamperes, and D is the distortion (or harmonic) voltamperes. Show that for a highly inductive load the distortion voltamperes is given by

$$D^2 = E^2 I_d^2 \left(\frac{\pi - \alpha}{\alpha} - \frac{8}{\pi^2} \cos^2 \frac{\alpha}{2} \right)$$

3.43 A single-phase bridge circuit (Fig. 3.21) is supplied from an ideal supply $e = E_m \sin \omega t$. If the load is highly inductive, deduce the waveform of the supply current and calculate the Fourier components a_1 and b_1, of its fundamental component. Hence derive expressions for the distortion factor, displacement factor, and power factor.

3.44 Repeat Problem 3.40 for the circuit of Fig. 3.21.

3.45 Repeat Problem 3.41 for the circuit of Fig. 3.21.

3.46 Repeat Problem 3.42 for the circuit of Fig. 3.4d and hence show that

$$D^2 = E^2 I_d^2 \left[1 - \left(\frac{2\sqrt{2}}{\pi} \right)^2 \right]$$

3.47 Calculate the Fourier coefficients a_1 and b_1 for the load voltage of Fig. 3.19a. Hence show that the nth harmonic is given by Eq. (3.72) with the phase-angle given by Eq. (3.73).

3.48 In the single-phase, full-wave rectifier of Fig. 3.17d the load impedance elements are $R = 20 \ \Omega$ and $L = 100$ mH. The supply voltage is $e_s = 240 \ \sqrt{2} \ E_m \sin \omega t$ at 50 Hz. If the firing angle $\alpha = 30°$, calculate (a) values of average and rms currents, (b) average power dissipation, and (c) power factor. Use the Fourier series method.

3.49 A single-phase, full-wave bridge rectifier circuit (Fig. 3.21) has a series R-L load of phase angle $\phi = 60°$. If the thyristor switches are symmetrically triggered at $\alpha = 30°$, sketch the load current and load voltage waveforms, assuming ideal sinusoidal supply.

3.50 For Problem 3.49, how do the average and rms values of the load current waveform compare with corresponding values obtained from the half-wave circuit of Fig. 3.12?

4

Three-Phase, Half-Wave, Uncontrolled (Diode) Bridge Rectifier Circuits

Three-phase electricity supplies with balanced, sinusoidal voltages are widely available. It is found that the use of a three-phase rectifier system, in comparison with a single-phase system, provides smoother output voltage and higher rectifier efficiency. Also, the utilization of any supply transformers and associated equipment is better with polyphase circuits. If it is necessary to use an output filter this can be realized in a simpler and cheaper way with a polyphase rectifier.

A rectifier system with three-phase supply is illustrated by the general representation of Fig. 4.1. The instantaneous load voltage $e_L (\omega t)$ may have an amplitude ripple but is of fixed polarity. The output current $i_L (\omega t)$ is unidirectional but not necessarily continuous. In rectifier operation one seeks to obtain the maximum realizable average values of load voltage and current. This implies minimum load current ripple. An ideal rectifier circuit results in a continuous load current of constant amplitude and therefore constitutes an ideal dc supply. In order to illustrate the principles of polyphase rectifier operation some simple cases of three-phase diode rectifiers are chosen. The diode elements are assumed to be ideal voltage-actuated switches having zero conducting voltage drop.

4.1 RESISTIVE LOAD AND IDEAL SUPPLY

Figure 4.2a shows a three-phase, half-wave uncontrolled rectifier. The supply phase voltages are presumed to remain sinusoidal at all times which implies an

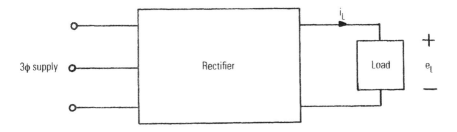

FIG. 1 General representation of a three-phase rectifier system.

ideal, impedanceless supply. The arrangement of Fig. 4.2a is electrically equivalent to three-single-phase, half-wave diode rectifiers in parallel, with a common
load R, in which the three supply voltages are mutually displaced in time phase
by 120°.

$$e_{aN} = E_m \sin \omega t \tag{4.1}$$

$$e_{bN} = E_m \sin(\omega t - 120°) \tag{4.2}$$

$$e_{cN} = E_m \sin(\omega t - 240°) \tag{4.3}$$

Since the circuit contains no inductance, it is not necessary to consider time
derivatives of the currents and the operating waveforms can be deduced by inspection.

Some waveforms for steady-state operation are given in Fig. 4.2. Consider
operation at a sequence of intervals of time. The terminology $t = 0 -$ means the
moment of time immediately prior to $t = 0$. Similarly, $t = 0 +$ refers to the
instant of time immediately following $t = 0$.

At $t = 0$, e_{aN} and e_{bN} are negative; $i_a = i_b = 0$. Diode D_c is conducting
i_c so that the common cathode has the potential of point c. Diodes D_a and D_b
are reverse-biased.

At $t = 0 +$, $e_{aN} = 0$ and e_{bN} remains negative. Voltage e_{cN} remains more
positive than e_{aN} so that the common cathode still has higher potential than points
a and b. Diodes D_a and D_b remain in extinction.

At $t = 30°$, $e_{aN} = e_{bN} =$ positive, and e_{bN} remains negative and so diode
D_b remains in extinction

At $t = 30° +$, e_{aN} is more positive than e_{cN}. Point a is higher in potential
than the common cathode so that D_a switches into conduction. The common
cathode then has the potential of point a. The cathode of D_c is then of higher
potential than the anode so that D_c switches off.

The switching sequence described above is cyclic, and a smooth transfer
is effected by which the three-phase supply lines are sequentially connected to

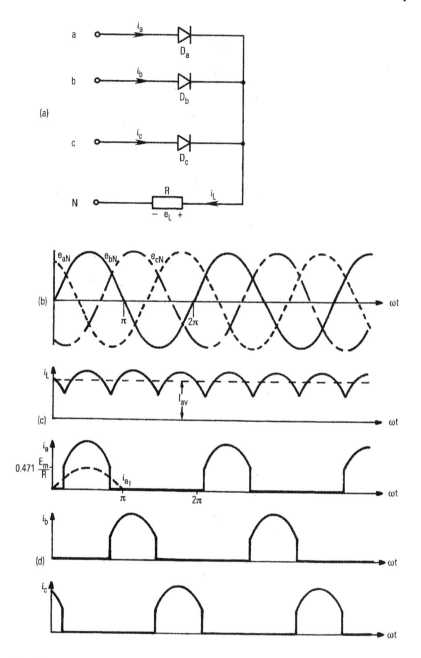

FIG. 2   Three-phase, half-wave diode rectifier with resistive load: (a) circuit connection, (b) phase voltages at the supply, (c) load current, and (d) supply currents.

the load. Each diode is, in turn, extinguished by natural commutation due to the cycling of the supply voltages. A supply current waveform consists of the middle pieces of the corresponding positive half-wave with a conduction angle of 120°. Instantaneous supply current $i_a(\omega t)$ (Fig. 4.2d), for example, is defined by

$$i_a = \frac{E_m}{R} \sin \omega t \left|_{\pi/6,\ \pi/6 + 2\pi, ...}^{5\pi/6,\ 5\pi/6 + 2\pi, ...}\right. \tag{4.4}$$

It can be seen in Fig. 4.2c that the load current contains a ripple of three times supply frequency, and consequently, this form of rectifier is sometimes known as a *three-pulse* system. The average value I_{av} of the load current may be obtained by taking the average value of any 120° interval in Fig. 4.2c or of certain 60° intervals.

$$\begin{aligned} I_{av} &= \text{average value of } i_L(\omega t) \\ &= \text{average value of } \frac{E_m}{R} \sin \omega t \left|_{30°}^{90°}\right. \\ &= \frac{6}{2\pi} \int_{30°}^{90°} \frac{E_m}{R} \sin \omega t \, d\omega t \\ &= \frac{3\sqrt{3}}{2\pi} \frac{E_m}{R} \\ &= 0.827 \frac{E_m}{R} \end{aligned} \tag{4.5}$$

Similarly, the average load voltage is seen to be

$$E_{av} = \frac{3\sqrt{3}}{2\pi} E_m = E_{av_0} = I_{av} R \tag{4.6}$$

It is of interest to compare the average current value with the corresponding values $0.318 \times (E_m/R)$ for single-phase, half-wave operation [Eq. (2.6)] and $0.637 \times (E_m/R)$ for single-phase, full-wave operation (Example 2.1).

The rms value of the load current is

$$\begin{aligned} I_L &= \sqrt{\frac{6}{2\pi} \int_{30°}^{90°} \frac{E_m^2}{R^2} \sin^2 \omega t \, d\omega t} \\ &= \frac{E_m}{R} \sqrt{\frac{4\pi + 3\sqrt{3}}{8\pi}} \\ &= 0.841 \frac{E_m}{R} \end{aligned} \tag{4.7}$$

The load dissipation is therefore

$$P_L = I_L^2 R = \frac{E_m^2}{R}\left(\frac{4\pi + 3\sqrt{3}}{8\pi}\right) = \frac{1}{\sqrt{2}}\frac{E_m^2}{R}$$

(4.8)

From the waveforms of Fig. 4.2d it can be seen that the rms value I_a of the supply current is given by

$$I_a^2 = \frac{1}{2\pi}\int_{30°}^{150°}\frac{E_m^2}{R^2}\sin^2\omega t \, d\omega t$$

this gives

$$I_a = 0.485\frac{E_m}{R}$$
$$= I_b = I_c = I_s$$

(4.9)

Comparison of Eqs. (4.7) and (4.9) shows that $I_s = I_L/\sqrt{3}$.

The power factor of the three-phase, half-wave rectifier is given by

$$PF = \frac{P_L}{3E_s I_s}$$

(4.10)

Substituting (4.8)(4.9) into (4.10) gives

$$PF = \frac{\left(1/\sqrt{2}\right)\times\left(E_m^2/R\right)}{3\left(E_m/\sqrt{2}\right)\times 0.485\times\left(E_m/R\right)} = 0.687$$

(4.11)

For the line current waveform i_a in Fig. 4.2d it is found that the Fourier coefficients of the fundamental component are given by

$$a_1 = 0$$
$$b_1 = 0.471\frac{E_m}{R}$$
$$c_1 = 0.471\frac{E_m}{R}$$
$$\psi_1 = 0°$$

(4.12)

The current displacement factor $\cos\psi_1$ of i_a (ωt) is therefore unity, and the power factor, Eq. (4.10), is entirely attributable to distortion effects rather than displacement effects. Part of the fundamental harmonic component i_{a_1} (ωt) is shown in Fig. 4.2d and is, because $\psi_1 = 0$, in time phase with its phase voltage

e_{aN}. This was also true in the single-phase, half-wave diode rectifier with resistive load described in Sec. 2.1.2. Because $\cos \psi_1 = 1$, no power factor correction of this displacement factor can be realized. Any power factor improvement would have to be sought by increasing the distortion factor (i.e., by reducing the degree of distortion of the supply current waveform).

The ripple factor for the load current waveform of Fig. 4.2c is obtained by substituting values of Eqs. (4.5) and (4.7) into Eq. (2.11).

$$
RF = \sqrt{\left(\frac{I_L}{I_{av}}\right)^2 - 1}
$$
$$
= \sqrt{1.034 - 1} = 0.185 \tag{4.13}
$$

The value 0.185 in Eq. (4.13) compares very favorably with values 1.21 for the single-phase, half-wave rectifier and 0.48 for the single-phase, full-wave rectifier, with resistive load.

4.1.1 Worked Examples

Example 4.1   A three-phase, half-wave, uncontrolled bridge circuit with resistive load $R = 25\ \Omega$ is supplied from an ideal, balanced three-phase source. If the rms value of the supply voltage is 240 V, calculate (1) average power dissipation, (2) average and rms load currents, and (3) rms supply current.

The circuit diagram is shown in Fig. 4.2a. It is customary to specify a three-phase voltage supply in terms of its line-to-line rms value. In the present case the rms voltage per phase is

$$
E_s = \frac{240}{\sqrt{3}} = 138.6\ \text{V}
$$

The peak supply voltage per-phase is therefore

$$
E_m = \sqrt{2}E_s = 240\sqrt{\frac{2}{3}} = 196\ \text{V}
$$

From Eq. (4.5), the average load current is

$$
I_{av} = \frac{0.827 \times 240 \times \sqrt{2}}{25 \times \sqrt{3}} = 6.48\ \text{A}
$$

Similarly, the rms load current is given by Eq. (4.7)

$$
I_L = \frac{0.841 \times 240 \times \sqrt{2}}{25 \times \sqrt{3}} = 6.59\ \text{A}
$$

The load power dissipation is

$$P_L = I_L^2 = 1.086\text{-kW}$$

In comparison, it is of interest to note that three 25-Ω resistors connected in parallel across a single-phase 240 V supply would dissipate 6.912 kW. The rms value of the supply current is obtained from Eq. (4.9).

$$I_a = \frac{0.485 \times 240 \times \sqrt{2}}{25 \times \sqrt{3}} = 3.8\,\text{A}$$

Example 4.2 For the three-phase, half-wave rectifier of Example 4.1, calculate the input voltamperes and the power factor. Is any correction of the power factor possible by energy storage devices connected at the supply terminals?
The rms supply current per phase is, from Eq. (4.9).

$$I_s = \frac{0.485 \times 240 \times \sqrt{2}}{25 \times \sqrt{3}} = 3.8\,\text{A}$$

The rms supply voltage per phase is given, in Example 4.1, by $E_s = 138.6$ V. For a three-phase load drawing symmetrical supply currents the input voltamperes is therefore

$$S = 3E_sI_s = 1.580\text{ kVA}$$

The load power dissipated is all drawn through the supply terminals and is seen, from Example 2.1, to be

$$P_L = P_s = 1.086\text{ kW}$$

The power factor is therefore

$$PF = \frac{P_s}{S} = \frac{1.086}{1.580} = 0.687$$

This value of power factor is independent of E_s and R and may also be obtained directly from Eq. (4.11).
It is shown in Sec. 4.1.1 that the displacement factor $\cos\psi_1$ is unity. This means that the displacement angle ψ_1 is zero and that the fundamental supply current is in time phase with its respective voltage. There is therefore no quadrature component of the fundamental current and no power factor correction can be accomplished by the use of energy storage devices (which draw compensating quadrature current).
Example 4.3 For the three-phase, half-wave rectifier circuit of Fig. 4.2 calculate the displacement factor and distortion factor at the supply point and hence calculate the input power factor

Take phase a as the reference phase. Current i_a (ωt) is given by Eq. (4.4). Coefficients a_1 and b_1 of the fundamental component of the Fourier Series are given by

$$a_1 = \frac{1}{\pi} \int_0^{2\pi} i_a(\omega t) \cos \omega t \, d\omega t$$

$$= \frac{E_m}{\pi R} \int_{30°}^{150°} \sin \omega t \cos \omega t \, d\omega t$$

$$= \frac{E_m}{4\pi R} \left[-\cos 2\omega t \right]_{30°}^{150°} = 0$$

$$b_1 = \frac{1}{\pi} \int_0^{2\pi} i_a(\omega t) \sin \omega t \, d\omega t$$

$$= \frac{E_m}{\pi R} \int_{30°}^{150°} \sin^2 \omega t \, d\omega t$$

$$= \frac{E_m}{2\pi R} \left[\omega t - \frac{1}{2} \sin 2\omega t \right]_{\pi/6}^{5\pi/6}$$

$$= \frac{E_m}{2\pi R} \left[\frac{2\pi}{3} + \frac{\sqrt{3}}{2} \right] = 0.471 \frac{E_m}{R}$$

The peak magnitude c_1 of the fundamental component i_{a_1} (ωt) of i_a (ωt) is therefore

$$c_1 = \sqrt{a_1^2 + b_1^2} = 0.471 \frac{E_m}{R}$$

Since $a_1 = 0$, the phase angle ψ_1 is also zero

$$\psi_1 = \tan^{-1} \left(\frac{a_1}{b_1} \right) = 0$$

Displacement factor $= \cos \psi_1 = 1.0$

Distortion factor $= \dfrac{I_{a_1}}{I_a}$

The rms value of the fundamental supply current component is given by

$$I_{a_1} = \frac{0.471}{\sqrt{2}} \frac{E_m}{R} = 0.333 \frac{E_m}{R}$$

The ms value of' the total supply current per phase is given by Eq. (4.8)

$$I_a = 0.485 \frac{E_m}{R}$$

Therefore, the distortion factor is given by

$$\text{Distortion factor} = \frac{0.333}{0.485} = 0.687$$

The system power factor is therefore

$$PF = (\text{distortion factor})(\text{displacement factor}) = 0.687$$

which agrees with Example 4.2

4.2 RESISTIVE LOAD WITH TRANSFORMER COUPLED SUPPLY

In some applications a three-phase rectifier circuit is fed from the star-connected secondary windings of a transformer, as in Fig. 4.3.

The ratio load power/transformer secondary voltamperes is sometimes referred to as the *secondary utilization factor*. But, for a star–star-connected transformer, the primary voltamperes is equal to the secondary voltamperes so that the above ratio is then more familiarly seen as the power factor of the trans-

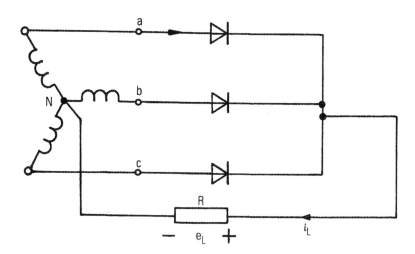

FIG. 3   Three-phase, half-wave diode rectifier fed from transformer source.

former–rectifier–load combination. The primary windings of a rectifier transformer also may be connected in delta to provide a path for triplen harmonic currents. Many different transformer connections have been used, and they are characterized by particular waveforms, requiring relevant ratings of the transformer windings. For example, two three-phase, half-wave secondary connections may be combined via an interphase transformer winding, as shown in Fig. 4.4. This results in a performance known as *six-pulse operation*, described in Chapter 6.

The great disadvantage of the circuit of Fig. 4.2a is that a direct current component is drawn through each supply line. With transformer coupling (Fig. 4.3) the dc components in the secondary windings could saturate the transformer cores. This may be avoided by the use of a zigzag connection in which the dc magnetomotive forces of two secondary windings on the same core cancel out.

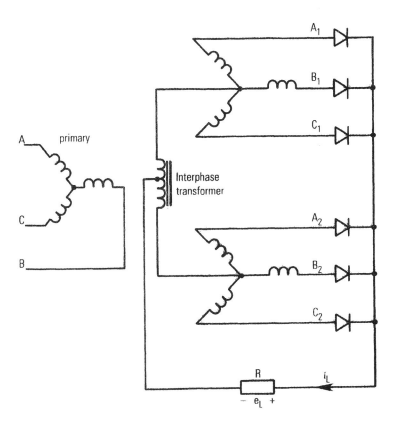

FIG. 4   Double-star or interphase transformer connection to produce six-pulse operation.

The half-wave bridge is of limited practical value in industry but is very useful as an educational aid in understanding the basic operation of polyphase bridge circuits.

4.3 HIGHLY INDUCTIVE LOAD AND IDEAL SUPPLY

If the load resistor R in the three-phase, half-wave circuit, has a highly inductive load impedance L in series (Fig. 4.5a), this smoothing reactor absorbs most of the load voltage ripple. Although this ripple can never be entirely eliminated, the load current and the voltage across the load resistor are virtually constant (Fig. 4.5d). The instantaneous load voltage retains its segmented sinusoidal form (Fig. 4.5c), as with resistive load.

For the load branch

$$e_L(\omega t) = e_R(\omega t) + e_l(\omega t)$$
$$= i_L R + L\, di_L / dt \tag{4.14}$$

It is seen from Fig. 4.5c that

$$e_L(\omega t) = E_m \sin\omega t \Big|_{30°}^{150°} + E_m \sin(\omega t - 120)\Big|_{150°}^{270°} + E_m \sin(\omega t - 240)\Big|_{0,\,270°}^{30°,\,360°}$$

$$\tag{4.15}$$

With a large series inductance the load current becomes very smooth with an almost constant value I given by

$$I_{av} = \frac{E_{av}}{R} = \frac{3\sqrt{3}}{2\pi}\frac{E_m}{R} = \frac{E_{av_0}}{R} \tag{4.16}$$

which is identical to Eq. (4.5) for resistive load. But the instantaneous load voltage $e_L(\omega t)$ may be thought of as consisting of its average value E_{av} plus a ripple component (i.e., a nonsinusoidal alternating component), $e_r(\omega t)$:

$$e_L(\omega t) = E_{av} + e_r(\omega t) \tag{4.17}$$

Now with a constant load current (Fig. 4.5d), its instantaneous, average, rms, and peak values are identical.

$$i_L(\omega t) = I_{av} = I_L = \frac{E_{av_0}}{R} \tag{4.18}$$

Comparing Eqs. (4.14) and (4.17), noting that $i_L = I_{av}$ shows that

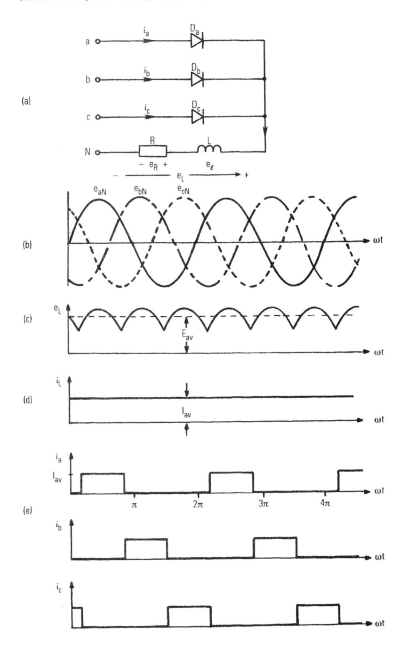

Fig. 5 Three-phase, half-wave diode rectifier with highly inductive load: (a) circuit connection, (b) supply phase voltages, (c) load voltage, (d) load current, and (e) supply currents.

$$e_r(\omega t) = L \frac{di_L}{dt} = \omega L \frac{di_L}{d\omega t}$$

(4.19)

The time variation of $e_L (\omega t)$ in Fig. 4.5c, above and below the E_{av} value, is the component $e_r (\omega t)$, and this falls entirely on the inductor L. Instantaneous ripple voltage $e_r (\omega t)$ can also be expressed in terms of the flux φ associated with the inductor

$$e_r(\omega t) = N \frac{d\varphi}{dt} = L \frac{di_L}{dt}$$

(4.20)

The flux is therefore given by

$$\varphi = \frac{1}{N} \int e_r \, dt$$

(4.21)

which is the net area shaded in Fig. 4.5d. The time average values of the inductor voltage $e_r (\omega t)$ and flux $\varphi(\omega t)$ are zero so that the two respective shaded portions represent equal areas and equal values in volt-seconds.

Note that the instantaneous inductor voltage $e_r (\omega t)$ is only zero at six instants in each supply voltage cycle and not at every instant as would be the case if a pure direct current was injected into a pure inductor.

As with resistive load, the supply current pulses (Fig. 4.5e) have a conduction angle of 120°. The rms value I_a of the supply currents is related to the average load current by

$$I_a = \sqrt{\frac{1}{2\pi} \int_{30^\circ}^{150^\circ} I_{av}^2 \, d\omega t} = \frac{I_{av}}{\sqrt{3}} = \frac{I_L}{\sqrt{3}}$$

(4.22)

Combining Eqs. (4.5) and (4.22) gives

$$I_a = \frac{3}{2\pi} \frac{E_m}{R} = 0.477 \frac{E_m}{R}$$

(4.23)

which is about 2% lower than the value $0.485 E_m/R$ for resistive load, Eq. (4.9). All the average power dissipation in the circuit is presumed to occur in the load resistor. Therefore,

$$P_L = I_L^2 R = I_{av}^2 R = \left(\frac{3\sqrt{3}}{2\pi} \right)^2 \frac{E_m^2}{R} = 0.684 \frac{E_m^2}{R}$$

(4.24)

The circuit power factor is therefore given by

$$PF = \frac{\left(3\sqrt{3}/2\pi\right)^2 \dfrac{E_m^2}{R}}{3\left(E_m/\sqrt{2}\right)\left(3/2\pi\right)\left(E_m/R\right)} = 0.675 \tag{4.25}$$

which is marginally lower than the corresponding value 0.687, Eq. (4.10), with resistive load. It should be noted that the value of the power factor is independent of voltage level and of load impedance values. A comparative summary of some of the properties of the three-phase, half-wave, uncontrolled bridge is given in the top half of Table 4.1

TABLE 4.1   Uncontrolled, Three-Phase Rectifier with Ideal Supply

		Resistive load	Highly inductive load
Three-pulse (half-wave) operation	Average load current	$0.\,27\dfrac{E_m}{R}$	$0.\,27\dfrac{E_m}{R}$
	RMS load current	$0.\,41\dfrac{E_m}{R}$	$0.\,27\dfrac{E_m}{R}$
	Load power	$0.\,07\dfrac{E_m}{R}$	$0.\,84\dfrac{E_m}{R}$
	RMS supply current	$0.\,85\dfrac{E_m}{R}$	$0.\,77\dfrac{E_m}{R}$
	Power factor	0.684	0.676
	Ripple factor Load voltage	0.185	0.185
	Load current	0.185	0
Six-pulse (full-wave) operation	Average load current	$0.\,55\left(\sqrt{3}\dfrac{E_m}{R}\right)$	$0.\,55\left(\sqrt{3}\dfrac{E_m}{R}\right)$
	RMS load current	$0.\,56\left(\sqrt{3}\dfrac{E_m}{R}\right)$	$0.\,55\left(\sqrt{3}\dfrac{E_m}{R}\right)$
	Load power	$2.74\dfrac{E_m^2}{R}$	$2.736\dfrac{E_m^2}{R}$
	RMS supply current	$1.352\dfrac{E_m}{R}$	$1.35\dfrac{E_m}{R}$
	Power factor	0.965	0.955
	Ripple factor Load voltage	0.185	0.185
	Load current	0.185	0

4.3.1 Worked Examples

Example 4.4   A three-phase, half-wave diode bridge is supplied from an ideal three-phase voltage source. Calculate the ripple factor of the load voltage, the supply current and load current with (1) resistive load and (2) highly inductive load

The load voltage waveform for an ideal bridge with ideal supply voltages is shown by Fig. 4.5c for both resistive and inductive loads. For this waveform, from Eq. (4.16),

$$E_{av} = I_{av}R = 0.827\frac{E_m}{R}$$

Also, from Eq. (4.6), the rms value E_L of the load voltage is

$$E_L = 0.841\frac{E_m}{R}$$

The ripple factor of the load voltage is therefore

$$RF = \sqrt{\left(\frac{I_L}{I_{av}}\right)^2 - 1} = 0.185$$

This low value is consistent with the relatively smooth waveform of Fig. 4.5c.
   1. For resistive load, the load current ripple factor is also equal to 0.185. For the supply current with resistive load, the rms value, Eq. (4.8) is

$$I_a = 0.485\frac{E_m}{R}$$

The average value of the supply current waveform (Fig. 4.2d), is one third the average value of the load current waveform (Fig. 4.2c). From Eq. (4.5),

$$I_{a_{av}} = \frac{0.827}{3}\frac{E_m}{R} = 0.276\frac{E_m}{R}$$

With resistive load the supply current ripple factor is therefore

$$RF = \sqrt{\left(\frac{I_L}{I_{a_{av}}}\right)^2 - 1} = 1.44$$

This high value is consistent with the high value ($0.471E_{m/R}$) of the fundamental ac component of $i_s(\omega t)$ shown in Fig. 4.2d.

2. With highly inductive load (Fig. 4.5), the rms supply current [Eq. (4.23)] is

$$I_a = 0.477 \frac{E_m}{R}$$

The average value of the supply current is one third the average value of the load current [Eq. (4.16)]:

$$I_{av} = \frac{\sqrt{3}}{2\pi} \cdot \frac{E_m}{R} = \frac{E_{av_0}}{3R} = 0.276 \frac{E_m}{R}$$

With highly inductive load the supply current ripple factor is therefore

$$RF = \sqrt{\left(\frac{0.477}{0.276}\right)^2 - 1} = 1.41$$

The effect of the load inductance is seen to reduce the supply current ripple factor only marginally, because the discontinuous current waveform is mainly attributable to natural switching by the diode elements.

Example 4.5   A three-phase, half-wave diode bridge with a highly inductive load is supplied from an ideal, three-phase source. Determine the waveform of a diode voltage and calculate the diode rms voltage rating.

Consider the circuit of Fig. 4.5a. During the conduction of current i_a (ωt), the voltage drop on diode D_a is ideally zero. While D_a is in extinction, its anode is held at potential e_{aN} and its cathode is held at either e_{bN} or e_{cN} depending on whether D_b or D_c is conducting, respectively. In Fig. 4.5 the following voltages pertain to parts of the cycle

$$0 < \omega t < 30° \qquad e_{D_a} = e_{aN} - e_{cN} = e_{ac}$$

$$30° < \omega t < 150° \qquad e_{D_a} = 0$$

$$150° < \omega t < 270° \qquad e_{D_a} = e_{aN} - e_{bN} = e_{ab}$$

$$270° < \omega t < 360° \qquad e_{D_a} = e_{aN} - e_{cN} = e_{ac}$$

The line voltages e_{ac} and e_{ab} are $\sqrt{3}$ times the magnitude values of the phase voltages and result in the diode anode-cathode voltage waveform shown in Fig. 4.6, which can be deduced from the three-phase line voltage waveforms.

The rms value of waveform E_{D_a} (ωt) in Fig. 4.6c may be obtained from

$$E_{D_a} = \sqrt{\frac{2}{2\pi} \int_{150°}^{270°} \left[\sqrt{3} E_m \sin(\omega t - 30°)\right]^2 d\omega t}$$

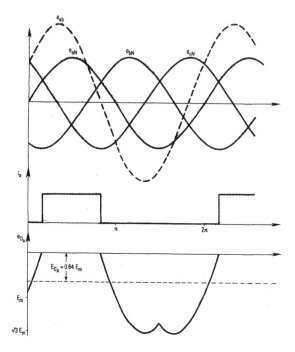

FIG. 6 Waveforms for operation of a three-phase, half-wave diode rectifier with highly inductive load.

$$E_{D_a}^2 = \sqrt{\frac{3}{\pi} E_m^2 \int_{150°}^{270°} \sin^2(\omega t - 30°) \, d\omega t}$$

$$= \frac{3}{\pi} E_m^2 \left[\frac{\omega t - 30°}{2} - \frac{\sin 2(\omega t - 30°)}{4} \right]_{150°}^{270°}$$

Using radian values for the angle terms, this becomes

$$E_{D_a}^2 = \frac{3}{\pi} E_m^2 \left[\frac{\sqrt{3}}{2} - \frac{1}{4} \left(\frac{\sqrt{3}}{2} + \frac{\sqrt{3}}{2} \right) \right]$$

From which,

$$E_{D_a} = 0.64 \, E_m$$

This value is sketched in Fig. 4.6c and looks reasonable. For design purposes a more relevant property of the diode is the peak reverse voltage (PRV) that it is required to withstand. In this case, it is clear that the PRV has a magnitude of $\sqrt{3}E_m$. The reverse (i.e., cathode–anode) voltage waveform of diode D_a will be the reverse of waveform E_{D_a} in Fig. 4.6.

Example 4.6   A three-phase, half-wave rectifier is fed from an ideal three-phase supply of value 400 V at 50 Hz. The load current is maintained constant at 50 A by the use of a suitable load side inductor. If the conducting voltage drop of the diodes is 1 V, calculate the required diode ratings and the value of the load resistor.

The diode peak reverse voltage rating is the peak value of the line voltage.

$$E(\text{PRV}) = 400 \sqrt{2} = 566 \text{ V}$$

The average load current is specified as

$$I_{av} = 50 \text{ A}$$

The supply current pulses in Fig. 4.5 have an rms value given by Eq. (4.22)

$$I_a = \frac{I_{av}}{\sqrt{3}} = 28.9 \text{ A}$$

The conducting voltage drop on the bridge diodes has no effect on their voltage or current ratings.

Now the average load voltage is given, from Eq. (4.16), by

$$E_{av} = \frac{3\sqrt{3}}{2\pi} E_m$$

where E_m is the peak phase voltage. In this case $E_m = 400 \sqrt{2}/\sqrt{3}$ so that

$$E_{av} = \frac{3}{2\pi} 400\sqrt{2} - 1 = 269 \text{ V}$$

From Eq. (4.16),

$$R = \frac{E_{av}}{I_{av}} = \frac{269}{50} = 5.38 \text{ } \Omega$$

4.4   HIGHLY INDUCTIVE LOAD IN THE PRESENCE OF SUPPLY IMPEDANCE

An electrical power supply is not usually a perfect voltage source because it contains series impedance. The action of drawing current from the supply into a resistive or inductive load causes the supply voltage at the terminals to reduce below its no-load value. In public electricity supply undertakings, the generator voltage level is usually automatically boosted to provide constant voltage at a consumer's terminals when load current is drawn from the supply. The series impedance of an electricity supply system is usually resistive-inductive, being created by transformers, cables, and transmission supply lines. In transformer-supplied bridge circuits the supply inductance is mainly the transformer leakage inductance.

The magnitude of the supply inductance is typically such that not more than about 5% reduction would occur in an unregulated supply voltage at full-load current. Because of this inductance the instantaneous commutation of current from one diode to another that occurs in resistive circuits, described in Sec. 4.1.1, for example, cannot occur. When switching closure occurs in an open inductive circuit a definite time is required for a current to build up from zero to its final steady-state value. The instantaneous transitions in the value of the supply currents in Fig. 4.2, for example, no longer take place.

Consider operation of the half-wave diode bridge (Fig. 4.7). The balanced sinusoidal voltages of the generator are given by e_{AN}, e_{BN}, and e_{CN}, where

$$e_{AN} = E_m \sin \omega t \tag{4.26}$$

$$e_{BN} = E_M \sin (\omega t - 120°) \tag{4.27}$$

$$e_{CN} = E_m \sin (\omega t - 240°) \tag{4.28}$$

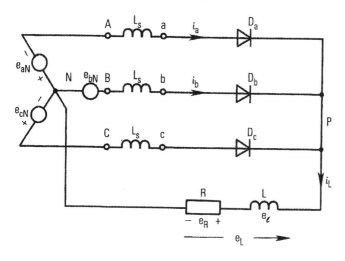

FIG. 7 Three-phase, half-wave diode rectifier with highly inductive load inductance.

When each supply line contains an effective series inductance L_s; Fig. 4.7, the bridge terminal voltages e_{AN}; e_{BN}, and e_{CN} do not remain sinusoidal on load but are given by

$$e_{aN} = e_{AN} - L_s \frac{di_a}{dt} \tag{4.29}$$

$$e_{bN} = e_{BN} - L_s \frac{di_b}{dt} \tag{4.30}$$

$$e_{cN} = e_{CN} - L_s \frac{di_c}{dt} \tag{4.31}$$

Compared with operation with an ideal supply, as in Fig. 4.5a, the waveforms of both the terminal voltage and current and the load voltage are affected by the presence of supply reactance. Now the magnitude of the supply inductance L_s is usually small compared with the value of the load inductance L. The presence of supply inductance therefore does not significantly affect the magnitude I_{av} of the load current nor the maximum value I_{av} and average value $I_{av}/3$ of the supply currents in Fig. 4.5. The magnitude of the load current at fixed supply voltage is determined almost entirely by the value of the load resistance because the load inductor offers no net impedance to the (hopefully predominant) direct current component. It is also of interest that the waveform of the supply currents, at fixed supply voltage, varies according to the level of the load current and the value of the supply inductance. Several different modes of supply current behavior are identifiable, depending on the particular application.

Figure 4.8 gives some detail of the waveforms due to operation of the circuit of Fig. 4.7. Diode D_a carries the rectified current of peak magnitude I_{av} up to a point x. Since the anode voltages e_{AN} and e_{BN} of D_a and D_b are then equal, diode D_b starts to conduct current i_L. Due to the effect of the supply inductance, the current i_a cannot extinguish immediately as it does with ideal supply, shown in Fig. 4.8b. Due to the supply line inductance, diodes D_a and D_b then conduct simultaneously, which short circuits terminals a and b of the supply. During this interval of simultaneous conduction, called the *overlap period* or *commutation angle u*, the common cathode has a potential $(e_{AN} + e_{BN})/2$. In the overlap period xy, e_{BN} is greater than e_{AN} and the difference voltage can be considered to cause a circulating current in loop aPbNa (Fig. 4.7), which increases i_b and diminishes i_a. When current i_a falls below its holding value diode D switches into extinction and the load voltage then jumps to the corresponding point z on wave e_{BN}. Each supply current has the characteristic waveform of Fig. 4.8c, but the load current is continuous and smooth at the value I_{av}, (Fig. 4.8e). Because the presence of supply inductance does not affect the maximum value I_{av} nor the

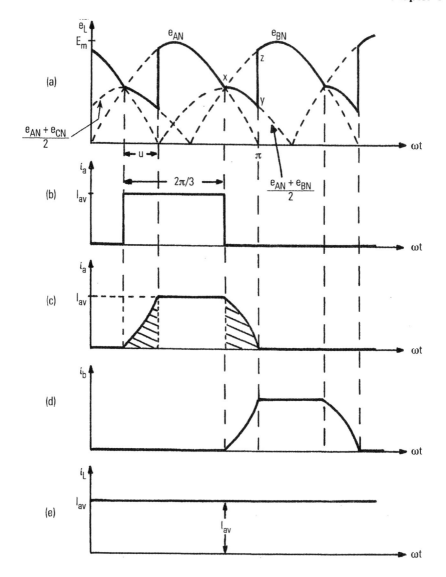

FIG. 8 Waveforms for operation of a three-phase, half-wave diode rectifier with highly inductive load and supply inductance: (a)load voltage, (b) supply line current (with ideal supply), (c) and (d) supply line currents i_a (ωt) and i_b (ωt), and (e) load current.

average value $I_{av}/3$ of the supply current the total area under the current pulse (Fig. 4.8c) is unchanged, compared with ideal supply.

Comparison of Fig. 4.8a with Fig. 4.5c shows, however, that an effect of supply reactance is to reduce the average value E_{av} of the load voltage. From Fig. 4.8a it is seen that, with overlap angle u,

$$
\begin{aligned}
E_{av} &= \frac{3}{2\pi} \left[\int_{30°}^{u+30°} \frac{e_{AN} + e_{CN}}{2} \, d\omega t + \int_{u+30°}^{150°} e_{AN} \, d\omega t \right] \\
&= \frac{3\sqrt{3}}{2\pi} E_m \cos^2 \frac{u}{2} \\
&= \frac{3\sqrt{3}E_m}{4\pi} (1 + \cos u)
\end{aligned}
\tag{4.32}
$$

This may also be expressed

$$
E_{av} = E_{av_0} \cos^2 \frac{u}{2} = \frac{E_{av_0}}{2} (1 + \cos u)
\tag{4.33}
$$

where E_{av_0} is the average load voltage with zero overlap, or ideal ac supply, defined in Eq. (4.7).

In the circuit of Fig. 4.7, during the overlap created by the simultaneous conduction of diodes D_a and D_b, there is no current in supply line c, and

$$
e_{AN} - L_s \frac{di_a}{dt} = e_{BN} - L_s \frac{di_b}{dt}
\tag{4.34}
$$

But the load current $I_{av} = i_a + i_b$ is not affected by the presence of L_s. Therefore,

$$
e_{AN} - L_s \frac{di_a}{dt} = e_{BN} - L_s \frac{d}{dt} (I_{av} - i_a)
\tag{4.35}
$$

and

$$
e_{AN} - e_{BN} = 2L_s \frac{di_a}{dt}
\tag{4.36}
$$

Substituting Eqs. (4.29) and (4.30) into Eq. (4.36) and integrating from $150°$ to $150° + u$, noting that $i_a = I_{av}$ at $\omega t = 150°$, gives

$$
\cos u = 1 - \frac{2\omega L_s I_{av}}{\sqrt{3}E_m}
\tag{4.37}
$$

Combining Eqs. (4.6) and (4.37) permits u to be expressed in terms of impedance parameters, utilizing the fact that $I_{av} = E_{av}/R$,

$$\cos u = 1 - \frac{3\omega L_s I_{av}}{\pi E_{av_0}} = 1 - \frac{3}{\pi} \frac{\omega L_s}{R} \frac{E_{av}}{E_{av_0}} \tag{4.38}$$

Combining Eqs. (4.35) and (4.38) permits cos u to be expressed in terms of impedance parameters

$$\cos u = \frac{1 - (3\omega L_s / 2\pi R)}{1 + (3\omega L_s / 2\pi R)} \tag{4.39}$$

Provided that the load inductance L is large, the actual value of L does not occur in the relevant circuit equations. With a good electrical supply the ratio $\omega L_s/R$ is about 0.05 at full load and the value of u is then about 18°. For a poor (i.e., relatively high inductance) supply, or with reduced load resistance such that $\omega L_s/ R = 0.2$, then u is about 34°. A value $u = 18°$ results in a reduction of E_{av} of less than 3%, while $u = 34°$ results in about 9% reduction. The reduction of average load voltage can be expressed in terms of impedance parameters by combining Eqs. (4.33) and (4.39).

$$E_{av} = \frac{E_{av_0}}{1 + (3\omega L_s / 2\pi R)} \tag{4.40}$$

The supply inductance is found to modify the previously appropriate expression Eq. (4.22), for rms supply current to

$$I_a = \frac{I_{av}}{\sqrt{3}} \sqrt{1 - 3\psi(u)} \tag{4.41}$$

where

$$\psi(u) = \frac{(2 + \cos u)\sin u - (1 + 2\cos u)u}{2\pi(1 - \cos u)^2} \tag{4.42}$$

Function $\psi(u)$ varies almost linearly with u for values up to $u = 60°$. At $u = 34°$, for example, the effect of supply reactance is found to reduce I_a by about 3.5%.

The effect of gradually increased overlap, with fixed supply voltage, is demonstrated sequentially in Figs. 4.9–4.15. Note that for values of supply inductance such that $u > 90°$, Figs. 4.13–4.15, the load resistance is here modified to give the same peak value of supply current. For values of $u < 90°$ the performance is usually described as mode I operation. With $u > 90°$ the load phase voltages become discontinuous and the performance is referred to as mode II operation.

It is seen that the conduction angle of the supply currents progressively increases with overlap.

The boundary between mode I operation and mode II operation occurs at $u = 90°$. Under that condition $\cos u$ is zero and it is seen from equation (4.37) that

$$I_{av}\bigg|_{u=90°} = \frac{\sqrt{3}E_m}{2\omega L_s} \tag{4.43}$$

But from Fig. 4.7, the right hand side of Eq. (4.43) is seen to be the peak value of the short-circuit current I_{sc} in (say) loop APBN.

Therefore,

$$\hat{I}_{sc} = \frac{\sqrt{3}E_m}{2\omega L_s} = I_{av}\bigg|_{u=90°} \tag{4.44}$$

Combining Eqs. (4.37), (4.43), and (4.44) results in

$$\frac{I_{av}}{\hat{I}_{sc}} = 1 - \cos u \tag{4.45}$$

The short circuit current can be expressed in terms of the average load voltage by eliminating $\cos u$ between Eqs. (4.38), (4.40), and (4.45).

$$\frac{E_{av}}{E_{av_0}} = 1 - \frac{I_{av}}{2\hat{I}_{sc}} \tag{4.46}$$

For $u > 90°$, which occurs in mode II operation, $I_{av} > \hat{I}_{sc}$ and the average load voltage, E_{av} becomes less than one half of the value E_{av_0} with ideal supply.

All the energy dissipation in the circuit of Fig. 4.7 is presumed to occur in the load resistor R. The power rating of the circuit in mode I is given in terms of the constant value I_{av} of the load current.

$$P = I_{av}^2 R \tag{4.47}$$

In mode I the supply current has the rms value denoted in Eq. (4.41). The supply voltage $e_{aN}(\omega t)$ is seen from Figs. 4.10–12 to be given by

$$e_{aN}(\omega t) = E_m \sin \omega t \bigg|_{0,\, u+30°,\, u+150°}^{30°,\, 150°,\, 360°} + \frac{1}{2}(e_{AN} + e_{CN})\bigg|_{30°}^{u+30°} + \frac{1}{2}(e_{AN} + e_{BN})\bigg|_{150°}^{u+150°} \tag{4.48}$$

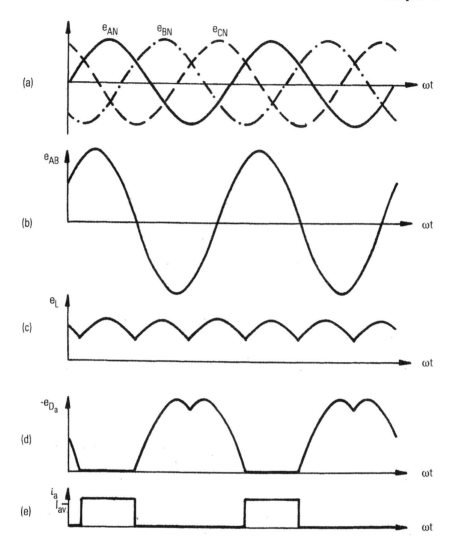

FIG. 9 Waveforms for three-phase, half-wave diode bridge with highly inductive load. Ideal supply $u = 0$.

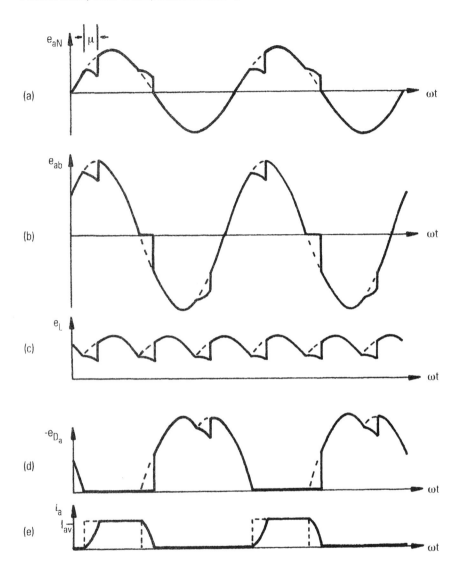

Fig. 10   Waveforms for three-phase, half-wave diode bridge with highly inductive load. Mode I, $\mu = 30°$, ideal supply.

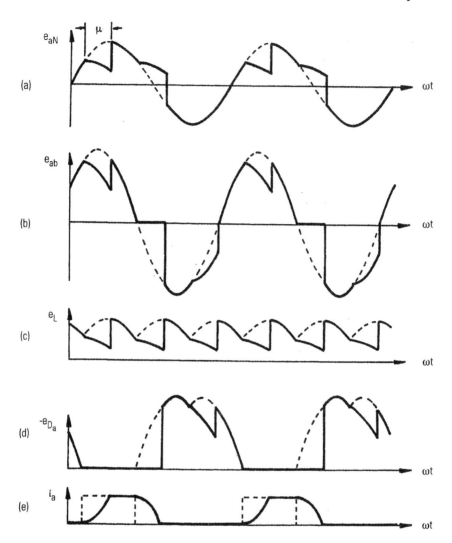

FIG. 11 Waveforms for three-phase, half-wave diode bridge with highly inductive load. Mode I, $\mu = 60°$, ideal supply.

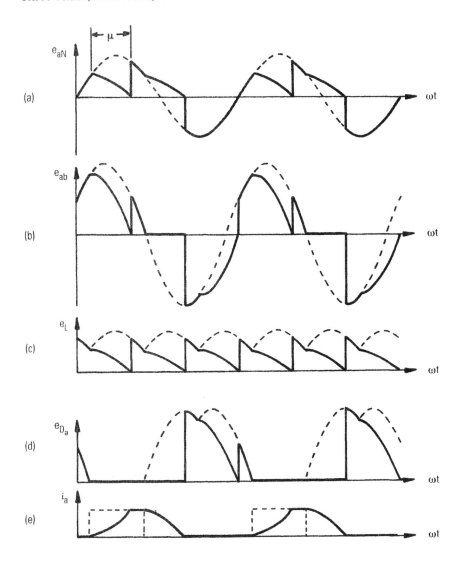

FIG. 12 Waveforms for three-phase, half-wave, diode bridge with highly inductive load. Limit mode I, $\mu = 90°$, ideal supply.

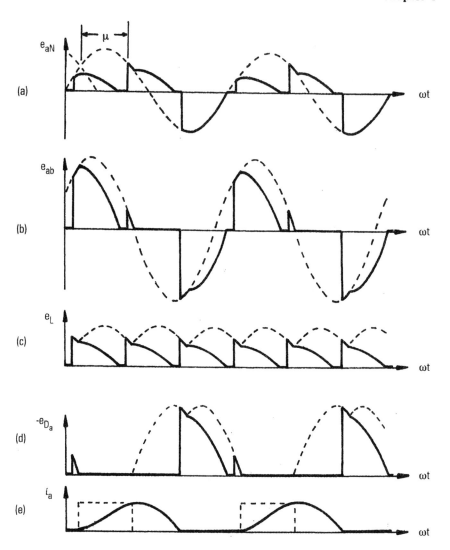

FIG. 13 Waveforms for three-phase, half-wave diode bridge with highly inductive load. Mode II, $\mu = 105°$, ideal supply.

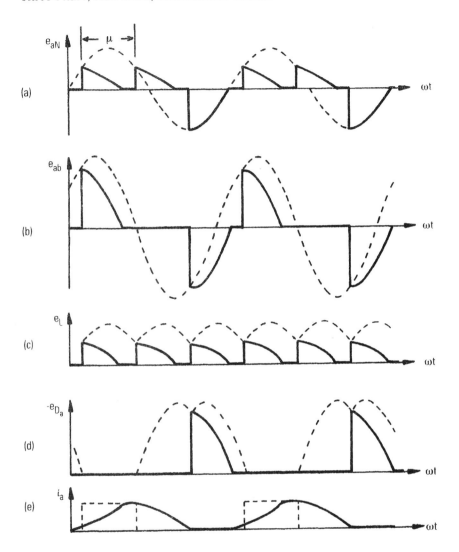

FIG. 14   Waveforms for three-phase, half-wave diode bridge with highly inductive load. Mode II, $\mu = 120°$, ideal supply.

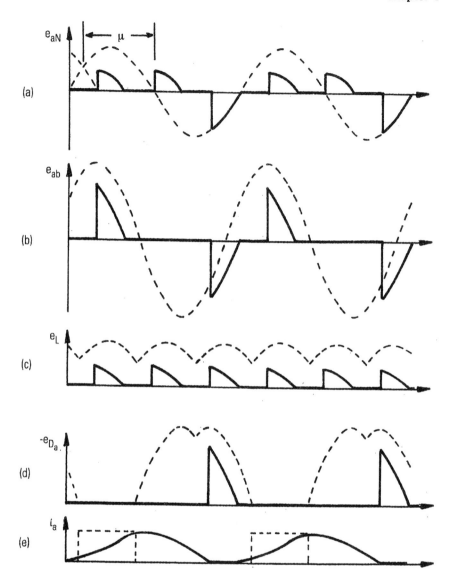

FIG. 15 Waveforms for three-phase, half-wave diode bridge with highly inductive load. μ = 150°, ideal supply.

which has the rms value E_{aN}, where

$$E_{aN}^2 = \frac{1}{2\pi} \int_0^{2\pi} e_{aN}^2 \, d\omega t \qquad (4.49)$$

The substitution of Eq. (4.48) into Eq. (4.49) gives

$$E_{aN} = E_m \sqrt{\frac{1}{2\pi} \left(\pi - \frac{3u}{4} + \frac{1}{4} \sin 2u \right)} \qquad (4.50)$$

The power factor of the three-phase, half-wave bridge is given by

$$PF = \frac{P}{3E_{aN}I_a} \qquad (4.51)$$

Supply reactance causes both the rms voltage E_{aN} and rms current I_a to be reduced below their respective levels with ideal supply. The power factor is therefore increased.

4.4.1 Worked Examples

Example 4.7   A three-phase, half-wave uncontrolled bridge circuit transfers energy from a three-phase supply to a highly inductive load consisting of a resistor R in series with inductor L. Each supply line may be considered to have a series inductance L_s. Show that the average load voltage is given by

$$E_{av} = \frac{3\sqrt{3}}{2} - \frac{3\omega L_s I_{av}}{2\pi}$$

where E_m is the peak phase voltage.

The circuit diagram is shown in Fig. 4.7. From Eq. (4.33)

$$E_{av} = \frac{E_{av_0}}{2} (1 + \cos u) = \frac{3\sqrt{3} E_m}{4\pi} (1 + \cos u)$$

Substituting $\cos u$ from Eq. (4.37) gives

$$E_{av} = \frac{3\sqrt{3} E_m}{4\pi} \left(2 - \frac{2L_s \omega I_{av}}{\sqrt{3} E_m} \right)$$

Therefore,

$$E_{av} = \frac{3\sqrt{3} E_m}{2\pi} - \frac{3L_s \omega I_{av}}{2\pi}$$

$$= \frac{3}{2\pi} \left(\sqrt{3} E_m - \omega L_s I_{av} \right)$$

The final form of E_{av} above shows that this incorporates the peak line-to-line voltage $\sqrt{3}\,E_m$ less the line voltage drop due to the supply reactance.

Example 4.8   A three-phase, half-wave diode bridge supplies power to a load consisting of resistor R and series inductor L. Each phase of the supply has a series inductance L_s where $L_s \ll L$. Sketch waveforms of the per-phase voltage and current of the supply for mode I operation when $u = 30°$. Derive an expression for the instantaneous supply current i_a (ωt) for the overlap period $150° < \omega t < 150° + u$.

Waveforms of e_{aN} (ωt) and i_a (ωt) for $u = 30°$ are given in Figs. 4.8 and 4.10. Instantaneous current i_a (ωt) is defined by Eq. (4.36),

$$e_{AN} - e_{BN} = 2L_s \frac{di_a}{dt}$$

where

$$e_{AN} = E_m \sin \omega t$$

$$e_{BN} = E_m \sin \left(\omega t - \frac{2\pi}{3} \right)$$

Therefore,

$$\frac{di_a}{dt} = \frac{e_{AN} - e_{BN}}{2L_s} = \frac{E_m}{2L_s} 2 \sin \frac{\pi}{3} \cos \left(\omega t - \frac{\pi}{3} \right)$$

$$= \frac{\sqrt{3}}{2} \frac{E_m}{L_s} \cos \left(\omega t - \frac{\pi}{3} \right)$$

Integrating both sides of the differential equation gives

$$i_a(t) = \frac{\sqrt{3}}{2} \frac{E_m}{\omega L_s} \sin \left(\omega t - \frac{\pi}{3} \right) + K$$

where K is a constant of integration.

Now (1) at $\omega t = 150°$, $i_a = I_{av}$; (2) at $\omega t = 150° + u, i_a = 0$. Under condition (1),

$$K = I_{av} - \frac{\sqrt{3}}{2} \frac{E_m}{\omega L_s} = I_{av} - \hat{I}_{sc}$$

which is negative because $\hat{I}_{sc} > I_{av}$ for mode I operation.

Under condition (2),

$$K = -\frac{\sqrt{3}}{2}\frac{E_m}{\omega L_s}\cos u = -\hat{I}_{sc}\cos u$$

Since $u = 30°$,

$$K = -\frac{\sqrt{3}}{2}\hat{I}_{sc}$$

$$= -\frac{3}{4}\frac{E_m}{\omega L_s}$$

Equating these two values of K between the two consistent conditions shows that

$$I_{av} = \frac{\sqrt{3}E_m}{2\omega L_s}\left(1 - \frac{\sqrt{3}}{2}\right)$$

$$= \hat{I}_{sc}\left(1 - \frac{\sqrt{3}}{2}\right)$$

which is seen to be consistent with Eq. (4.45)

Using the value of K from condition (2) in the equation for i_a gives

$$i_a = \frac{\sqrt{3}}{2}\frac{E_m}{\omega L_s}\left[\sin\left(\omega t - \frac{\pi}{3}\right) - \frac{\sqrt{3}}{2}\right]$$

for $150° < \omega t < 150° + u$

for $150° < \omega t < 150° + u$

Example 4.9   A three-phase, half-wave diode rectifier has a load resistance $R = 10\ \Omega$ in series with a large inductor. Each supply line contains a series inductance L_s such that the inductive reactance in the line is 10% of the load resistance. The generator three-phase voltages have an rms line value of 400 V. Calculate the power factor of operation and compare this with the case of ideal supply.

The equivalent circuit is given in Fig. 4.7. The first step of the solution is to calculate the overlap angle u. Since

$$\frac{\omega L_s}{R} = 0.1$$

then from Eq. (4.39),

    $\cos u = 0.9086$

Therefore, $u = 24.7° = 0.431$ rad, which is mode I operation. Function $\psi(u)$, Eq. (4.42), is found to have the value

$$\psi(u) = \frac{(2.9086)(0.418) - (1+1.817)(0.431)}{2\pi(0.0915)^2}$$

$$= \frac{1.2158 - 1.2141}{0.0526}$$

$$= \frac{0.0017}{0.0526} = 0.032$$

The average load voltage in the presence of L is given by Eq. (4.32) in which E_m is the peak phase voltage. In the present case,

$$E_m = \frac{400\sqrt{2}}{\sqrt{3}} = 326.6V$$

Therefore,

$$E_{av} = \frac{3\sqrt{3}E_m}{4\pi}(1+\cos u)$$

$$= 135(1.9085)$$

$$= 257.7 \text{ V}$$

This compares with the value $E_{av_o} = (3\sqrt{3}/2\pi) E_m = 270$ V obtainable with an ideal supply.

The average load current is unchanged by the presence of the supply inductance.

$$I_{av} = \frac{E_{av_o}}{R} = \frac{270}{10} = 27 \text{ A}$$

From Eq. (4.42) the rms supply current is

$$I_a = \frac{I_{av}}{\sqrt{3}}\sqrt{1-3\psi(u)}$$

$$= \frac{27}{\sqrt{3}}\sqrt{1-0.096} = 14.82 \text{ A}$$

This compares with the value $I_a = I_{av}/\sqrt{3} = 15.6$ A with ideal supply. The power into the bridge circuit is presumed to be dissipated entirely in the load resistor:

$$P = I_{av}^2 R = (27)^2 \, 10 = 7290 \text{ W}$$

The power factor, seen from the supply terminals, is given by Eq. (4.51), which incorporates the rms value E_{an} of the terminal voltage. For a value $u = 24.7°$

the waveform $e_{aN}(\omega t)$ is very similar to that given in Fig. 4.10a and is defined by Eq. (4.48). Inspection of the more detailed diagram (Fig. 4.8a) shows that

$$e_{aN}(\omega t) = \frac{E_m}{2}\sin(\omega t + 60°)\Big|_{30°}^{30°+u} + \frac{E_m}{2}\sin(\omega t - 60°)\Big|_{150°}^{150°+u}$$

$$+ \frac{E_m}{2}\sin\omega t\Big|_{0,\,30°+u,\,150°+u}^{30°,\,150°,\,360°}$$

Therefore,

$$E_L^2 = \frac{E_m^2}{2\pi}\left[\frac{1}{4}\int_{30°}^{30°+u}\sin^2(\omega t + 60°)\,d\omega t + \frac{1}{4}\int_{150°}^{150°+u}\sin^2(\omega t - 60°)\,d\omega t\right.$$

$$\left.+ \int_{0,30°+u}^{30°,150°,360°}\sin^2\omega t\,d\omega t\right] = \frac{E_m^2}{2\pi}\left\{\frac{1}{4}\left[\frac{\omega t}{2} - \frac{\sin^2(\omega t + 60°)}{4}\right]_{30°}^{30°+u}\right.$$

$$+ \frac{1}{4}\left[\frac{\omega t}{2} - \frac{\sin^2(\omega t - 60°)}{4}\right]_{150°}^{150°+u}$$

$$+ \left(\frac{\omega t}{2} - \frac{\sin 2\omega t}{4}\right)_{0,\ 30°+u,\ 150°+u}^{30°,\ 150°,\ 360°}\right\}$$

$$= \frac{E_m^2}{2\pi}\left\{\frac{1}{4}\left[\frac{u}{2} - \frac{1}{4}\sin(\pi + 2u)\right] + \frac{1}{4}\left[\frac{u}{2} - \frac{1}{4}\sin(\pi - 2u)\right] + \frac{2\pi - 2u}{2} - \frac{1}{4}\frac{\sqrt{3}}{2} - 0\right.$$

$$\left.- \frac{\sqrt{3}}{2} - \sin(60° + 2u) + 0 - \sin(300° + 2u)\right\}$$

$$= \frac{E_m^2}{2\pi}\left[\frac{1}{4}\left(u + \frac{1}{4}\sin 2u - \frac{1}{4}\sin 2u\right) + (\pi - u) + \frac{1}{4}\left(\frac{\sqrt{3}}{2}\cos 2u + \frac{1}{2}\sin 2u\right.\right.$$

$$\left.\left.- \frac{\sqrt{3}}{2}\cos 2u + \frac{1}{2}\sin 2u\right)\right]$$

$$= \frac{E_m^2}{2\pi}\left(\pi - \frac{3u}{4} + \frac{1}{4}\sin 2u\right)$$

for $u = 24.7°$,

$$E_L = E_m\sqrt{\frac{3}{2\pi}} = 0.691E_m$$

This compares with the value $E_L = 0.707 E_m$ for sinusoidal supply.
The power factor is therefore

$$PF = \frac{P}{3E_{aN}I_a} = \frac{7290}{3 \times 0.691 \times 400/\sqrt{3} \times \sqrt{2} \times 14.92} = 0.722$$

This value is about 7% higher than the value 0.676 obtained with ideal supply.

PROBLEMS

Three-Phase, Half-Wave Bridge Circuit with Resistive Load and Ideal Supply

4.1 A set of balanced, three-phase sinusoidal voltages from an ideal supply is applied to a three-phase, half-wave bridge of ideal diodes with a resistive load (Fig. 4.2a). Sketch waveforms of the load current and supply current and show that the average load current is given by

$$I_{av} = \frac{3\sqrt{3}}{2\pi} \frac{E_m}{R}$$

where E_m is the peak value of the supply phase voltage.

4.2 For the bridge of Problem 4.1, calculate the rms values I_L and I_a of the load and supply currents, respectively. Hence show that

$$I_a = \frac{I_L}{\sqrt{3}}$$

4.3 Calculate the operating power factor of the resistively loaded bridge of Problem 4.1 and show that it is independent of load resistance and supply voltage level.

4.4 Calculate values for the Fourier coefficients a_1 and b_1 of the fundamental (supply frequency) component of the line current in the bridge of Fig. 4.2a. Hence calculate the displacement factor.

4.5 For the bridge circuit of Fig. 4.2a calculate the rms value of the input current and also the rms value of its fundamental component. Hence calculate the supply current distortion factor. Use the value of the distortion together with the displacement factor (Problem 4.4) to calculate the power factor.

4.6 Derive expressions for the nth harmonic components a_n and b_n of the Fourier series representing the line current waveform $i_a (\omega t)$ of Fig. 4.2. Show that $a_n = 0$ for n odd and $b_n = 0$ for n even.

4.7 A set of three-phase voltages of rms value 400 V at 50 Hz is applied to a half-wave diode bridge with a resistive load $R = 40 \, \Omega$. Calculate the power transferred to the load.

4.8 In the three-phase, half-wave bridge of Fig. 4.2a deduce and sketch waveforms of the three supply currents and the load current if diode D_a fails to an open circuit.

4.9 Calculate the ripple factor for the supply current waveform $i_a \, (\omega t)$, (Fig. 4.2d), obtained by resistive loading of a three-phase, half-wave diode bridge.

Three-Phase Half-Wave Bridge Circuit with Highly Inductive Load and Ideal Supply

4.10 A set of balanced, three-phase, sinusoidal voltages from an ideal supply is applied to a three-phase, half-wave diode bridge with a highly inductive load (Fig. 4.5a). Sketch waveforms of the supply voltages and currents and the load voltage and current. Show that the average load current retains the same value as with resistive load.

4.11 For the ideal, three-phase bridge of Problem 4.10, show that the rms value of the supply current I_s is related to the rms value of the load current I_L by

$$I_s = \frac{I_L}{\sqrt{3}}$$

4.12 Calculate the operating power factor of the inductively loaded bridge of Fig. 4.5a. Show that with a highly inductive load, the power factor is constant and independent of the values of L and R. Compare the value of the power factor with that obtained when the load is purely resistive.

4.13 For a three-phase, half-wave bridge with highly inductive load calculate the average and rms values of the supply current. Hence show that the ripple factor of the supply current has a value $RF = 1.41$. How does this compare with the corresponding value for resistive load?

4.14 A set of three-phase voltages of rms line value 240 V at 50 Hz is applied to a three-phase, half-wave bridge of ideal diodes. The load consists of a resistor $R = 10 \, \Omega$ in series with a large inductor. Calculate the power dissipation and compare this with the corresponding value in the absence of the (inductor) choke.

4.15 Calculate values for the Fourier coefficients a_1 and b_1 of the fundamental (supply frequency) component of the line current in the inductively loaded bridge of Fig. 4.5a. Hence calculate the displacement factor.

4.16   Calculate the magnitude and phase angle of the fundamental component of the line current for the inductively loaded bridge of Fig. 4.5a. Sketch this component for phase a together with the corresponding phase voltage and current.

4.17   For the three-phase bridge circuit of Fig. 4.5a calculate the rms value of the input current and also the rms value of its fundamental component. Hence calculate the supply current distortion factor. Use this value of the distortion factor together with the value of the displacement factor obtained from Problem 4.15 to determine the input power factor.

4.18   Derive expressions for the nth harmonic components of the Fourier series representing the line current $i_a (\omega t)$ of Fig. 4.5e. Compare the component terms, for particular values of n, with corresponding values obtained for resistive load in Problem 4.6.

4.19   For an inductively loaded, three-phase, half-wave bridge the load voltage waveform is given in Fig. 4.5c. Calculate the Fourier series for this periodic waveform and show that its lowest ripple frequency is three times the supply frequency.

4.20   In the three-phase, half-wave bridge circuit of Fig. 4.5a, deduce and sketch waveforms of the three supply currents and the load current if diode D_a fails to an open circuit.

4.21   For the three-phase bridge in Problem 4.14 calculate the rms current and peak reverse voltage ratings required of the diodes.

Three-Phase, Half-Wave Bridge Circuit with Highly Inductive Load in the Presence of Supply Inductance

4.22   A set of balanced three-phase voltages is applied to an uncontrolled, three-phase, half-wave bridge with a highly inductive load. Each supply line has a series inductance L_s such that ωL_s is about 10% of load resistor R. Sketch the waveform of the load voltage and derive an expression for the average value in terms of the peak supply voltage per phase E_m and the overlap angle u.

4.23   For the three-phase, half-wave diode bridge circuit with supply reactance, state an expression for the waveform of the supply voltage per phase. Show that the rms value of this is given by Eq. (4.50) and sketch its variation with u for $0 \le u \le 90°$.

4.24   The overlap function $\psi(u)$ for a three-phase, half-wave, diode bridge circuit is defined by Eq. (4.42). Sketch the variation of $\psi(u)$ versus u in the range $0 \le u \le 90°$.

4.25   The rms value I_a of the supply current to a three-phase, half-wave, uncontrolled bridge rectifier with highly inductive load is related to the average load current I_{av} by Eq. (4.41). Calculate the variation of I_a, assuming fixed I_{av}, for a range of values of overlap function $\psi(u)$. Sketch the per unit variation of I_a with overlap angle u for the range $0 \leq u \leq 90°$.

4.26   A set of three-phase voltages of rms line value 240 V at 50 Hz is applied to a three-phase, uncontrolled, half-wave bridge. The load consists of a resistor $R = 10\ \Omega$ in series with a large choke. Each supply line contains a series inductor L_s of such value that $\omega L_s = 0.2R$. Calculate values of the rms voltage and current per phase at the bridge terminals and compare these with the values obtained with ideal supply.

4.27   For the bridge circuit of Problem 4.26 calculate the power dissipation and power factor. Compare the values with the respective values obtained with ideal supply.

4.28   The average load voltage E_{av} for a three-phase, half-wave uncontrolled bridge is defined by Eq. (4.33). Calculate and sketch the variation of E_{av} versus u for the range $0 \leq u \leq 90°$. Extend the sketch of E_{av} for $u > 90°$, using an appropriate relationship. Does the variation of E_{av} versus u indicate the change of mode of operation?

4.29   A three-phase, half-wave diode bridge circuit supplies power to a load resistor R in series with a large inductor. The open-circuit supply voltages are a balanced set of sinusoidal voltages. Each supply line contains a series reactance ωL_s, where $L_s << L$. Sketch waveforms of the supply phase, assuming mode I operation with an overlap angle $u \approx 20°$. Show that during overlap the supply current varies sinusoidally.

5

Three-Phase, Half-Wave Controlled Bridge Rectifier Circuits

In order to achieve controlled variation of the load voltage the three-phase circuits of Chapter 4 may be modified by replacing the diodes by controlled switches such as the silicon controlled rectifier, as shown in Fig. 5.1.

With a supply of zero impedance the three supply voltages in Fig. 5.1 retain a balanced sinusoidal form for any load condition. Equations (4.1) to (4.3) still apply and are reproduced below for convenience.

$$e_{aN} = E_m \sin \omega t \tag{5.1}$$

$$e_{bN} = E_m \sin\left(\omega t - \frac{2\pi}{3} \right) \tag{5.2}$$

$$e_{cN} = E_m \sin\left(\omega t - \frac{4\pi}{3} \right) \tag{5.3}$$

5.1 RESISTIVE LOAD AND IDEAL SUPPLY

The onset of conduction in any phase may be delayed by retarding the switching angle of the switch connected in that phase. The initiation of conduction also requires that the anode voltage of an SCR switch be positive with respect to its cathode. For this reason thyristor Th_a in line a of Fig. 5.1 cannot be successfully

136

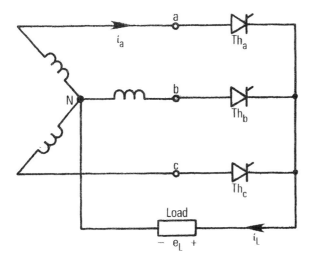

FIG. 1   Three-phase, half-wave controlled rectifier circuit using silicon controlled rectifier (SCR) switches.

fired until $\omega t = \pi/6 = 30°$. Prior to the instant $\omega t = 30°$, shown in Fig. 5.2, voltage e_{aN} (ωt) is less positive than $e_{cN}(\omega t)$, so that a reverse voltage exists across Th_a. The crossover point of successive phase voltages (i.e., $\omega t = 30°$) is therefore taken as the zero or datum from which switching-angle retardation is measured. The most usual form of control is to switch on each device at an identical point on wave of its respective anode voltage. This causes equal currents in the supply lines.

Let the three thyristors in Fig. 5.1 be fired at firing angle $\alpha = 30°$. Each phase current then begins to flow at the instant $\omega t = \alpha + 30° = 60°$ after its positive going anode voltage zero. Phase current $i_a(\omega t)$, for example, then has the form shown in Fig. 5.2e. Because of the delayed firing of thyristor Th_b the potential e_{aN} (ωt) at the anode of thyristor Th_a remains the most positive potential in the circuit until $\omega t = 180° = \pi$. Thyristor Th_a therefore conducts from $\omega t = \alpha + 30°$ to $\omega t = 180°$, at which point thyristor Th_b switches on. The cathode of Th_a then acquires the potential of point b in Fig. 5.1, so that Th_a is reverse biased. The combination of reverse anode voltage on Th_a, combined with zero current, causes the commutation or switch-off of thyristor Th_a.

If the ignition or switch-on of the three thyristors is delayed until $\alpha = 60° = \pi/3$ (i.e., $90°$ after their respective voltage zeros) the load current becomes discontinuous, as shown in Fig. 5.2f. At a firing angle $\alpha = 150° = 5\pi/6$ conduction will cease.

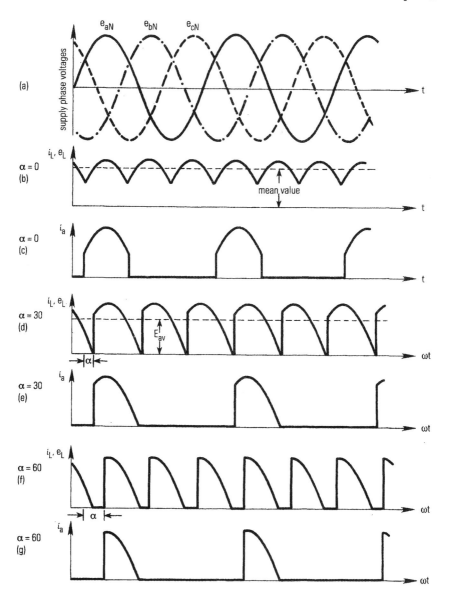

FIG. 2 Waveforms of three-phase, half-wave controlled rectifier, R load: (a) supply phase voltages, (b) load voltage and current ($\alpha = 0°$), (c) supply current i_a (ωt) ($\alpha = 0°$), (d) load voltage and current ($\alpha = 30°$), (e) supply current i_a (ωt) ($\alpha = 30°$), (f) load voltage and current ($\alpha = 60°$), and (g) supply current i_a (ωt) ($\alpha = 60°$).

The load voltage waveforms of Fig. 5.2 cannot be represented by a single mathematical expression that is true for all values of α. It turns out to be necessary to consider separately the cases for intervals $0 \leq \alpha \leq \pi/6$ and $\pi/6 \leq \alpha \leq 5\pi/6$. These two cases distinguish between continuous and discontinuous conduction of the load current and are often described as *modes* of operation

For $0 \leq \alpha \leq \pi/6$,

$$e_L = E_m \sin\left(\omega t - \frac{4\pi}{3}\right)\Big|_{0,\,\alpha+3\pi/2}^{\alpha+\pi/6,\,2\pi} + E_m \sin\omega t\Big|_{\alpha+\pi/6}^{\alpha+5\pi/6} + E_m \sin\left(\omega t - \frac{2\pi}{3}\right)\Big|_{\alpha+5\pi/6}^{\alpha+3\pi/2} \tag{5.4}$$

For $\pi/6 \leq \alpha \leq 5\pi/6$,

$$e_L = E_m \sin\left(\omega t - \frac{4\pi}{3}\right)\Big|_{0,\,\alpha+3\pi/2}^{\pi/3,\,2\pi} + E_m \sin\omega t\Big|_{\alpha+\pi/6}^{\pi} + E_m \sin\left(\omega t - \frac{2\pi}{3}\right)\Big|_{\alpha+5\pi/6}^{5\pi/3} \tag{5.5}$$

The load voltage (Fig. 5.2) is seen to have a repetitive period of one-third of the supply voltage periodicity. This means that the ripple frequency of the load voltage is three times that of the supply. It also means that the average value of the load voltage can be obtained by taking the average value of any $2\pi/3$ slice of it. Consider the section of $e_L(\omega t)$ contributed by phase a

For $0 \leq \alpha \leq \pi/6$,

$$E_{av} = \frac{3}{2\pi} \int_{\alpha+\pi/6}^{\alpha+5\pi/6} E_m \sin\omega t \, d\omega t$$

$$= \frac{3\sqrt{3}}{2\pi} E_m \cos\alpha = E_{av_0} \cos\alpha \tag{5.6}$$

For $\pi/6 \leq \alpha \leq 5\pi/6$,

$$E_{av} = \frac{3}{2\pi} \int_{\alpha+\pi/6}^{\pi} E_m \sin\omega t \, d\omega t$$

$$= \frac{3E_m}{2\pi}\left[1 + \cos\left(\alpha + \frac{\pi}{6}\right)\right] \tag{5.7}$$

The corresponding average load currents are given by:

For $0 \leq \alpha \leq \dfrac{\pi}{6}$,

$$I_{av} = \frac{E_{av}}{R} = \frac{3\sqrt{3}E_m}{2\pi R}\cos\alpha = \frac{E_{av_0}}{R}\cos\alpha \tag{5.8}$$

For $\pi/6 \leq \alpha \leq 5\pi/6$,

$$I_{av} = \frac{E_{av}}{R} = \frac{3E_m}{2\pi R}\left[1 + \cos\left(\alpha + \frac{\pi}{6}\right)\right] \tag{5.9}$$

When $\alpha = 0$, Eq. (5.8) reduces to the corresponding expression Eq. (4.5) for a diode circuit. At $\alpha = \pi/6 = 30°$, Eqs. (5.8) and (5.9) are seen to be identical.

It is seen in Fig. 5.2 that the area under the curve of a load current $i_L(\omega t)$ is three times the area under the curve of the corresponding supply current $i_a(\omega t)$. The average value of the load current is therefore three times the average value of the supply current.

The rms value I_a of the supply current is obtained in the usual way from its defining integral

$$I_a = \sqrt{\frac{1}{2\pi} \int_0^{2\pi} i_a^2(\omega t)\, d\omega t} \tag{5.10}$$

The instantaneous supply current $i_a(\omega t)$ depends on the mode of operation.

For $0 \le \alpha \le \pi/6$,

$$i_a(\omega t) = \frac{E_m}{R} \sin \omega t \Big|_{\alpha+\pi/6}^{\alpha+5\pi/6} \tag{5.11}$$

For $\pi/6 \le \alpha \le 5\pi/6$,

$$i_a(\omega t) = \frac{E_m}{R} \sin \omega t \Big|_{\alpha+\pi/6}^{\pi} \tag{5.12}$$

The substitution of Eqs. (5.11) and (5.12), respectively, into Eq. (5.10) gives for $0 \le \alpha \le \pi/6$,

$$I_a = \frac{E_m}{R} \sqrt{\frac{4\pi + 3\sqrt{3} \cos 2\alpha}{24\pi}} \tag{5.13}$$

When $\alpha = 0$, Eq. (5.13) reduces to Eq. (4.9). For $\pi/6 \le \alpha \le 5\pi/6$,

$$I_a = \frac{E_m}{R} \sqrt{\frac{(5\pi/3) - 2\alpha + \sin(2\alpha + \pi/3)}{8\pi}} \tag{5.14}$$

At $\alpha = \pi/6$, Eqs. (5.13)(5.14) give identical results.

The supply current $i_a(\omega t)$ flows only through thyristor Th_a and the load resistor R. Its rms value I_a can therefore be used to define the power dissipation per phase

$$P_a = I_a^2 R = P_b = P_c \tag{5.15}$$

Therefore, for $0 \le \alpha \le \pi/6$,

$$P_a = \frac{E_m^2}{2R} \frac{4\pi + 3\sqrt{3} \cos 2\alpha}{12\pi} \tag{5.16}$$

It is seen that (5,16) is consistent with (4.7) for the case $\alpha = 0$. For $\pi/6 \le \alpha \le 5\pi/6$,

$$P_a = \frac{E_m^2}{2R} \frac{(5\pi/3) - 2\alpha + \sin(2\alpha + \pi/3)}{4\pi} \qquad (5.17)$$

The total power dissipation is the sum of the per-phase power dissipations and may be used to obtain the rms load current.

$$P_L = I_L^2 R = P_a + P_b + P_c \qquad (5.18)$$

The power factor of the half-wave controlled bridge may be obtained from Eqs. (5.13)–(5.18).

$$PF = \frac{P_a}{E_a I_a} = \frac{P_a}{\left(E_m/\sqrt{2}\right) I_a} \qquad (5.19)$$

For $0 \le \alpha \le \pi/6$,

$$PF = \sqrt{\frac{4\pi + 3\sqrt{3}\cos 2\alpha}{12\pi}} \qquad (5.20)$$

For $\pi/6 \le \alpha \le 5\pi/6$,

$$PF = \sqrt{\frac{(5\pi/3) - 2\alpha + \sin(2\alpha + \pi/3)}{4\pi}} \qquad (5.21)$$

Expressions for the related properties of distortion factor and displacement factor are derived in Example 5.3. Certain properties of this rectifier are given in Table 5.1.

5.1.1 Worked Examples

Example 5.1   A three-phase, half-wave controlled rectifier with ideal three-phase supply provides power for a resistive load. Calculate an expression for the average load voltage and plot the variation of this over the entire possible range of firing angle α.

A circuit diagram is given as Fig. 5.1. The average load voltage is given by Eqs. (5.6) and (5.7).

$$E_{av} = \begin{cases} \dfrac{3\sqrt{3}}{2\pi} E_m \cos\alpha & 0 \le \alpha \le \dfrac{\pi}{6} \\[3mm] \dfrac{3E_m}{2\pi}\left[1 + \cos\left(\alpha + \dfrac{\pi}{6}\right)\right] & \dfrac{\pi}{6} \le \alpha \le \dfrac{5\pi}{6} \end{cases}$$

TABLE 5.1 Some Properties of the Three-Phase, Half-Wave Controlled Bridge Rectifier with Resistive Load and Ideal Supply

Circuit property	Mode I $0 < \alpha < 30°$	Mode II $30° < \alpha < 150°$
Instantaneous supply current	$\dfrac{E_m}{R}\sin\omega t \Big|_{\alpha+\pi/6}^{\alpha+5\pi/6}$	$i_a = \dfrac{E_m}{R}\sin\omega t \Big|_{\alpha+\pi/6}^{\pi}$
RMS supply current	$\dfrac{E_m}{R}\sqrt{\dfrac{4\pi+3\sqrt{3}\cos 2\alpha}{24\pi}}$	$\dfrac{E_m}{R}\sqrt{\dfrac{5\pi/3-2\alpha+\sin(2a+\pi/3)}{8\pi}}$
Average load current	$\dfrac{3\sqrt{3}}{2\pi}\dfrac{E_m}{R}\cos\alpha$	$\dfrac{3}{2\pi}\dfrac{E_m}{R}\big[1+\cos(\alpha+\pi/6)\big]$
Average load power	$\dfrac{E_m^2}{R}\dfrac{4\pi+3\sqrt{3}\cos 2\alpha}{8\pi}$	$\dfrac{E_m^2}{R}\left[\dfrac{5\pi/3-2\alpha+\sin(2\alpha+\pi/3)}{8\pi/3}\right]$
Power factor	$\sqrt{\dfrac{4\pi+3\sqrt{3}\cos 2\alpha}{12\pi}}$	$\sqrt{\dfrac{5\pi/3-2\alpha+\sin(2\alpha+\pi/3)}{4\pi}}$
RMS load current	$\dfrac{E_m}{R}\sqrt{\dfrac{4\pi+3\sqrt{3}\cos 2\alpha}{8\pi}}$	$\dfrac{E_m}{R}\sqrt{\dfrac{5\pi/3-2\alpha+\sin(2\alpha+\pi/3)}{\dfrac{8\pi}{3}}}$
a_1	$-\dfrac{E_m\sqrt{3}}{4\pi R}\sin 2\alpha$	$-\dfrac{E_m}{4\pi R}\big[1-\cos(2\alpha+\pi/3)\big]$
b_1	$\dfrac{E_m}{R}\dfrac{4\pi+3\sqrt{3}\cos 2\alpha}{12\pi}$	$\dfrac{E_m}{R}\sin\omega t \Big|_{\alpha+\pi/6}^{\alpha+5\pi/6}$

It was shown in Chapter 4 that the average load voltage with a three-phase, half-wave diode bridge is

$$E_{av_0} = \frac{3\sqrt{3}}{2\pi} E_m$$

Therefore,

$$E_{ar} = \begin{cases} E_{av_0}\cos\alpha & 0 \le \alpha \le \dfrac{\pi}{6} \\[2ex] \dfrac{E_{av_0}}{\sqrt{3}}\left[1+\cos\left(\alpha+\dfrac{\pi}{6}\right)\right] & \dfrac{\pi}{6} \le \alpha \le \dfrac{5\pi}{6} \end{cases}$$

The variation of E_{av}/E_{avO} is shown in Fig. 5.4.

Example 5.2 For the three-phase, half-wave controlled rectifier circuit of Fig. 5.1 a 240-V/phase, 50-Hz source supplies power to a load of 100 ω. Calculate the average and rms load currents and hence the load current ripple factor for (1) α = 30° and (2) α = 90°.

For a firing angle α = 30° either set of modal equations can be used. From Eq. (5.8),

$$I_{av} = \frac{3\sqrt{3}\sqrt{2}240}{2\pi100} \cos 30°$$
$$= 2.43 \text{ A}$$

From Eq. (5.13),

$$I_a = \frac{240\sqrt{2}}{100} \sqrt{\frac{4\pi + 3\sqrt{3}/2}{24\pi}}$$
$$= \frac{240\sqrt{2}}{100}(0.476)$$
$$= 1.522 \text{ A}$$

Now it may be deduced from Eq. (5.18) that

$$P_L = I_L^2 R = 3I_a^2 R$$

so that

$$I_L = \sqrt{3}I_a$$

Therefore, in this case,

$$I_L = 2.64 \text{ A}$$

In Eq. (4.13), therefore,

$$RF = \sqrt{\left(\frac{I_L}{I_{av}}\right)^2 - 1}$$
$$= \sqrt{\left(\frac{2.64}{2.43}\right)^2 - 1} = 0.425$$

The above value for the ripple factor compares with the value 0.185 obtained with α = 0°. The higher value is consistent with the waveforms, whereby one would expect the waveform of Fig. 5.2d to have a greater ac ripple content than the waveform of Fig. 5.2b.

For α = 90°, from Eq. (5.9),

$$I_{av} = \frac{3\sqrt{2}\,240}{2\pi 100}\left(1 - \frac{1}{2}\right)$$

$$= 0.81 \text{ A}$$

From Eq. (5.14)

$$I_a = \frac{240\sqrt{2}}{100}\sqrt{\frac{(5\pi/3) - \pi - \sqrt{3}/2}{8\pi}}$$

$$= \frac{240\sqrt{2}}{100}(0.221)$$

$$= 0.75 \text{ A}$$

But

$$I_L = \sqrt{3}I_a = 1.3 \text{ A}$$

The ripple factor at $\alpha = 90°$ therefore becomes

$$RF = \sqrt{\left(\frac{1.3}{0.81}\right)^2 - 1} = 1.254$$

Example 5.3 A three-phase, half-wave controlled rectifier is fed from an ideal three-phase supply. Calculate the displacement factor and the distortion factor with resistive load and show that these are consistent with the power factor Eqs. (5.20) and (5.21)

In order to calculate the displacement factor and distortion factor, it is necessary to determine the Fourier coefficients of the fundamental component of the supply current. For current $i_a(\omega t)$, in general,

$$a_1 = \frac{1}{\pi}\int_0^{2\pi} i_a(\omega t)\cos\omega t\, d\omega t$$

$$b_1 = \frac{1}{\pi}\int_0^{2\pi} i_a(\omega t)\sin\omega t\, d\omega t$$

For $0 \le \alpha \le \pi/6$, the substitution of Eq. (5.11) into the two equations above yields

$$a_1 = \frac{-E_m\sqrt{3}}{4\pi R}\sin 2\alpha$$

$$b_1 = \frac{E_m}{4\pi R}\frac{4\pi + 3\sqrt{3}\cos 2\alpha}{3}$$

For $\pi/6 \le \alpha \le 5\pi/6$, the substitution of Eq. (5.12) into the defining equation for a_1 and b_1 gives

$$a_1 = \frac{-E_m}{4\pi R}\left[1 - \cos\left(2\alpha + \frac{\pi}{3}\right)\right]$$

$$b_1 = \frac{E_m}{4\pi R}\left[\frac{5\pi}{3} - 2\alpha + \sin\left(2\alpha + \frac{\pi}{3}\right)\right]$$

The phase angle ψ_1 which represents the phase difference between the supply-phase voltage and the fundamental component of the supply current is defined by

$$\psi_1 = \tan^{-1}\frac{a_1}{b_1}$$

The displacement factor, $\cos\psi_1$, may therefore be expressed in terms of components a_1 and b_1

$$\text{Displacement factor} = \cos\psi_1 = \cos\left(\tan^{-1}\frac{a_1}{b_1}\right) = \frac{a_1}{\sqrt{a_1^2 + b_1^2}}$$

For $0 \le \alpha \le \pi/6$,

$$\text{Displacement factor} = \frac{4\pi + 3\sqrt{3}\cos 2\alpha}{\sqrt{27 + 24\pi\sqrt{3}\cos 2\alpha + 16\pi^2}}$$

For $\pi/6 \le \alpha \le 5\pi/6$,

$$\text{Displacement factor} = \frac{(5\pi/3) - 2\alpha + \sin(2\alpha + \pi/3)}{\sqrt{\left[1 - \cos(2\alpha + \pi/3)\right]^2 + \left[(5\pi/3) - 2\alpha + \sin(2\alpha + \pi/3)\right]^2}}$$

The distortion factor may also be expressed in terms of the Fourier coefficients a_1 and b_1.

$$\begin{aligned}\text{Distortion factor} &= \frac{I_1}{I_a} \\ &= \frac{1}{\sqrt{2}}\frac{c_1}{I_a} \\ &= \frac{1}{\sqrt{2}}\frac{\sqrt{a_1^2 + b_1^2}}{I_a}\end{aligned}$$

The rms line current I_a is obtained from Eqs. (5.13) and (5.14). Substituting for a_1 and b_1, and I_a gives
For $0 \le \alpha \le \pi/6$

$$\text{Distortion factor} = \sqrt{\frac{(9/4\pi) + (4\pi/3) + 2\sqrt{3}\cos 2\alpha}{4\pi + 3\sqrt{3}\cos\alpha}}$$

For $\pi/6 \le \alpha \le 5\pi/6$

$$\text{Distortion factor} = \sqrt{\frac{\left[1 - \cos(2\alpha + \pi/3)\right]^2 + \left[(5\pi/3) - 2\alpha + \sin(2\alpha + \pi/3)\right]^2}{4\pi\left[(5\pi/3) - 2\alpha + \sin(2\alpha + 5\pi/3)\right]}}$$

Now

Power factor = (distortion factor)(displacement factor)

Substituting the respective expressions for displacement factor and distortion factor into the power factor equation above gives
For $0 \le \alpha \le \pi/6$

$$PF = \sqrt{\frac{4\pi + 3\sqrt{3}\cos 2\alpha}{12\pi}}$$

For $\pi/6 \le \alpha \le 5\pi/6$

$$PF = \sqrt{\frac{(5\pi/3) - 2\alpha + \sin(2\alpha + \pi/3)}{4\pi}}$$

These expressions for the power factor are seen to agree with those of Eqs. (5.20) and (5.21) which were obtained from the system power dissipation.

5.2 RESISTIVE LOAD WITH IDEAL SUPPLY AND SHUNT CAPACITOR COMPENSATION

Let the three-phase bridge circuit of Fig. 5.1 be now modified to incorporate identical capacitors at the supply point, as shown in Fig. 5.3. The supply is considered to be ideal so that the supply voltage is not affected by connection of the capacitance. If the three capacitors are of good quality they will not absorb any significant amount of power so that the expressions in Eqs. (5.16) and (5.17) remain unchanged. Because the supply voltage is unchanged there is no change in the thyristor switch currents nor in the load current, due to the capacitors.

The capacitor currents constitute a balanced three-phase set of sinusoidal currents.

$$i_{c_a} = \frac{E_m}{X_c}\sin\left(\omega t + \frac{\pi}{2}\right) = \frac{E_m}{X_c}\cos\omega t$$

$$(5.22)$$

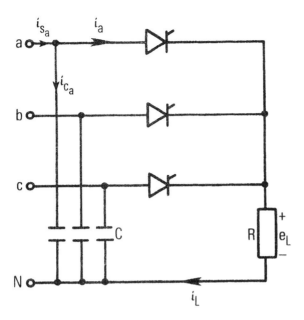

FIG. 3   Three-phase, half-wave, silicon controlled rectifier with supply point capacitance.

$$i_{c_b} = \frac{E_m}{X_c} \sin\left(\omega t - \frac{2\pi}{3} + \frac{\pi}{2}\right) \tag{5.23}$$

$$i_{c_c} = \frac{E_m}{X_c} \sin\left(\omega t - \frac{4\pi}{3} + \frac{\pi}{2}\right) \tag{5.24}$$

The instantaneous supply current $i_{s_a}(\omega t)$ entering phase a in Fig. 5.3 is therefore given by the sum of the switch current i_a, Eq. (5.11) or (5.12) and the capacitor current i_{ca}, Eq. (5.22).

For $0 \le \alpha \le \pi/6$,

$$i_{s_a}(\omega t) = \frac{E_m}{X_c} \cos\omega t + \frac{E_m}{R} \sin\omega t \Big|_{\pi/6+\alpha}^{5\pi/6+\alpha} \tag{5.25}$$

For $\pi/6 \le \alpha \le 5\pi/6$,

$$i_{s_a}(\omega t) = \frac{E_m}{X_c} \cos\omega t + \frac{E_m}{R} \sin\omega t \Big|_{\pi/6+\alpha}^{\pi} \tag{5.26}$$

When the above expressions are substituted into Eq. (5.10), corresponding expressions for the rms current are obtained as follows:

For $0 \leq \alpha \leq \pi/6$,

$$I_{s_a} = \frac{E_m}{\sqrt{2}R} \sqrt{\frac{R^2}{X_c^2} + \frac{4\pi + 3\sqrt{3}\cos 2\alpha}{12\pi} - \frac{\sqrt{3}R}{2\pi X_c}\sin 2\alpha} \tag{5.27}$$

For $\pi/6 \leq \alpha \leq 5\pi/6$,

$$I_{s_a} = \frac{E_m}{\sqrt{2}R} \sqrt{\frac{R^2}{X_c^2} + \frac{5\pi/3 - 2\alpha + \sin(2\alpha + \pi/3)}{4\pi} - \frac{R}{2\pi X_c}\left[1 - \cos\left(2\alpha + \frac{\pi}{3}\right)\right]} \tag{5.28}$$

When $C = 0$, $X_c = \infty$ and Eqs. (5.27) and (5,28) reduce to Eqs. (5.13) and (5.14), respectively. Now because the power and supply voltage are unchanged, compared with uncompensated operation, the power factor will increase (i.e., improve) if the rms value of the supply current decreases. The presence of capacitor compensation will therefore improve the power factor if the value of the current in Eq. (5.27) is less than the corresponding load current in Eq. (5.13).

For $0 \leq \alpha \leq \pi/6$, power factor compensation will occur if

$$\frac{R^2}{X_c^2} + \frac{4\pi + 3\sqrt{3}\cos 2\alpha}{12\pi} - \frac{\sqrt{3}R}{2\pi X_c}\sin 2\alpha < \frac{4\pi + 3\sqrt{3}\cos 2\alpha}{12\pi}$$

or if

$$\frac{R^2}{X_c^2} < \frac{\sqrt{3}R}{2\pi X_c}\sin 2\alpha \tag{5.29}$$

Rearranging relation (5.29) gives as the required criterion

$$X_c > \frac{2\pi}{\sqrt{3}R}\operatorname{cosec} 2\alpha \tag{5.30}$$

When $\alpha = 0$, the required inequality is that $X_c > \infty$. This is clearly impossible and demonstrates that no power factor improvement is possible at $\alpha = 0$, as was shown in Chapter 4.

For any fixed, nonzero value of α, with any fixed load R, Eq. (5.30) shows that some power factor improvement may be realized if an appropriate value of C is used. The power factor in the presence of terminal point capacitance can be calculated by the substitution of Eqs. (5.22), (5.23), (5.27), and (5.28) into Eq. (5.25). The appropriate expressions are shown in Table 5.1 Maximum power factor will occur when the rms supply current is a minimum. Let the expressions for rms current I_{s_a} be differentiated with respect to C and equated to zero.

For $0 \leq \alpha \leq \pi/6$,

$$\frac{dI_{s_a}}{dC} = \frac{\left(E_m/\sqrt{2R}\right)\left[2R^2\omega^2C - \left(\sqrt{3}R\omega/2\pi\right)\sin 2\alpha\right]}{R^2/X_c^2 + \left(4\pi + 3\sqrt{3}\cos 2\alpha\right)/12\pi - \left(\sqrt{3}R/2\pi X_c\right)\sin 2\alpha} = 0 \qquad (5.31)$$

The bracketed term in the numerator must be zero so that, for a maximum,

$$X_c = \frac{4\pi R}{\sqrt{3}\sin 2\alpha}$$

$$C = \frac{\sqrt{3}\sin 2\alpha}{4\pi\omega R} \qquad (5.32)$$

With this value of compensating capacitance substituted into (5.27)

$$I_{s_{a_{min}}} = \frac{E_m}{\sqrt{2R}}\sqrt{\frac{4\pi + 3\sqrt{3}\cos 2\alpha}{12\pi} - \frac{3\sin^2 2\alpha}{16\pi^2}} \qquad (5.33)$$

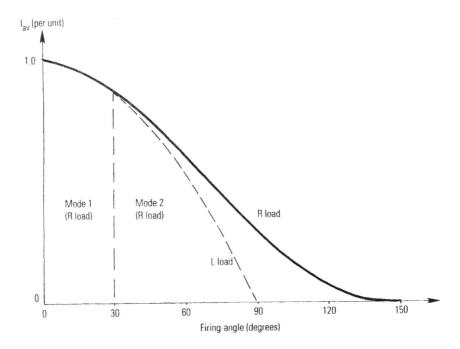

FIG. 4 Average load current versus firing angle for the three-phase, half-wave controlled rectifier.

The uncompensated rms supply current in Eq. (5.13) is therefore reduced by the second term of Eq. (5.33) and the power factor is therefore increased,

For $0 \leq \alpha \leq \pi/6$, the differentiation technique results in the corresponding expression,

$$I_{S_{a\,min}} = \frac{E_m}{\sqrt{2R}} \sqrt{\frac{\frac{5\pi}{3} - 2\alpha + \sin\left(2\alpha + \frac{\pi}{3}\right)}{4\pi} - \frac{\left[1 - \cos\left(2\alpha + \frac{\pi}{3}\right)\right]^2}{16\pi^2}} \qquad (5.34)$$

5.2.1 Worked Example

Example 5.4   A resistive load $R = 100\ \Omega$ is supplied from an ideal three-phase 50-Hz supply via a three-phase, half-wave controlled bridge rectifier. Equal capacitors C are connected across each phase of the supply. Calculate the value of capacitance C that would result in maximum power factor at $\alpha = 30°$. What is the percentage improvement of power factor then realized by capacitance compensation?

At $\alpha = 30°$, the optimum value of the compensation capacitance is given by Eq. (5.26)

$$C = \frac{\sqrt{3}\sin 2\alpha}{4\pi\omega R}$$

$$= \frac{\sqrt{3}\left(\sqrt{3}/2\right)}{4\pi(2\pi)(50)(100)}$$

$$= 38\mu F$$

The new value of the rms supply current is given by Eq. (5.27)

$$I_{S_{a\,min}} = \frac{240}{100} \sqrt{\frac{4\pi + 3\sqrt{3}/2}{12\pi} - \frac{9/4}{16\pi^2}}$$

$$= \frac{240}{100}\sqrt{0.402 - 0.0142}$$

$$= 1.495\ A$$

In the absence of the capacitors only the first term of Eq. (5.27) is valid, so that

$$I_{S_a} = \frac{240}{100}\sqrt{0.402} = 1.522\ A$$

The ratio of compensated to uncompensated power factor is the inverse ratio of the respective rms currents

$$\frac{PF_c}{PF} = \frac{I_{s_a}}{I_{s_{a_{min}}}} = \frac{1.522}{1.495} = 1.018$$

The improvement of power factor at $\alpha = 30°$ is therefore 1.8%

5.3 HIGHLY INDUCTIVE LOAD AND IDEAL SUPPLY

A three-phase, half-wave controlled bridge rectifier with a highly inductive load is shown in Fig. 5.5. The purpose of the load inductor is to smooth the load current to, as nearly as possible, an ideal direct current with no ripple at all.

The supply voltages defined by Eqs. (5.1)–(5.3) are still valid. Circuit power dissipation is presumed to take place only in the load resistor.

If the load inductance is sufficiently high the load current becomes very smooth as shown in Fig. 5.6c–g. The instantaneous load voltage retains the same form as for resistive load (Fig. 5.2) until $\alpha = 30°$. For $\alpha = 30°$, the load voltage now goes negative for part of its cycle as in, for example, Fig. 5.6f. At $\alpha = 90° =$

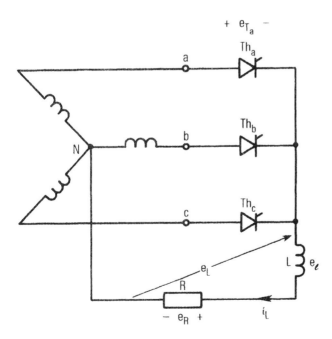

FIG. 5 Three-phase, half-wave controlled rectifier with highly inductive load.

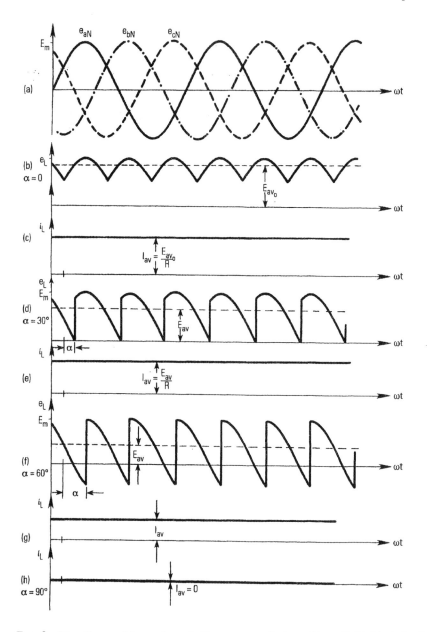

Fig. 6 Waveforms of three-phase, half-wave controlled rectifier with highly inductive load: (a) supply phase voltages, (b) load voltage ($\alpha = 0°$), (c) load current ($\alpha = 0°$), (d) load voltage ($\alpha = 30°$), (e) load current ($\alpha = 30°$), (f) load voltage ($\alpha = 60°$), (g) load current ($\alpha = 60°$), and (h) load current ($\alpha = 90°$).

$\pi/2$, the negative and positive alternations (not shown) are equal so that the average load voltage and current become zero (Fig. 5.6h).

The load voltage e_L (ωt) in Fig. 5.6 is now described by Eq. (5.4) for both α < 30° and α > 30°. Unlike the case with resistive load (Fig. 5.2), only one mode of operation now occurs and this is equal to the mode $0 \le \alpha \le \pi/6$ that occurred with resistive load. The average load voltage with highly inductive load is therefore given by Eq. (5.6), repeated below.

For $0 \le \alpha \le \pi/2$,

$$E_{av} = \frac{3\sqrt{3}}{2\pi} E_m \cos\alpha = E_{av_0} \cos\alpha \qquad (5.35)$$

The average load voltage is given in Fig. 5.4 and is seen to be less than the corresponding value with resistive load for all values of α in the working range. The average load current is, once again, given by

$$I_{av} = \frac{E_{av}}{R} = \frac{E_{av_0}}{R} \cos\alpha \qquad (5.36)$$

With highly inductive load the smooth load current waveform satisfies (very nearly) the current relationship

$$i_L(\omega t) = I_{av} = I_L = I_m \qquad (5.37)$$

The load current ripple factor is therefore zero.

The supply current now consists of rectangular pulses of conduction angle $2\pi/3$. Each phase carries identical pulses with a phase difference of $2\pi/3$ or 120° from the other two phases. In phase a, Fig. 5.7, the supply current, in the first supply voltage cycle, is given by

$$i_a(\omega t) = I_{av} \Big|_{(5\pi/6)+\alpha}^{(5\pi/6)+\alpha} \qquad (5.38)$$

The rms value of this supply current is therefore

$$I_a = \sqrt{\frac{1}{2\pi} \int_{(\pi/6)+\alpha}^{(5\pi/6)+\alpha} i_a^2(\omega t)\, d\omega t}$$

$$= \frac{I_{av}}{\sqrt{3}} \qquad (5.39)$$

Combining Eqs. (5.36) and (5.39) gives

$$I_a = \frac{E_{av_0}}{\sqrt{3}R} \cos\alpha \qquad (5.40)$$

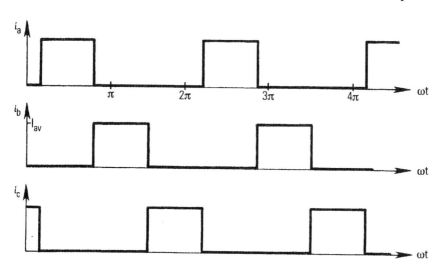

FIG. 7   Supply currents of three-phase, half-wave controlled rectifier with highly inductive load

The average value of the supply current is seen, by inspection of Fig. 5.6, to be one-third of the average load current. From Eq. (5.36), therefore,

$$I_{a_{av}} = \frac{1}{3} I_{av} = \frac{E_{av_0}}{\sqrt{3}R} \cos \alpha \tag{5.41}$$

The total load power dissipation may be obtained from Eqs. (5.36) and (5.37),

$$P_L = I_L^2 R = \frac{E_{av_0}^2}{R} \cos^2 \alpha \tag{5.42}$$

It may be assumed that the power is transferred from the supply to the load equally by the three phases, so that

$$P_a = \frac{E_{av_0}^2}{3R} \cos^2 \alpha = P_b = P_c \tag{5.43}$$

The power factor may be obtained by the substitution of Eqs. (5.40) and (5.43) into Eq. (5.19)

$$PF = \frac{P_a}{E_a I_a}$$

$$= \frac{P_a}{\dfrac{E_m}{\sqrt{2}} I_a}$$

$$= \frac{3}{\sqrt{2\pi}} \cos\alpha$$

$$= 0.675 \cos\alpha \tag{5.44}$$

When $\alpha = 0$, Eq. (5.44) reduces to Eq. (4.25) which was derived for the uncontrolled (diode), half-wave bridge. Some properties of the three-phase, half-wave bridge circuit with highly inductive load are given in Table 5.2.

If the bridge is compensated by equal capacitors at the supply point in the manner of Fig. 5.4, the supply current in phase a becomes

$$i_{s_a}(\omega t) = \frac{E_m}{X_c} \cos\omega t \bigg|_0^{2\pi} + I_{av} \bigg|_{(\pi/6)+\alpha}^{(5\pi/6)+\alpha} \tag{5.45}$$

The substitution of Eq. (5.45) into the defining integral gives an expression for the rms current

$$I_{s_a} = \sqrt{\frac{1}{2\pi} \int_0^{2\pi} i_{s_a}^2(\omega t)\, d\omega t}$$

$$= \sqrt{\frac{E_m^2}{2X_c^2} - \frac{\sqrt{3}}{\pi} \frac{E_m}{X_c} I_{av} \cos 2\alpha + \frac{I_{av}^2}{3}} \tag{5.46}$$

It is seen that (5.46) reduces to (5.39) when the bridge is uncompensated, and, in effect, $X_c = \infty$. The rms supply current may be expressed in terms of the rms thyristor current $I_{av}/\sqrt{3}$ by combining Eqs. (5.39) and (5.46).

$$I_{s_a} = \sqrt{\frac{E_m^2}{2X_c^2} - \frac{3}{\pi} \frac{E_m}{X_c} I_a \cos 2\alpha + I_a^2} \tag{5.47}$$

The terminal voltage and load power are not affected by the connection of shunt capacitance at the supply. The power factor is therefore improved, by reduction of the rms supply current, if

$$\frac{3}{\pi} \frac{E_m}{X_c} I_a \cos 2\alpha > \frac{E_m^2}{2X_c^2} \tag{5.48}$$

TABLE 5.2   Some Properties of the Three-Phase, Half-Wave
Controlled Bridge Rectifier with Highly Inductive Load and Ideal
Supply ($0 \leq \alpha \leq 90°$)

Circuit property	Expression
Instantaneous supply current	$i_a(\omega t) = \begin{vmatrix} (5\pi/6) + \alpha \\ (\pi/6) + \alpha \end{vmatrix}$
RMS supply current (I_s)	$\dfrac{I_{av}}{\sqrt{3}} = \dfrac{3}{2\pi}\dfrac{E_m}{R}\cos\alpha$
Average load current	$I_{av} = \dfrac{3\sqrt{3}}{2\pi}\dfrac{E_m}{R}\cos\alpha$
Average load power	$\left(\dfrac{3\sqrt{3}}{2\pi}\right)^2 \dfrac{E_m^2}{R}\cos^2\alpha$
Power factor	$\dfrac{3}{\sqrt{2\pi}}\cos\alpha$
RMS load current (I_L)	$I_L = I_{av} = \dfrac{3\sqrt{3}}{2\pi}\dfrac{E_m}{R}\cos\alpha$
a_1	$-\dfrac{\sqrt{3}}{\pi}I_{av}\sin\alpha$
b_1	$+\dfrac{\sqrt{3}}{\pi}I_{av}\cos\alpha$
Displacement factor ($\cos\psi_1$)	$\cos\alpha$
Distortion factor	$\dfrac{3}{\sqrt{2}\,\pi}$

Rearranging Eq. (5.48) gives as the required criterion

$$X_c > \frac{\pi}{6}\frac{E_m}{I_a}\sec 2\alpha \tag{5.49}$$

The minimum value of rms supply current obtainable by capacitance compensation
may be obtained, as before, by differentiating Eq. (5.47) with respect to C and equating to zero. The appropriate optimum values of X_c and C then prove to be

$$X_c > \frac{\pi}{3} \frac{E_m}{I_a} \sec 2\alpha$$

$$C > \frac{3}{\pi} \frac{I_a}{\omega E_m} \cos 2\alpha \tag{5.50}$$

With this value of capacitance, the value of the rms supply current minimum value is

$$I_{S_{a_{min}}} = I_a \sqrt{1 - \frac{9}{2\pi^2} \cos^2 2\alpha} \tag{5.51}$$

Both the power and the supply phase voltage are unaffected by the connection of good quality capacitors across the terminals. The power factor with optimum capacitor compensation PF_c is therefore obtained by substituting Eq. (5.51) into Eq. (5.44).

$$\begin{aligned}
PF_c &= \frac{P_a}{E_a I_{S_{a_{min}}}} \\
&= \frac{P_a}{E_a I_a \sqrt{1 - (9/2\pi^2)(\cos^2 2\alpha)}} \\
&= \frac{PF}{\sqrt{1 - (9/2\pi^2)(\cos^2 2\alpha)}} \\
&= \frac{0.675 \cos\alpha}{\sqrt{1 - (9/2\pi^2)(\cos^2 2\alpha)}}
\end{aligned} \tag{5.52}$$

5.3.1 Worked Examples

Example 5.5   A three-phase, half-wave controlled rectifier supplies a highly inductive load from an ideal three-phase supply. Derive an expression for the average load voltage and compare this with the corresponding case for resistive load at $\alpha = 60°$.

The circuit diagram is shown in Fig. 5.5. Although $\omega L \gg R$ at supply frequency, the inductor offers negligible impedance at the load frequency (ideally zero). Resistor R may well represent the only load device that is being powered.

The load voltage $e_L(\omega t)$ shown in Fig. 5.6 is a continuous function for all values of α and is seen to be described by Eq. (5.4), which has an average value

$$E_{av} = \frac{3\sqrt{3}}{2\pi} E_m \cos\alpha = E_{av_0} \cos\alpha \tag{Ex. 5.5a}$$

This is precisely the same expression that is pertinent to resistive loads in the mode when $0 \le \alpha \le \pi/6$. For resistive loads where $\alpha > \pi/6$, the current becomes discontinuous, the modal behavior changes, and the average load voltage is represented by Eq. (5.7), reproduced below.

$$E_{av} = \frac{3E_m}{2\pi}\left[1+\cos\left(\alpha+\frac{\pi}{6}\right)\right]$$

$$= \frac{E_{av_0}}{\sqrt{3}}\left[1+\cos\left(\alpha+\frac{\pi}{6}\right)\right]$$

(Ex. 5.5b)

At $\alpha = 60°$, the average voltage with inductive load, from (Ex. 5.5a), is

$$E_{av}(\text{inductive load}) = \frac{1}{2}E_{av_0}$$

Similarly, at $\alpha = 60$, the average load voltage with resistive load, from (Ex. 5.5b), is

$$E_{av}(\text{resistive load}) = \frac{E_{av_0}}{\sqrt{3}}(1+0) = \frac{E_{av_0}}{\sqrt{3}}$$

Therefore, at $\alpha = 60°$,

$$\frac{E_{av}(\text{inductive load})}{E_{av}(\text{resistive load})} = \frac{\sqrt{3}}{2} = 0.866$$

This result is seen to be consistent with the data of Fig. 5.4.

Example 5.6 In the three-phase rectifier bridge of Fig. 5.5 a three-phase, 415-V, 50-Hz, supply transfers power to a 100-Ω load resistor. What is the effect on the power transferred at (a) $\alpha = 30°$, (b) $\alpha = 60°$, if a large inductance is connected in series with the load resistor.

With resistive load the input power per phase is given by Eqs. (5.16) and (5.17). Assuming that each phase provides one-third of the load power, then $P_L = 3P_a$. At $\alpha = 30°$, from Eq. (5.16), assuming that 415 V is the line voltage,

$$P_L = \frac{3E_m^2}{2R}\frac{4\pi+3\sqrt{3}\cos2\alpha}{12\pi}$$

$$= \frac{3\left(415\sqrt{2}/\sqrt{3}\right)^2}{200}\left(\frac{4\pi+3\sqrt{3}/2}{12\pi}\right)$$

$$= 1722(0.4022) = 692.7 \text{ W}$$

At $\alpha = 60°$, from Eq. (5.17),

$$P_L = \frac{3E_m^2}{2R}\left[\frac{\dfrac{5\pi}{3} - 2\alpha + \sin\left(2\alpha + \dfrac{\pi}{3}\right)}{4\pi}\right]$$

$$= 1722(0.25) = 430.5 \text{ W}$$

With highly inductive load the load power is given by Eq. (5.42) for all α.

$$P_L = \frac{E_{av_0}^2}{R}\cos^2\alpha$$

$$= \left(\frac{3\sqrt{3}}{2\pi}E_m\right)^2\frac{1}{R}\cos^2\alpha$$

$$= 785\cos^2\alpha$$

At $\alpha = 30°$,

$$P_L = 785\times\frac{3}{4} = 589 \text{ W}$$

At $\alpha = 60°$,

$$P_L = 785\times\frac{1}{4} = 196 \text{ W}$$

The effect of the load inductor is therefore to reduce the rms load current and thereby the load power dissipation below the value obtained with the load resistor acting alone. If the same load power dissipation is required in the presence of a high inductor filter, then the value of the load resistor must be reduced to permit more load current to flow.

   Example 5.7   A three-phase, half-wave controlled rectifier has ideal thyristor elements and is fed from a lossless supply. The rectifier feeds a load consisting of a resistor in series with a large filter inductor. Sketch the waveform of the reverse voltage across a thyristor at $\alpha = 60°$ and specify its maximum value.

   The circuit is shown in Fig. 5.5 with some typical load voltage and current waveforms given in Fig. 5.6. Let the reverse voltage on thyristor Th_a be designated e_{T_a} as shown in Fig. 5.5. When thyristor Th_c is conducting, for example, the instantaneous value of this reverse voltage can be written

$$-e_{T_a}(\omega t) = e_{cN} - e_{aN} \tag{Ex. 5.7a}$$

From Eqs. (5.1) and (5.3) it is seen that

$$-e_{T_a}(\omega t) = E_m \sin\left(\omega t - \frac{4\pi}{3}\right) - \sin \omega t$$

$$= -\sqrt{3}E_m \cos\left(\omega t - \frac{2\pi}{3}\right)$$

$$= -\sqrt{3}E_m \sin\left(\frac{\pi}{2} - \omega t + \frac{2\pi}{3}\right)$$

$$= \sqrt{3}E_m \sin\left(\omega t - \frac{7\pi}{6}\right)$$

Similarly, if thyristor Th_b is conducting, the instantaneous reverse voltage across Th_a is

$$-e_{T_a}(\omega t) = e_{bN} - e_{aN}$$

$$= E_m \sin\left(\omega t - \frac{2\pi}{3}\right) - \sin \omega t$$

$$= \sqrt{3}E_m \sin\left(\omega t - \frac{5\pi}{6}\right) \qquad \text{(Ex. 5.7b)}$$

But it can be seen from the three-phase voltage waves of Fig. 5.8 that

$$\sqrt{3}E_m \sin\left(\omega t - \frac{7\pi}{6}\right) = e_{ca}(\omega t)$$

$$\sqrt{3}E_m \sin\left(\omega t - \frac{5\pi}{6}\right) = e_{ba}(\omega t)$$

The reverse voltage $-e_{T_a}$ can therefore be defined in terms of the line-to-line supply voltages, which is obvious from the basic expressions in Eqs. (Ex. 5.7a) and (Ex. 5.7b).

The waveform of the reverse blocking voltage at thyristor Th_a, for $\alpha = 60°$, is given in Fig. 5.8. Its peak value is clearly equal to $\sqrt{3}E_m$, being the peak value of the line-to-line supply voltage.

Example 5.8   Equal capacitors C are to be used across the supply to neutral terminals to compensate the power factor of a three-phase, half-wave, thyristor controlled bridge rectifier with a highly inductive load in which the load resistor $R = 100\ \Omega$. Calculate the value of capacitance that will give maximum power factor and the degree of power factor improvement at (1) $\alpha = 30°$ and (2) $\alpha = 60°$ if the supply voltage is 240 V/phase at 50 Hz.

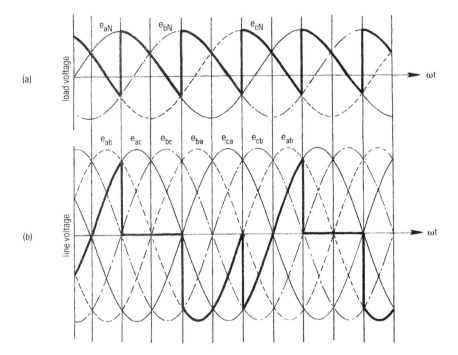

(a)

(b)

FIG. 8   Voltage waveforms of three-phase, half-wave controlled rectifier with highly inductive load, $\alpha = 60°$: (a) load voltage and (b) voltage across thyristor Th_a.

The appropriate circuit diagram is given in Fig. 5.3. For both of the specified values of thyristor firing angle the optimum capacitance is given by Eq. (5.50). The rms supply current I_a in the absence of capacitance is, from Eq. (5.40),

$$I_a = \frac{E_{av_0}}{\sqrt{3}R}\cos\alpha$$

$$= \frac{3E_m}{2\pi R}\cos\alpha$$

At $\alpha = 30°$,

$$I_a = \frac{3\sqrt{2}\,(240)}{2\pi 100}\frac{\sqrt{3}}{2} = 1.4\ A$$

At $\alpha = 60°$,

$$I_a = \frac{3\sqrt{2}\,(240)}{2\pi(100)}\frac{1}{2} = 0.81\text{ A}$$

Substituting into Eq. (5.50) gives

at $\alpha = 30°$

$$C = \frac{3}{\pi}\frac{1.4}{2\pi 50}\frac{10^6}{\sqrt{2}240}\frac{1}{2} = 0.63\ \mu\text{F}$$

at $\alpha = 60°$

$$C = \frac{3}{\pi}\frac{0.81}{2\pi 50}\frac{10^6}{\sqrt{2}240}\frac{1}{2} = 0.36\ \mu\text{F}$$

The supply current reduction is given in Eq. (5.51).

$$\frac{I_{s_{a\min}}}{I_a} = \sqrt{1 - \frac{9}{2\pi^2}\cos^2 2\alpha}$$

At $\alpha = 30°$,

$$\frac{I_{s_{a\min}}}{I_a} = \sqrt{1 - 0.342} = 0.81$$

At $\alpha = 60°$,

$$\frac{I_{s_{a\min}}}{I_a} = \sqrt{1 - 0.114} = 0.941$$

The connection of the capacitors makes no difference to the transfer of power or to the supply voltages. The power factor is therefore improved by the inverse ratio of the supply current reduction. At $\alpha = 30°$,

$$\frac{PF_c}{PF} = \frac{1}{0.81} = 1.235$$

At $\alpha = 60°$,

$$\frac{PF_c}{PF} = \frac{1}{0.941} = 1.063$$

The actual values of power factor after compensation may be obtained from Eq. (5.52). At $\alpha = 30°$, $PF_c = 0.621$. At $\alpha = 60°$, $PF_c = 0.36$.

5.4 SERIES *R-L* LOAD AND IDEAL SUPPLY

A general case exists in which the load inductance is finite and cannot be considered as infinitely large. In such cases the circuit can be analyzed using the appropriate differential equations. If thyristor Th_a in Fig. 5.5 is conducting, for example, then supply voltage $e_a(\omega t)$ is impressed across the load so that

$$e_a(\omega t) = E_m \sin \omega t = i_L R + L \frac{di_L}{dt} \tag{5.53}$$

If $i_L = 0$ at the switching angle α, the load current consists of discontinuous, nonsinusoidal current pulses described by the equation

$$i_L(\omega t) = \frac{E_m}{|Z_L|} \left[\sin(\omega t - \phi_L) - \sin\left(\alpha + \frac{\pi}{6} - \phi_L\right) \varepsilon^{-\cot \phi_L [\omega t - \pi/6 - \alpha]} \right] \tag{5.54}$$

where

$$|Z_L| = \sqrt{R^2 + \omega^2 L^2} \qquad \phi_L = \tan^{-1} \frac{\omega L}{R} \tag{5.55}$$

The average value I_{av} and rms value I_L of the load current can be obtained by the use of the defining integrals

$$I_{av} = \frac{1}{2\pi} \int_0^{2\pi} i_L(\omega t)\, d\omega t \tag{5.56}$$

$$I_L = \sqrt{\frac{1}{2\pi} \int_0^{2\pi} i_L^2(\omega t)\, d\omega t} \tag{5.57}$$

If the load inductance is sufficiently large to maintain continuous conduction the load current waveform is of the general type shown in Fig. 5.9. At $\alpha = 60°$, the load voltage retains the same waveform as in Fig. 5.6f and its average value is still given by Eq. (5.35), with the average current given by Eq. (5.36).

For the load current waveform of Fig. 5.9c, the instantaneous value is given by Eq. (5.54) plus an additional term involving current value I_{min}. This is obtained by solving Eq. (5.52) for $i_L(\omega t) = I_{min}$ when $\omega t = \alpha + \pi/6$.

$$i_L(\omega t) = \frac{E_m}{|Z|} \left[\sin(\omega t - \phi_L) - \sin\left(\alpha + \frac{\pi}{6} - \phi_L\right) \varepsilon^{-\cot \phi_L [\omega t - \pi/6 - \alpha]} \right]$$
$$+ I_{min} \varepsilon^{-\cot \phi_L [\omega t - \pi/6 - \alpha]} \tag{5.58}$$

Equation (5.58) can be rearranged to give an explicit expression for I_{min} by noting that $i_L(\omega t) = I_{min}$ at $\omega t = \alpha + \pi/6 + 2\pi/3$.

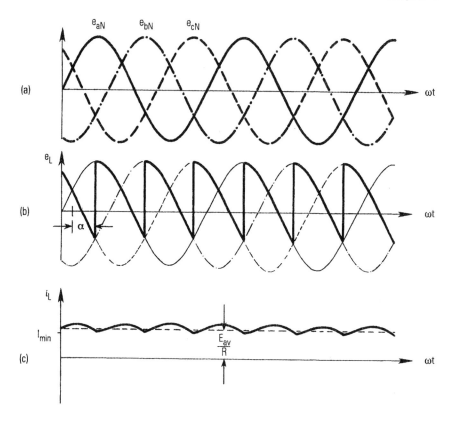

Fig. 9 Waveforms of three-phase, half-wave controlled rectifier with series R-L load of arbitrary value, $\alpha = 60°$.

If a bypass diode D is connected across the load impedance, as in Fig. 5.10, the load voltage $e_L(\omega t)$ cannot go negative. At $\alpha = 60°$, for example, the load voltage then has the form of Fig. 5.2f (which is valid for a resistive load) even when the load is highly inductive. The effect of the bypass diode is to chop off the negative part of $e_L(\omega t)$ in Fig. 5.6f and thereby to increase the average load voltage and current without affecting the supply current.

It was shown in Sec. 5.1 that average load voltage with resistive load, for $\alpha \leq 30°$, is given by Eq. (5.6), which compares with the corresponding expression Eq. (5.35) for highly inductive load.

The ratio of the two average voltage expressions represents a net increase

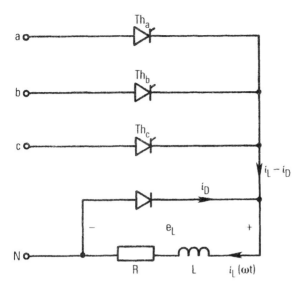

Fig. 10   Three-phase, half-wave controlled rectifier with bypass diode.

$$\frac{E_{av_R}}{E_{av_L}} = \left(\begin{array}{c}\text{increase of load voltage}\\\text{due to bypass diode}\end{array}\right) = \frac{1+\cos(\alpha+\pi/6)}{\sqrt{3}\cos\alpha} \tag{5.59}$$

At $\alpha = 60°$, for example, the ratio in Eq. (5.59) has the value 1.155. In other words, the effect of the bypass diode is to increase the average load voltage by 15.5%.

5.5   SERIES R-L LOAD PLUS A CONSTANT EMF WITH IDEAL SUPPLY

The most general form of burden on a three-phase, half-wave controlled bridge rectifier is shown in Fig. 5.11. In addition to a series R-L impedance of arbitrary phase angle, the load current is opposed by an emf E, which is constant or which changes in a known manner, This might arise, for example, in battery charging or in the speed control of a separately excited dc motor. With the polarity shown in Fig. 5.11 the average load current is reduced because emf E is in opposition to the mean driving voltage E_{av} created by rectification:

$$I_{av} = \frac{E_{av}-E}{R} = \frac{E_{av_0}\cos\alpha - E}{R} \tag{5.60}$$

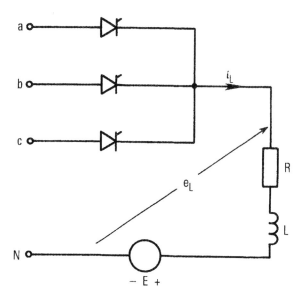

Fig. 11   Three-phase, half-wave controlled rectifier with series R-L load, incorporating a load-side emf.

Further accounts of the applications of bridge rectifiers to dc motor control are given in Refs. 10, 11, 15, 20, 25, and 27.

5.6   HIGHLY INDUCTIVE LOAD IN THE PRESENCE OF SUPPLY IMPEDANCE

Consider that a series inductance L_s now exists in each phase of the supply. The terminal voltages e_{aN}, e_{bN}, and e_{cN} at the bridge terminals (Fig. 5.12) are not sinusoidal while the bridge draws current from the supply but are given by Eqs. (4.26)–(4.28). The supply voltages at the generation point are still given by Eqs. (4.29)–(4.31).

In a controlled rectifier the effect of delayed triggering is to delay the start of conduction in the conductor containing the switch. Once conduction has begun the supply reactance delays the buildup of current for a period known as the *overlap period*. Much of the discussion of Sec. 4.4, relating to the three-phase, half-wave diode bridge rectifier is also relevant here. The effects of delayed triggering plus overlap are illustrated in Fig. 5.13. At a firing angle $\alpha = 30°$ conduction commences. The phase current does not reach its final value E_{av}/R until the overlap period μ is completed, although the load current is smooth and continuous. The

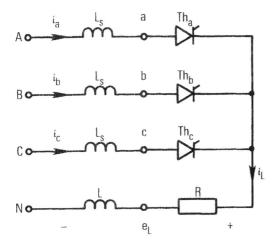

FIG. 12 Three-phase, half-wave controlled rectifier circuit, including supply inductance.

effect of overlap removes a portion (shown dotted) from the bridge terminal voltages that would be present with ideal supply. As a result, the area under the curve of the load voltage is reduced and so is its average value.

In Fig. 5.13 the average load voltage is given by

$$
\begin{aligned}
E_{av} &= \frac{3}{2\pi}\left[\int_{30°}^{30°+\alpha} e_{CN}\,d\omega t + \int_{30°+\alpha}^{30°+\alpha+\mu}\frac{e_{AN}+e_{CN}}{2}\,d\omega t + \int_{30°+\alpha+\mu}^{150°+\alpha} e_{aN}\,d\omega t\right] \\
&= \frac{3\sqrt{3}}{4\pi}E_m\left[\cos\alpha + \cos(\alpha+\mu)\right] \\
&= \frac{E_{av_0}}{2}\left[\cos\alpha + \cos(\alpha+\mu)\right]
\end{aligned}
\tag{5.61}
$$

When $\alpha = 0$, Eq. (5.61) reduces to Eq. (4.33), which was derived for diode operation. Since the time average value of the voltage on the load inductor is zero, then

$$
I_{av} = \frac{E_{av}}{R}
\tag{5.62}
$$

If it is required to supply the load with constant current, then since E_{av} varies with α and μ, resistor R most be varied at the same rate. During the first overlap interval in Fig. 5.13, thyristors Th_a and Th_c are conducting simultaneously. A complete circuit therefore exists such that, starting from point N in Fig. 5.12,

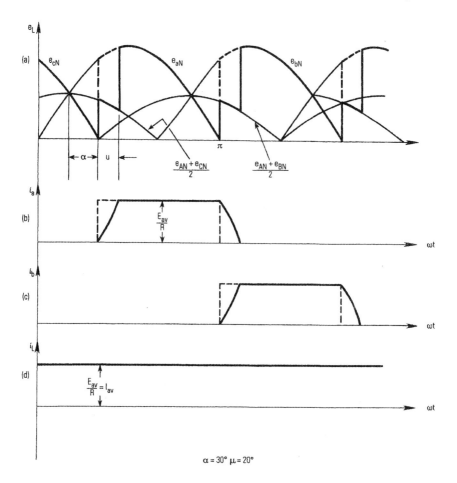

FIG. 13 Waveforms of three-phase, half-wave controlled rectifier with highly inductive load, in the presence of supply inductance, $\alpha = 30°$, $\mu = 20°$: (a) load voltage, (b) supply current i_a (ωt), (c) supply current i_b (ωt), and (d) load current.

$$-e_{AN} + L_s \frac{di_a}{dt} - L_s \frac{di_c}{dt} + e_{CN} = 0 \qquad (5.63)$$

Since thyristor Th_b is in extinction, current i_b is zero and

$$i_a + i_c = I_{av} \qquad (5.64)$$

Eliminating i_c from Eqs. (5.63) and (5.64) gives

$$e_{AN} - e_{CN} = 2L_s \frac{di_a}{dt} = e_{AC} \tag{5.65}$$

Note that this voltage e_{AC} (ωt), which falls across two line reactances in series, is not the same as the corresponding instantaneous load voltage $(e_{AN} + e_{CN})/2$ (Fig. 5.13), which is also valid during the same period of overlap.

The instantaneous time variation of current i_a (ωt) in the overlap period $\alpha + 30° \leq \omega t \leq \alpha + 30° + \mu$ may be obtained from Eq. (5.65).

$$i_a(t) = \frac{1}{L_s} \int e_{AC}(t) \, dt$$

or

$$i_a(\omega t) = \frac{1}{2\omega L_s} \int e_{AC}(\omega t) \, d\omega t$$

$$= \frac{1}{2\omega L_s} \int \sqrt{3} E_m \sin(\omega t - 30°) \, d\omega t$$

$$= -\frac{\sqrt{3} E_m}{2\omega L_s} \cos(\omega t - 30°) + K$$

But (1) $i_a = I_{av}$ at $\omega t = 30° + \alpha + \mu$

(2) and $i_a = 0$ at $\omega t = 30° + \alpha.$ $\tag{5.66}$

But (1) $i_a = I_{av}$ at $\omega t = 30° + \alpha + \mu$
(2) and $i_a = 0$ at $\omega t = 30° + \alpha.$ From condition (1)

$$K = I_{av} + \frac{\sqrt{3} E_m}{2\omega L_s} \cos(\alpha + \mu) \tag{5.67}$$

From condition (2)

$$K = \frac{\sqrt{3} E_m}{2\omega L_s} \cos \alpha \tag{5.68}$$

The instantaneous time variation i_a (ωt) may be written by combining Eqs. (5.66) and (5.67).

$$i_a(\omega t) \bigg|_{\omega t = 30° + \alpha}^{\omega t = 30° + \alpha + \mu} = I_{av} + \frac{\sqrt{3} E_m}{2\omega L_s} \cos\left[(\alpha + \mu) - \cos(\omega t - 30°) \right] \tag{5.69}$$

The two expressions Eqs. (5.67) and (5.68) for constant of integration K may be equated to give

$$I_{av} = \frac{\sqrt{3}E_m}{2\omega L_s}\left[\cos\alpha - \cos(\alpha+\mu)\right]$$

(5.70)

Note that the sign in the bracketed term of Eq. (5.70) is different from the corresponding sign in the similar expression Eq. (5.61) for the average load voltage. Equations (5.61) and (5.70) may be combined to give

$$E_{av} = E_{av_0}\cos\alpha - \frac{3\omega L_s}{2\pi}I_{av}$$

(5.71)

The first term of Eq. (5.71) is seen to be equal to Eq. (5.35) and represents the average load voltage with ideal supply. The second term in Eq. (5.71) represents the effect of voltage drop in the supply inductances.

Since $I_{av} = E_{av}/R$, Eq. (5.71) can be rearranged to give

$$E_{av} = \frac{E_{av_0}\cos\alpha}{1+\frac{3\omega L_s}{2\pi R}}$$

(5.72)

It is seen that Eq. (5.72) reduces to Eq. (4.40) when $\alpha = 0$.

A discussion on the effects of delayed firing and of overlap on bridge operation is given at the end of Chapter 7.

5.6.1 Worked Examples

Example 5.9   A three-phase, half-wave controlled rectifier contains inductance L_s in each supply line. The rectifier supplies a highly inductive load with a load current of average value I_{av}. Devise an equivalent circuit to show the effect on the average load voltage E_{av} of the switch firing angle α and the supply inductance.

From Eq. (5.71) it can be inferred that there is an effective open-circuit voltage $E_{av0}\cos\alpha$ on the load side. This constitutes a dc driving voltage that is reduced due to the effect of supply side inductance when current flows. A possible equivalent circuit is given in Fig. 5.14, which also shows the effect of supply line resistance.

$$E_{av} = E_{av_0}\cos\alpha - \frac{3\omega L_s}{2\pi}I_{av} - R_sI_{av}$$

It is often permissible to assume that resistance $R_s = 0$.

Example 5.10   A three-phase, half-wave rectifier uses silicon controlled rectifier switches and supplies a highly inductive load circuit in which the current is maintained constant at 50 A (by adjustment of the load resistance). The three-phase supply has a line-to-line voltage of 230 V at 50 Hz, and each supply line contains an effective series inductance $L_s = 1.3$ mH plus a series resistance $R_s = 0.05$ ω. Calculate the average load voltage for firing angles $\alpha = 0°$, 30°, and 60°.

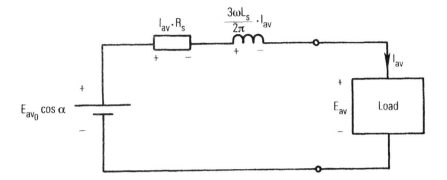

FIG. 14   Load-side equivalent circuit for a three-phase, half-wave controlled rectifier with supply impedance.

The average load-side voltage on open circuit is, from Eq. (5.35),

$$E_{av_0} = \frac{3\sqrt{3}}{2\pi} \frac{230\sqrt{2}}{\sqrt{3}} = 155.3 \text{ V}$$

Without the effects of voltage drop in the supply lines the average load voltage would be

$$E_{av} = E_{av0} \cos \alpha = 155.3 \cos \alpha$$

At a constant load current of 50 A there is a constant voltage drop $50 \times 0.05 = 2.5$ V in the supply line resistor. Also at a load current of 50 A the supply line inductor results in a voltage drop, Eq. (5.71), of

$$\frac{3\omega L_s}{2\pi} I_{av}$$
$$= \frac{3 \times 2\pi 50 \times 1.3}{2\pi 1000} 50$$
$$= 9.75 \text{ V}$$

Because this calculation is performed in equivalent dc side terms, no question of phase relationship arises, and the voltage drops on the supply line resistor and inductor can be added algebraically. For all values of firing angle the voltage drop along a supply line is $2.5 + 9.75 = 12.25$ V.
At $\alpha = 0°$,

$$E_{av} = 155.3 - 12.25 = 143 \text{ V}$$

At $\alpha = 30°$,

$\qquad E_{av} = 155.3 \times 0.866 - 12.25 = 122.2$ V

At $\alpha = 0°$,

$\qquad E_{av} = 155.3 \times 0.5 - 12.25 = 65.4$ V

Example 5.11 A three-phase, half-wave controlled rectifier using SCR switches supplies power from a 415-V, 50-Hz bus with a short-circuit reactance of 0.415 Ω/phase to a highly inductive load. The load current is 50 A when the load voltage is maximum. Calculate the communication time of the switches at $\alpha = 30°$.

The effect of commutation or simultaneous conduction between two phases is illustrated by the waveforms in Fig. 5.13, for the case of $\alpha = 30°$. The per-phase supply reactance, (sometimes called the *short-circuit reactance* or *commutation reactance*) is given as

$\qquad X_{sc} = \omega L_s = 2\pi 50 L_s = 0.415$ Ω

It can be assumed that the given value 415 V represents the rms line voltage.
Therefore,

$\qquad \sqrt{3}\,E_m = $ peak line voltage $= \sqrt{2} \times 415 = 587$ V

Now maximum average load current occurs when $\alpha = 0$. In Eq. (5.70)

$$50 = \frac{587}{2 \times 0.415}(1 - \cos\mu)$$

from which

$\qquad \mu = \cos^{-1} 0.929 = 21.7° = 0.378$ radian

At $\alpha = 0$, the average load voltage is, from (5.71),

$$E_{av} = \frac{3\sqrt{3}}{2\pi}E_m - \frac{3}{2\pi} \times 0.415 \times 50$$

$$= \frac{3}{2\pi}(587 - 20.75)$$

$$= \frac{3}{2\pi} \times 566.25 = 270 \text{ V}$$

Constant Load Resistance. If the load resistance is kept constant at the value R $= E_{av}/I_{av} = 270/50 = 5.4$ Ω, from Eq. (5.62), then the average voltage with $\alpha = 30°$, from Eq. (5.72), is

$$E_{av} = \frac{(3/2\pi) \times 587 \times \sqrt{3}/2}{1 + (3/2\pi)(0.415/5.4)}$$

$$= \frac{242.5}{1.0367} = 234.1 \text{ V}$$

The load current at $\alpha = 30°$ will then have fallen from Eq. (5.62), to

$$I_{av} = \frac{E_{av}}{R} = \frac{234.1}{5.4} = 43.35 \text{ A}$$

Substituting into Eq. (5.70),

$$43.35 = \frac{587}{0.83}\left[0.866 - \cos(30° + \mu)\right]$$

from which

$$\mu = 6.42° = 0.112 \text{ radian}$$

But $\mu = \omega t_{com}$, where t_{com} is the commutation time.
Therefore,

$$t_{com} = \frac{0.112}{2\pi 50} = 0.357 \text{ mS}$$

Constant Load Current. If the load current is kept constant at 50 A then from Eq. (5.70), at $\alpha = 30°$

$$50 = \frac{587}{0.83}\left[0.866 - \cos(30° + \mu)\right]$$

which gives

$$\mu = 7.32° = 0.128 \text{ radian}$$

Therefore,

$$t_{com} = \frac{0.128}{2\pi 50} = 0.41 \text{ mS.}$$

With constant current of 50 A the average voltage at $\alpha = 30°$, from Eq. (5.61), is

$$E_{av} = \frac{1}{2} \cdot \frac{3\sqrt{3}}{2\pi}(0.866 + 0.795)$$

$$= 140.14(1.661)$$

$$= 232.8 \text{ V}$$

which is achieved by reducing the load resistor from 5.4 Ω to 232.8/50 = 4.660 Ω.

PROBLEMS

Three-Phase, Half-Wave Controlled Rectifier with Ideal Supply and Resistive Load

5.1   A three-phase, half-wave controlled rectifier is supplied from an ideal three-phase, 415-V, 50-Hz source. The load is purely resistive and of value 75 Ω. Calculate the average and ms load voltages and the load voltage ripple factor at α = 45°.

5.2   For the circuit of Problem 5.1 calculate the average load current at (a) α = 30°, (b) α = 60°, (c) α = 90°, and (d) α = 120°.

5.3   Calculate the power dissipation in the load for the circuit of Problem 5.1 when α = 60°.

5.4   Calculate the rms values of the supply currents for the circuit of Problem 5.1.

5.5   Calculate the value of the load current ripple factor for the circuit of Problem 5.1 at α = 0°, 30°, 60°, 90°, 120°, and 150°. Plot this function versus α.

5.6   For a three-phase, half-wave controlled bridge rectifier with R load show that the Fourier coefficients of the fundamental component of the supply currents are given by

$$a_1 = \frac{-E_m\sqrt{3}}{4\pi R}\sin 2\alpha$$

$$b_1 = \frac{E_m}{4\pi R}\frac{4\pi + 3\sqrt{3}\cos 2\alpha}{3}$$

5.7   Use the expressions for the Fourier coefficients a_1 and b_1 from Problem 5.6 to obtain an expression for the displacement angle ψ_1 of the fundamental component of the supply currents. Use this angle to obtain an expression for the input power per phase. Calculate the value of this power for the circuit of Problem 5.1 at α = 60° and check that the result is consistent with the value from Problem 5.3.

5.8   For the circuit of Problem 5.1 calculate the power factor for α = 30°, 60°, 90°, 120°, and 150° Plot the results against α. Would it be correct to refer to the results obtained as describing a "lagging" power factor?

5.9   Three equal ideal capacitors C are connected across the supply point of a three-phase, half-wave bridge, as shown in Fig. 5.3. If α = 30° and the peak values of the load and capacitor currents are equal, sketch the waveform of the corresponding supply current. Is it possible to tell if the supply current is ''improved'' by examining its waveform?

5.10   Show that the expressions given in Eqs. (5.27) and (5.28) for the rms current I_{s_a} are correct.

5.11   A three-phase, half-wave rectifier circuit is compensated by terminal shunt capacitance C as shown in Fig. 5.3. For $\pi/6 < \alpha < 5\pi/6$, derive a criterion for the relationship between C and load resistor R that would result in power factor improvement. [Hint: consider the derivation of Eq. (5.30).]

5.12   For the capacitance-compensated bridge circuit of Fig. 5.3 show that the minimum possible value of the rms supply current is given by Eq. (5.34) if $\alpha > 30°$. What is the percentage reduction of rms supply current due to optimal capacitance compensation at $\alpha = 60°$, $90°$, and $120°$?

5.13   Show that the reactive voltamperes of the capacitance compensated bridge circuit of Fig. 5.3 is given by

$$\frac{E_m^2}{2R}\left(\frac{R}{X_c} - \frac{\sqrt{3}\sin 2\alpha}{4\pi} \right) \quad \text{if } 0 \le \alpha \le \frac{\pi}{6}$$

$$\frac{E_m^2}{2R}\left[\frac{R}{X_c} - \frac{1 - \cos\left(2\alpha + \dfrac{\pi}{3} \right)}{4\pi} \right] \quad \text{if } \frac{\pi}{6} \le \alpha \le 5\frac{\pi}{6}$$

Is this reactive voltamperes associated with energy storage as would be the case in a linear, sinusoidal circuit?

5.14   Derive an expression for the dc or average value of the supply current for the circuit of Fig. 5.1. Is this value affected by the connection of compensating capacitors across the terminals?

Three-Phase, Half-Wave Controlled Rectifier with Ideal Supply and Highly Inductive Load

5.15   A three-phase, half-wave, silicon controlled rectifier bridge circuit is supplied from an ideal three-phase source of 415 V, 50 Hz. The load consists of a resistor $R = 75 \; \Omega$ in series with a high inductance filter. Calculate the average values of the load current and the supply current at $\alpha = 0°$, $30°$, $60°$, and $90°$.

5.16 For the circuit of Problem 5.15, calculate the ms supply current, the power dissipation and the power factor,

5.17 Derive expressions for the fundamental components a_1 and b_1 of the Fourier series describing the supply current to the three-phase bridge of Problem 5.15. Also, derive expressions for the displacement angle ψ_1 and the displacement factor $\cos \psi_1$, and calculate the values for the data of Problem 5.15.

5.18 Show that the rms value of the supply currents to a three-phase, half-wave controlled bridge rectifier with a high inductance in the load circuit is given by

$$I_a = \frac{E_{av_0}}{\sqrt{3R}} \cos \alpha$$

5.19 Show that the power factor of the three-phase bridge of Problem 5.15 is given by equation (5.44)

5.20 Calculate the operating power factor of the load bridge of Problem 5.15 at $\alpha = 0°$, $30°$, $60°$ and $90°$ and compare the results with the corresponding operation with resistive load.

5.21 A three-phase, half-wave, controlled bridge rectifier supplies power to a load consisting of resistor R in series with a large inductor. Power factor improvement is sought by the connection of equal capacitors C between the supply voltage terminals and the neutral point N. Show that the rms value of the resulting supply current is given by (5.47)

5.22 Show that the value of capacitance C that results in maximum power factor operation for the bridge of Fig. 5.3, with load filter inductance, is given by equation (5.50). Also show that the corresponding minimum value of the supply current is given by equation (5.51)

5.23 For the circuit of Problem 5.15 calculate the maximum improvement of operating power factor that can be realised by shunt capacitor compensation. Sketch the uncompensated and compensated power factors as functions of thyristor firing-angle.

5.24 Derive an expression for the reactive voltamperes at the terminals of the bridge of Problem 5.15. How is this expression modified by the connection of equal capacitors C of arbitrary value at the terminals? What is the value of C that will make the reactive voltamperes equal to zero?

5.25 Derive an expression for the average value of the supply current to a three-phase, half-wave, controlled bridge rectifier with highly inductive load. Is this value affected by the connection of compensating capacitors across the terminals? Compare the result with that obtained for resistive load, in Problem 5.15.

5.26 Use the Fourier coefficients $a_1 b_1$ derived in Problem 5.17 to obtain an expression for the rms fundamental component I_1 of the supply current in Problem 5.25. Hence obtain an expression of the distortion factor seen from the supply point.

Three-Phase, Half-Wave Controlled Rectifier with Series R-L Load Supplied from an Ideal Three-Phase Source

5.27 A three-phase, half-wave, silicon controlled rectifier bridge has a series $R\text{-}L$ load of phase angle ϕ. For the case when the load current is continuous, show that its instantaneous value is given by Eq. (5.58) in which the parameter $I_{\min}$ is given by

$$I_{\min} = \frac{\left(E_m / |Z|\right)\left[\sin\left(\alpha + 5\pi/6 - \phi\right) - \sin\left(\alpha + \pi/6 - \phi\right)\varepsilon^{-\cot\phi(2\pi/3)}\right]}{1 - \varepsilon^{-\cot\phi\,(2\pi/3)}}$$

5.28 For the three-phase, half-wave bridge of Problem 5.27 show that the average value of the load current at a firing angle α is given by

$$I_{av} = \frac{3}{2\pi}\frac{E_m}{|Z|}\left\{\left(1 - \varepsilon^{-\cot\phi\,(2\pi/3)}\right)\left[-\cos\left(\alpha + \frac{\pi}{6} - \phi\right) + \cos\left(\alpha + \frac{5\pi}{6} - \phi\right)\right.\right.$$

$$\left.\left.\tan\phi\,\sin\left(\alpha + \frac{\pi}{6} - \phi\right)\right]\right\}$$

$$+ \frac{3}{2\pi}I_{\min}\tan\phi\left(1 - \varepsilon^{-\cot\phi(2\pi/3)}\right)$$

5.29 In the three-phase bridge circuit of Fig. 5.10 calculate the increase of the average load voltage due to the presence of the bypass diode D at (a) $\alpha = 30°$, (b) $\alpha = 45°$, (c) $\alpha = 60°$, and (d) $\alpha = 75°$.

Three-Phase, Half-Wave Controlled Bridge Rectifier with Highly Inductive Load in the Presence of Supply Inductance

5.30 A three-phase, half-wave, silicon controlled bridge rectifier supplies power to a load containing a large series filter inductance. Each supply line contains series inductance that causes an overlap angle μ of about

20°. Sketch compatible waveforms of the phase voltage and current at the bridge terminals. Explain how the reduction of the average load voltage due to μ is related to the area (in voltseconds) under the phase voltage curve.

5.31 From the load voltage waveforms of Fig. 5.13, show that the average load voltage is given by

$$E_{av} = \frac{E_{av_0}}{2}[\cos\alpha + \cos(\alpha + \mu)]$$

5.32 A three-phase supply 415 V, 50 Hz transfers power to a load resistor R via a three-phase, half-wave controlled bridge rectifier. The load current is smoothed by a filter inductor and maintained at a constant level of 40 A. Each supply line contains an effective series inductance of 1 mH. Calculate the average load voltage and the value of the load resistance for firing angles $\alpha = 30°, 60°, 90°$.

5.33 For the three-phase, half-wave bridge of Problem 5.30, sketch compatible waveforms of the load voltage, SCR (line) current I_a, the voltage across switch Th_a and the voltage across the source inductance in line a for $\alpha = 15°$, assuming that overlap angle $\mu = 30°$.

5.34 A three-phase, half-wave, SCR bridge rectifier is operating with firing angle α (where $\alpha < 30°$) and supplying a highly inductive load. The effect of supply line inductance L_s is to cause an overlap angle $\mu \cong 20°$. Show that during the overlap period when $30° + \alpha \leq \omega t \leq 30° + \alpha + \mu$, the supply current in phase a is given by

$$i_a(\omega t) = \frac{\sqrt{3}E_m}{2\omega L_s}\left[\cos\alpha - \cos(\omega t - 30°)\right]$$

6

Three-Phase, Full-Wave Uncontrolled Bridge Rectifier Circuits

The basic full-wave uncontrolled (diode) rectifier circuit is shown in Fig. 6.1. Diodes D_1, D_3, and D_5 are sometimes referred to as the *upper half* of the bridge, while diodes D_2, D_4, and D_6 constitute the *lower half* of the bridge. As with half-wave operation the voltages at the anodes of the diode valves vary periodically as the supply voltages undergo cyclic excursions. Commutation or switch-off of a conducting diode is therefore accomplished by natural cycling of the supply voltages and is known as *natural commutation*. The load current i_L is unidirectional, but the supply currents are now bidirectional. In order to permit load current to flow, at least one diode must conduct in each half of the bridge. When this happens, the appropriate line-to-line supply point voltage is applied across the load. In comparison with the half-wave bridge (Fig. 4.2a), in which the supply-phase voltage is applied across the load, the full-wave bridge has the immediate advantage that the peak load voltage is $\sqrt{3}$ times as great.

6.1 RESISTIVE LOAD AND IDEAL SUPPLY

With resistive load and an ideal supply of three-phase, balanced sinusoidal voltages, there is an instantaneous transfer of anode voltage on each diode in sequence. Because of the resistive load, the load current and appropriate supply current map the waveform of the corresponding line voltage. Each diode conducts for one-third (120°) of each 360° supply voltage cycle. With the numbering notation

179

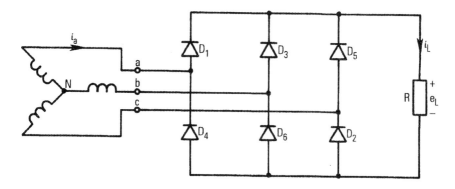

Fɪɢ. 1   Three-phase, full-wave diode rectifier circuit with resistive load.

of Fig. 6.1, which is standard for the three-phase, full-wave bridge in both its rectifier and inverter modes of operation, the conduction pattern of the rectifier circuit diodes is given in Fig. 6.2a in which the upper tier represents the upper half of the bridge. In the period from $30° < \omega t < 90°$ of the phase voltage waveform, for example, it is seen that diodes D_1 and D_6 are conducting. From Fig. 6.1 it is seen that line voltage e_{ab} is then applied across the load. An alternative representation of the diode conduction pattern is shown in Fig. 6.2b in which the upper and lower tiers do not represent the upper and lower halves of the bridge. The pattern of Fig. 6.2b shows that the diodes conduct in chronological order.

The supply phase voltages in Fig. 6.2 are given, once again, by Eqs. (4.1)–(4.3). The corresponding line-to-line voltages are

$$e_{ab} = e_{aN} + e_{Nb} = e_{aN} - e_{bN} = \sqrt{3}E_m \sin(\omega t + 30°) \tag{6.1}$$
$$e_{bc} = \sqrt{3}E_m \sin(\omega t - 90°) \tag{6.2}$$
$$e_{ca} = \sqrt{3}E_m \sin(\omega t - 120°) \tag{6.3}$$

The load current and voltage profile (Fig. 6.2e) contains a ripple component of six times supply frequency (i.e., the repetitive period of the ripple is one-sixth of the periodicity of the supply-phase voltage).

The average value of the load current is the same for every 60° interval in Fig. 6.2e

$$I_{av} = \text{average value of } i_L(\omega t)$$

$$= \text{average value of } \left. \frac{E_m}{R} \sin(\omega t + 30°) \right|_{30°}^{90°}$$

$$= \frac{6}{2\pi} \int_{30°}^{90°} \frac{E_m}{R} \sin(\omega t + 30°) \, d\omega t$$

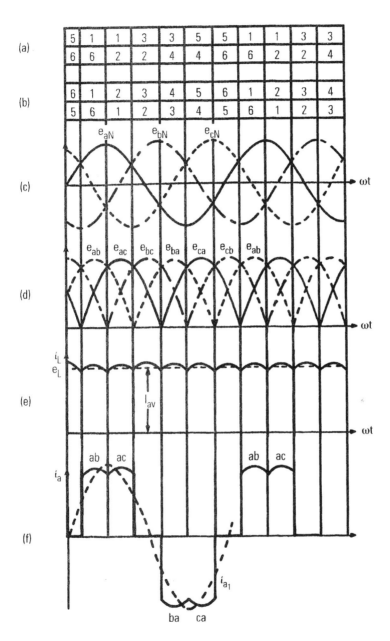

FIG. 2 Waveforms for three-phase, full-wave diode rectifier circuit with resistive load: (a) and (b) switching sequences, (c) supply phase voltages, (d) supply line voltages, (e) load current and voltage, and (f) line current.

$$= \frac{3}{\pi} \frac{\sqrt{3}E_m}{R}$$

$$= 1.654 \frac{E_m}{R} \tag{6.4}$$

Comparison of Eq. (6.4) with Eq. (4.5) shows that the average load current is doubled compared with corresponding half-wave operation. One of the principal functions of a bridge circuit is to produce the maximum possible value of the average or mean output voltage.

From Eqs. (6.4) it is seen that

$$E_{av} = \frac{3\sqrt{3}}{\pi} E_m = E_{av_o} \tag{6.5}$$

The value E_{av} in (6.5) refers to a zero impedance supply and is hereafter referred to as E_{av_o} to distinguish it from the more general case when the supply impedance has to be considered. The value E_{av_o} for a full-wave diode bridge Eqs. (6.5) is seen to be twice the value for the corresponding half-wave bridge, Eq. (4.6). The rms value of the load current is also equal for every 60° interval of the waveform of Fig. 6.2e

$$i_L(\omega t) = \frac{E_m}{R} \sin(\omega t + 30°) \Big|_{30°}^{90°} \tag{6.6}$$

so that

$$I_L = \sqrt{\frac{6}{2\pi} \int_{30°}^{90°} \frac{E_m^2}{R^2} \sin^2(\omega t + 30°)}$$

$$= 0.956 \frac{\sqrt{3}E_m}{R} = 1.66 \frac{E_m}{R} \tag{6.7}$$

This is seen to be almost double the corresponding value for half-wave operation, Eq. (4.7). The load power dissipation is therefore

$$P = I_L^2 R = 2.74 \frac{E_m^2}{R} = \frac{3E_m^2}{2R} \left(\frac{2\pi + 3\sqrt{3}}{2\pi} \right) \tag{6.8}$$

The value in Eqs. (6.8) is almost four times the corresponding value for half-wave operation given in Eq. (4.8).

The waveform of the supply current $i_a(\omega t)$ is given in Fig. 6.2f. This is a symmetrical function defined by the equation

$$i_a(\omega t) = \left.\frac{\sqrt{3}E_m}{R}\sin(\omega t + 30°)\right|_{30°}^{90°} + \left.\frac{\sqrt{3}E_m}{R}\sin(\omega t - 30°)\right|_{90°}^{150°} \qquad (6.9)$$

The rms value of every half wave is identical so that only a positive half wave need be considered. Proceeding by use of the defining integral for the rms line current I_a, it is found that

$$I_a = \sqrt{\frac{1}{\pi}\int_0^\pi i_a^2(\omega t)\,d\omega t}$$

$$= 0.78\frac{\sqrt{3}E_m}{R}$$

$$= 1.351\frac{E_m}{R}$$

$$= \frac{E_m}{R}\sqrt{\frac{2\pi + 3\sqrt{3}}{2\pi}} \qquad (6.10)$$

It may be deduced by inspection of Fig. 6.2f that the fundamental Fourier component $i_{a_1}(\omega t)$ of phase current $i_a(\omega t)$ is symmetrical with respect to the current waveform, as shown. This means that fundamental current $i_{a_1}(\omega t)$ is in time phase with its respective phase voltage $e_{aN}(\omega t)$. The current displacement angle ψ_1 is therefore zero, and the current displacement factor $\cos\psi_1 = 1$. The power factor seen from the bridge terminals is obtained from Eqs. (6.1), (6.9), and (6.10).

$$PF = \frac{P/3}{E_{aN}I_a} = \frac{(2.74/3)\times(E_m^2/R)}{(E_m/\sqrt{2})\times1.351\times(E_m/R)}$$

$$= 0.956 \qquad (6.11)$$

This compares with the corresponding value 0.686 for half-wave operation in Eq. (4.11). The Fourier coefficients of the fundamental component $i_{a_1}(\omega t)$ of the supply current waveform are found to be

$$a_1 = 0$$

$$b_1 = 1.055\frac{\sqrt{3}E_m}{R} = 1.827\frac{E_m}{R}$$

$$c_1 = 1.827\frac{E_m}{R}$$

$$\psi_1 = \tan^{-1}\frac{a_1}{b_1} = 0 \qquad (6.12)$$

The zero value of displacement angle in Eq. (6.12) obtained by Fourier series confirms the deduction made by inspection of Fig. 6.2f. The reduction of power factor below unity is therefore entirely due to distortion effects, and no power factor correction is possible by the connection of linear energy storage devices such as capacitors at the bridge input terminals.

The rms value of the fundamental component of the supply line current (Fig. 6.2f) is $1/\sqrt{2}$ of the peak value c_1 in Eq. (6.12)

$$
\begin{aligned}
I_{a_1} &= \frac{1}{\sqrt{2}} \times 1.055 \times \frac{\sqrt{3}E_m}{R} \\
&= 0.746 \times \frac{\sqrt{3}E_m}{R}
\end{aligned}
$$

(6.13)

The distortion factor of the supply current is given by

$$
\begin{aligned}
\text{Distortion factor} &= \frac{I_{a_1}}{I_a} \\
&= \frac{0.746}{0.78} \\
&= 0.956
\end{aligned}
$$

(6.14)

Since the displacement factor $\cos\psi_1 = 1.0$, Eq. (6.14) is seen to be consistent with Eq. (6.11). It can be seen from the waveform of Fig. 6.2f that the time average value of the supply current waveform is zero over any number of complete cycles. This eliminates the troublesome dc components that are present in the corresponding half-wave bridge circuit of Fig. 4.2.

Some circuit properties of the three-phase, full-wave diode bridge are given in Table 6.1

6.1.1 Worked Examples

Example 6.1   An ideal supply of balanced sinusoidal voltages is applied to the terminals of a three-phase, full-wave diode bridge with a pure resistance load $R = 25\ \Omega$. The supply voltages have an rms line value of 240 V at 50 Hz. Calculate the (1) average power dissipation, (2) average and rms load currents, and (3) the rms supply current. Compare the results with the corresponding values for the half-wave bridge of Example 4.1.

If the rms line-to-line voltage is 240 V, the peak phase voltage E_m is given by

$$
E_m = \frac{240\sqrt{2}}{\sqrt{3}} = 196\ \text{V}
$$

TABLE 6.1 Some Properties of the Three-Phase, Full-Wave Uncontrolled Bridge
Rectifier with Ideal Supply

Circuit property	Resistive load	Highly inductive load
Average load current	$\dfrac{3\sqrt{3}}{\pi}\dfrac{E_m}{R}$	$\dfrac{3\sqrt{3}}{\pi}\dfrac{E_m}{R}$
RMS load current	$\sqrt{\dfrac{3}{2}}\sqrt{\dfrac{2\pi+3\sqrt{3}}{2\pi}}\dfrac{E_m}{R}$	$\dfrac{3\sqrt{3}}{\pi}\dfrac{E_m}{R}$
Power dissipation	$\dfrac{3}{2}\left(\dfrac{2\pi+3\sqrt{3}}{2\pi}\right)\dfrac{E_m^2}{R}$	$\dfrac{27}{\pi^2}\dfrac{E_m}{R}$
RMS supply current	$\sqrt{\dfrac{2\pi+3\sqrt{3}}{2\pi}}\left(\dfrac{E_m}{R}\right)$	$\sqrt{\dfrac{2}{3}}\,I_{av}=\dfrac{3\sqrt{2}}{\pi}\dfrac{E_m}{R}$
RMS value of fundamental supply current	$1.292\dfrac{E_m}{R}$	$\dfrac{1}{\sqrt{2}}\dfrac{18}{\pi^2}\dfrac{E_m}{R}=1.290\dfrac{E_m}{R}$
Fourier coefficients of fundamental supply current $\quad a_1$	0	0
$\quad b_1$	$1.827\dfrac{E_m}{R}$	$\dfrac{3\sqrt{3}}{\pi}\dfrac{E_m}{R}$
Displacement factor	1.0	1.0
Distortion factor	0.956	0.955
Power factor	0.956	0.955

1. From Eq. (6.7) the rms load current is found to be

$$I_L = 1.66\frac{E_m}{R}$$
$$= 1.66 \times \frac{196}{25}$$
$$= 13.01 \text{ A}$$

The corresponding value for half-wave operation is 6.59A.
    The average load power is therefore

$$P = I_L^2 R = 4.234 \text{ kW}$$

which compares with 1.086 kW for half-wave operation.

2. The average load current is obtained from Eq. (6.4)

$$I_{av} = 1.654 \frac{E_m}{R}$$
$$= 12.967 \text{ A}$$

For the half-wave bridge the average load current is one half of this value being 6.48 A.

3. The rms supply current is given in Eq. (6.10)

$$I_a = 1.351 \frac{E_m}{R}$$
$$= 10.59 \text{ A}$$

This compares with a corresponding value 3.8 A for half-wave operation.

Example 6.2   A full-wave, three-phase diode bridge rectifier supplies power to a resistive load. With an ideal, three-phase power supply, the waveform of the supply line current is shown in Fig. 6.2f. Use Fourier analysis to calculate the magnitude of the fundamental component of this waveform. Sketch Fig. 6.2f and also draw, on the same axes, the fundamental component of current correct in magnitude and phase displacement. What is the time average value of the waveform over a supply cycle?

A sketch of the current waveform is reproduced as Fig. 6.3. With the terminology of this figure it is seen that

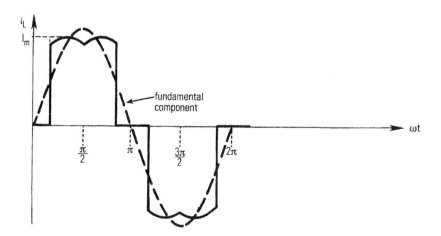

FIG. 3   Line current of three-phase, full-wave bridge rectifier with resistive load.

$$i_L(\omega t) = I_m \sin(\omega t + 30°)\Big|_{30°}^{90°} + I_m \sin(\omega t - 30°)\Big|_{90°}^{150°}$$

Since the waveform is symmetrical about the time axis

$$a_1 = \frac{1}{\pi}\int_0^{2\pi} i_L(\omega t)\cos\omega t \, d\omega t = 0$$

Therefore,

$$\psi_1 = \tan^{-1}\frac{a_1}{b_1} = 0$$

Now, in general,

$$b_1 = \frac{1}{\pi}\int_0^{2\pi} i_L(\omega t)\sin\omega t \, d\omega t$$

$$= \frac{2}{\pi}\int_0^{\pi} i_L(\omega t)\sin\omega t \, d\omega t$$

In this case,

$$b_1 = \frac{2I_m}{\pi}\left[\int_{30°}^{90°}\sin\omega t \, \sin(30° + \omega t)\, d\omega t + \int_{90°}^{150°}\sin\omega t \, \sin(\omega t - 30°)\, d\omega t\right]$$

Now

$$\sin\omega t \sin(30° + \omega t) = \frac{1}{2}\left[\cos(-30°) - \cos(2\omega t + 30°)\right]$$

$$\sin\omega t \sin(\omega t - 30°) = \frac{1}{2}\left[\cos 30° - \cos(2\omega t - 30°)\right]$$

Therefore

$$b_1 = \frac{2I_m}{\pi}\left\{\frac{1}{2}\left[\frac{\sqrt{3}}{2}\omega t - \frac{1}{2}\sin(2\omega t + 30°)\right]_{30°}^{90°} + \frac{1}{2}\left[\frac{\sqrt{3}}{2}\omega t - \frac{1}{2}\sin(2\omega t - 30°)\right]_{90°}^{150°}\right\}$$

$$= \frac{I_m}{\pi}\left[\sqrt{3}\left(\frac{\pi}{2} - \frac{\pi}{6} + \frac{5\pi}{6} - \frac{\pi}{2}\right) - \sin 210° + \sin 90° - \sin 270° + \sin 150°\right]$$

$$= \frac{I_m}{\pi}\left(\frac{2\sqrt{3}\pi}{3} + \frac{1}{2} + 1 + 1 + \frac{1}{2}\right)$$

$$= \frac{I_m}{\pi}(3.63 + 3) = 1.055 I_m = 1.055\frac{\sqrt{3}E_m}{R}$$

Since $a_1 = 0$, $c_1 = b_1$. The value $1.055 I_m$ therefore represents the peak value of the fundamental component of the supply current.

A sinusoid having this peak value and zero phase displacement is shown in Fig. 6.3 and is correct relative to the waveform itself. The time average value over any number of complete cycles is zero since the area under the positive half waves is canceled out by the area under the negative half waves.

Example 6.3   The load current waveform for a three-phase, full-wave diode bridge with resistive load is shown in Fig. 6.2e. Calculate a ripple factor for this waveform and compare it with the corresponding value for a half-wave, diode bridge

The use of Eq. (6.4) and (6.7) gives a value for the load current ripple factor

$$RF = \sqrt{\left(\frac{I_L}{I_{av}}\right)^2 - 1}$$

$$= \sqrt{\left(\frac{1.66}{1.654}\right)^2 - 1}$$

$$= 0.085$$

This low value compares with the value 0.185 for the corresponding half-wave bridge, Eq. (4.13).

It should be noted that the above value for the ripple factor is determined by the ratio of two similar figures. For this reason a small change of either I_{av} or I_L makes a comparatively large change in the value of the ripple factor. The actual value $RF = 0.085$ is important only with regard to its smallness compared with corresponding values for the single-phase bridge and the three-phase, half-wave bridge.

Example 6.4.   A three-phase, full-wave diode bridge is supplied from an ideal three-phase source and is resistively loaded. Show that the rms value of the supply line current is given by

$$I_a = 0.78 \frac{\sqrt{3} E_m}{R}$$

where E_m is the peak value of the supply phase voltage.

The circuit is shown in Fig. 6.1 and the waveform of supply current $i_a(\omega t)$ is shown in Fig. 6.2e. Over a complete cycle the current $i_a(\omega t)$ is defined by

$$i_a(\omega t) = i_{ab}(\omega t)\Big|_{30°, 210°}^{90°, 270°} + i_{ac}(\omega t)\Big|_{90°, 270°}^{150°, 330°}$$

$$= I_m \sin(\omega t + 30°)\Big|_{30°, 210°}^{90°, 270°} + I_m \sin(\omega t - 30°)\Big|_{90°, 270°}^{150°, 330°}$$

The peak current I_m is seen to have the value

$$I_m = \frac{\sqrt{3}E_m}{R}$$

Now, in general, the rms value of a function $i(\omega t)$ that is periodic in 2π is found from the defining expression

$$I = \sqrt{\frac{1}{2\pi} \int_0^{2\pi} i^2(\omega t)\, d\omega t}$$

If a function is symmetric so that the negative alternation is identical to the positive alternation, then

$$I = \sqrt{\frac{1}{\pi} \int_0^{\pi} i^2(\omega t)\, d\omega t}$$

In this present case,

$$I_a = \sqrt{\frac{I_m^2}{\pi}\left[\int_{30°}^{90°} \sin^2(\omega t + 30°)\, d\omega t + \int_{90°}^{150°} \sin^2(\omega t - 30°)\, d\omega t\right]}$$

Now,

$$\int \sin^2(\omega t \pm 30°)\, d\omega t = \frac{\omega t}{2} - \frac{1}{4}\sin 2(\omega t \pm 30°)$$

Therefore,

$$I_a^2 = \frac{I_m^2}{\pi}\left[\frac{\omega t}{2} - \frac{1}{4}\sin 2(\omega t + 30°)\right]_{30°}^{90°} + \frac{I_m^2}{\pi}\left[\frac{\omega t}{2} - \frac{1}{4}\sin 2(\omega t - 30°)\right]_{90°}^{150°}$$

$$= \frac{I_m^2}{\pi}\left(\frac{\pi}{4} - \frac{\pi}{12} - \frac{1}{4}\sin 240° + \frac{1}{4}\sin 120°\right) + \frac{I_m^2}{\pi}\left(\frac{5\pi}{12} - \frac{\pi}{4} - \frac{1}{4}\sin 240° + \frac{1}{4}\sin 120°\right)$$

$$= \frac{I_m^2}{\pi}\left[\frac{\pi}{6} + \frac{1}{4}\left(\frac{\sqrt{3}}{2}\right) + \frac{1}{4}\left(\frac{\sqrt{3}}{2}\right)\right]$$

$$= I_m^2 \frac{2}{\pi}(0.523 + 0.433)$$

$$= 0.6086 I_m^2$$

Therefore,

$$I_a = 0.78 I_m = 0.78 \frac{\sqrt{3}E_m}{R}$$

6.2 HIGHLY INDUCTIVE LOAD AND IDEAL SUPPLY

In some bridge applications a large inductance is included on the load side to smooth the output current. The bridge circuit then assumes the form shown in Fig. 6.4a. The instantaneous load voltage e_L (ωt) is still identical to the form for resistive load (Fig. 6.2f), but the load current is now constant.

$$i_L(\omega t) = I_L = I_{av} = \frac{E_{av}}{R} = \frac{3\sqrt{3}}{\pi} \frac{E_m}{R}$$

(6.15)

which is twice the value for half-wave operation, given in Eq. (4.16).

It should be noted that the average value of the load current is not changed by including the series load inductance. The inductor now absorbs any load side voltage ripple to result in a smooth, continuous current of fixed value through the load resistor. As with any purely direct current waveform, the instantaneous, average, rms, and peak values are identical. When the load current is completely smoothed, the load current ripple factor is, by definition, zero.

The average load voltage, E_{av}, across the output terminals in Fig. 6.4(a) is equal to the average voltage across the load resistor R since the average inductor voltage is zero.

$$E_{av} = E_{av_o} = \frac{3\sqrt{3}}{\pi} E_m$$

(6.16)

The load current is restrained largely by the load resistor not by the load inductor.

The average load current I_{av} is related to the average load voltage, as in the half-wave case, by

$$I_{av} = \frac{E_{av}}{R}$$

(6.17)

If the load impedance was entirely inductive, the load current would become destructively large and the overcurrent protection (fuses or circuit breakers) on the supply side would operate to isolate the bridge circuit from the supply.

The load power dissipation is now 4 times the value for corresponding half-wave operation, Eq. (4.24).

$$P_L = I_L^2 R = \frac{27}{\pi^2} \frac{E_m}{R}$$

(6.18)

Supply current i_a (ωt) (Fig. 6.4e) has the rms value

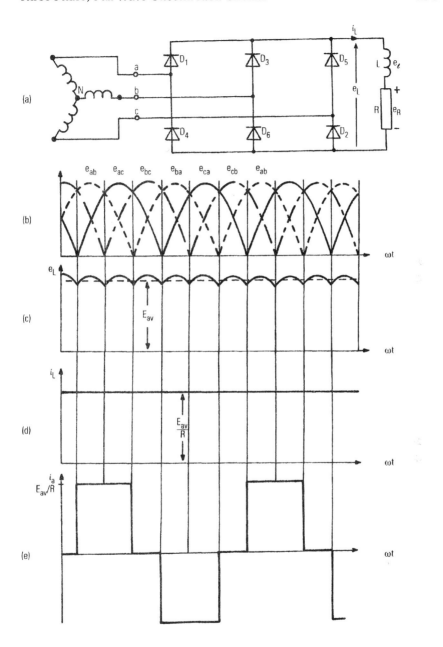

FIG. 4   Three-phase, full-wave diode bridge with highly inductive load (a) circuit connection: (b) supply line voltages, (c) load voltage, (d) load current, and (e) supply line current.

$$I_a = \sqrt{\frac{1}{\pi} \int_{30°}^{150°} I_{av}^2 \, d\omega t} = \sqrt{\frac{2}{3}} I_{av}$$

$$= \frac{3\sqrt{2}}{\pi} \frac{E_m}{R} = 1.350 \frac{E_m}{R} \tag{6.19}$$

which is $2\sqrt{2}$ times the corresponding half-wave value in Eq. (4.23).

A comparison of Eq. (6.10) and (6.19) shows that the rms supply current is, to a very good approximation, equal to the value for resistive load. The power factor for the full-wave bridge with highly inductive loading is found to be

$$PF = \frac{3}{\pi} = 0.955 \tag{6.20}$$

Comparison of Eqs. (6.20) with Eqs. (6.11) shows that the use of inductive loading makes virtually no difference to the operating power factor of a six-pulse bridge. The fundamental Fourier harmonic $i_{a_1} (\omega t)$ of $i_a (\omega t)$ is, as with resistive load, in time phase with supply voltage $e_{aN} (\omega t)$ so that the current displacement factor is again unity. Hence the reduction of the operating power factor below unity is due entirely to distortion of the current waveform rather than displacement or phase difference. The higher harmonics of the supply current combine with the supply frequency voltage to produce voltamperes but zero average power. Various features and parameters of the uncontrolled, three-phase, full-wave rectifier are listed in Table 6.1. Values of the Fourier components of the fundamental component of the supply current are given in Example 6.5 and in Table 6.1.

With series R-L load of fixed phase angle less than $90°$, the supply current waveform of a full-wave bridge assumes a shape intermediate between that of Fig. 6.2f and the flat topped waveform of Fig. 6.4e. In a rectifier circuit the ratings of the circuit components are an important technical and economic consideration. It can be seen in Fig. 6.4 that the maximum reverse voltage across a diode will be the peak value of the line-to-line voltage. Each diode in Fig. 6.4a conducts one pulse of current each supply voltage cycle. Diode D_1, for example, conducts only the positive pulses associated with current i_a, as in Fig. 4.22. The negative pulses of current i_a are conducted by diode D_4.

The rms current rating of diode D_1 (Fig. 6.4a) is therefore given by

$$I_{D_1} = \sqrt{\frac{1}{2\pi} \int_{30°}^{150°} I_{av}^2 \, d\omega t} \tag{6.21}$$

Comparison of Eq. (6.19) and (6.21) shows that

$$I_{D_1} = \frac{1}{\sqrt{2}} I_a \tag{6.22}$$

The rms current rating of the diodes in a full-wave, three-phase bridge is therefore $1/\sqrt{2}$ of the rms supply current rating.

6.2.1 Worked Examples

Example 6.5 Sketch the supply current waveforms, in correct time-phase relation to their respective phase voltages, for a full-wave diode bridge with highly inductive load. The three-phase supply can be considered an ideal voltage source. Calculate the fundamental component of the supply current and compare this with the respective value for the case of resistive load.

The waveform of the supply phase current $i_a (\omega t)$ is given in Fig. 6.4e. It is seen, by inspection, that this waveform is symmetrical with respect to the phase voltage waveform $e_{aN} (\omega t) = E_m \sin \omega t$, so that the fundamental displacement angle ψ is

$$\psi_1 = \tan^{-1} \frac{a_1}{b_1} = 0$$

Because $\psi_1 = 0$, then

$$a_1 = \frac{2}{\pi} \int_0^\pi i_a(\omega t) \cos \omega t \, d\omega t = 0$$

In Fig. 6.4, utilizing (6.16), the current $i_a (\omega t)$ can be defined as

$$i_a(\omega t) = \frac{3\sqrt{3}}{\pi} \frac{E_m}{R} \bigg|_{30°}^{90°} - \frac{3\sqrt{3}}{\pi} \frac{E_m}{R} \bigg|_{210°}^{330°}$$

Fundamental Fourier coefficient b_1 is therefore

$$b_1 = \frac{2}{\pi} \int_0^\pi i_a(\omega t) \sin \omega t \, d\omega t$$

$$= \frac{2}{\pi} \times \frac{3\sqrt{3}}{\pi} \times \frac{E_m}{R} \int_{30°}^{150°} \sin \omega t \, d\omega t$$

$$= \frac{6\sqrt{3}}{\pi^2} \frac{E_m}{R} (-\cos \omega t) \bigg|_{30°}^{150°}$$

$$= \frac{18}{\pi^2} \frac{E_m}{R} = 1.053 \frac{\sqrt{3} E_m}{R}$$

Comparison of this expression with the corresponding expression (6.12) for resistive load shows that the magnitude of the fundamental supply current component has decreased very slightly due to the inductive load.

Because $a_1 = 0$ and $\psi_1 = 0$ the fundamental component of current may be written

$$i_{a_1}(\omega t) = 1.053 \frac{\sqrt{3}E_m}{R} \sin \omega t$$

Since $a_1 = 0$, the rms value of the fundamental component of the line

$$I_1 = \frac{c_1}{\sqrt{2}} = \frac{1}{2} \frac{18}{\pi^2} \frac{E_m}{R}$$

but

$$I_{av} = \frac{E_{av}}{R} = \frac{3\sqrt{3}E_m}{\pi R}$$

Therefore,

$$I_1 = \frac{\sqrt{6}}{\pi} I_{av}$$

Example 6.6   Derive a general expression for the average load voltage of a p pulse uncontrolled rectifier with ideal voltage supply.

One can see an obvious similarity of wave shape between the load voltage waveforms for three-phase, half-wave rectification (Fig. 4.5c) and three-phase, full-wave rectification (Fig. 6.4c).

If one takes the peak point of a voltage excursion as the origin, the waveform can be represented in general by the diagram of Fig. 6.5. The pulse width of a section is $2\pi/p$, where p is the effective pulse number. For the interval $0 < \omega t < \pi/p$, the waveform is given by

$$e(\omega t) = E_m \cos \omega t$$

Its average value is therefore given by

$$E_{av} = \frac{1}{2\pi/p} \int_{-\pi/p}^{\pi/p} E_m \cos \omega t \, d\omega t$$

$$= \frac{pE_m}{2\pi} \left[\sin \frac{\pi}{p} - \sin \left(-\frac{\pi}{p} \right) \right]$$

Therefore,

$$E_{av} = \frac{pE_m}{\pi} \sin \frac{\pi}{p}$$

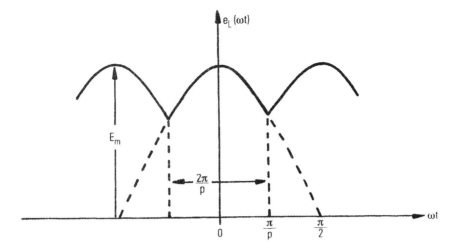

Fɪɢ. 5   General representation of the load voltage waveform for a three-phase, full-wave p pulse diode bridge rectifier.

When $p = 3$,

$$E_{av} = \frac{3\sqrt{3}}{2\pi} E_m$$

which agrees with Eq. (4.6), where E_m is the peak value of the line to neutral or phase voltage. When $p = 6$, in the general expression, then

$$E_{av} = \frac{3E_m}{\pi}$$

But E_m now refers to the peak value of the line-to-line voltage, as in Fig. 6.4c).

To be consistent with half-wave operation, the term E_m is reserved for the peak value of the phase voltage and, when $p = 6$,

$$E_{av} = \frac{3}{\pi}\sqrt{3}E_m$$

which is consistent with Eq. (4.56).

Example 6.7   A three-phase, full-wave diode bridge rectifier supplies a load current of 50 A to a load resistor R in the presence of a highly inductive series filter. If the ideal three-phase supply provides balanced sinusoidal voltages of 415 V, 50 Hz, calculate the required ratings of the bridge diodes and the value of R.

The appropriate circuit and waveforms are given in Fig. 6.4. The average load voltage is given by Eq. (6.16)

$$E_{av} = \frac{3\sqrt{3}E_m}{\pi}$$

in which $\sqrt{3}\,E_m$ is the peak value of the line voltage. It can be assumed in that the specified value 415 V is the rms value of the line voltage so that

$$E_{av} = \frac{3}{\pi} \times \sqrt{2} \times 415 = 560 \text{ V}$$

The value of the load resistor to sustain a current of 50 A is, from Eqs. (6.17),

$$R = \frac{560}{50} = 11.2 \ \Omega$$

The rms supply current may be given in terms of the average load current, as in Eqs. (6.19),

$$I_a = \sqrt{\frac{2}{3}} I_{av} = \frac{50\sqrt{2}}{\sqrt{3}} = 40.82 \text{ A}$$

From Eqs. (6.22), the rms current rating of each diode is

$$I_D = \frac{I_a}{\sqrt{2}} = 28.9 \text{ A}$$

The maximum voltage across a diode is the peak reverse value of the line voltage.

Diode PRV rating $= 415 \sqrt{2} = 587$ V.

Example 6.8   A three-phase, full-wave uncontrolled bridge is supplied from an ideal three-phase voltage source. The load current is filtered by a large inductor to produce negligible ripple. Derive expressions for the power dissipation on the dc side and on the ac side and show that these are equivalent, assuming ideal rectifier switches.

If the bridge diodes are ideal and therefore lossless, all of the power supplied to the ac terminals of the bridge must be dissipated in the load resistor.

Average power can only be transferred by combinations of voltage and current of the same frequency. Since the supply voltages are of supply frequency, the harmonic component of the line current that provides real power in watts (as opposed to voltamperes) is the supply frequency component. The input power is

$$P_{in} = 3\,EI_1\,\cos\psi_1$$

where E is the ms phase voltage, I_1 is the rms value of the fundamental component of phase (i.e., supply line) current and ψ_1 is the time-phase angle between them. From the calculations in Example 4.14 it is seen that

$$\psi_1 = 0 \quad \cos\psi_1 = 1.0$$

$$a_1 = 0 \qquad b_1 = c_1 = \frac{18}{\pi^2}\frac{E_m}{R}$$

The rms value of I_1 is therefore

$$I_1 = \frac{c_1}{\sqrt{2}} = \frac{1}{\sqrt{2}}\frac{18}{\pi^2}\frac{E_m}{R}$$

On the ac side, the input power is therefore

$$P_{ac} = 3\times\frac{E_m}{\sqrt{2}}\times\frac{1}{\sqrt{2}}\times\frac{18}{\pi^2}\frac{E_m}{R} = \frac{27}{\pi^2}\frac{E_m^2}{R}$$

This is seen to agree with Eqs. (6.18), which defines the power on the dc side.
Q.E.D.

6.3 HIGHLY INDUCTIVE LOAD IN THE PRESENCE OF SUPPLY IMPEDANCE

Let the full-wave bridge circuit now contain an inductance L_s in each supply line. Because of the nonsinusoidal currents drawn from the supply the voltages at the bridge input terminals a, b, and c of Fig. 6.6 are not sinusoidal but are given by Eqs. (4.29) to (4.31).

If the value of the source reactance is small the theory of operation is very similar to that previously described in Sec. 6.2 and a typical group of waveforms are shown in Fig. 6.7. With a significant amount of source inductance, the transfer of current from one phase to the next now occupies a finite time. For example, suppose that current is about to be transferred from phase a to phase b in the circuit of Fig. 6.6 This transfer will begin when voltages e_{aN} and e_{bN} are equal. Conduction begins through rectifier D_3. As the difference between the two instantaneous phase voltages increases the current increases in phase b and decreases in phase a. Since $i_a + i_b = i_L = I_{av}$, current i_a is zero with $i_b = i_L$, at which point the transfer of current, or commutation, is complete. While the commutation condition exists there is a short circuit between phases a and b via diodes D_1 and D_3. The phase voltage at the common point ab (neglecting diode voltage drops) is the average of voltages e_{AN} and e_{BN}

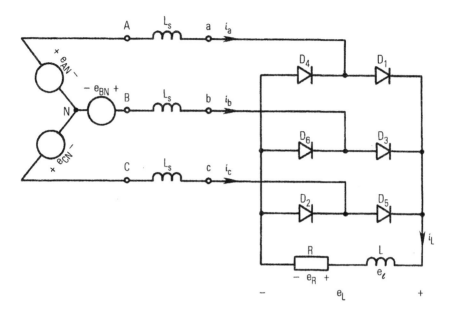

FIG. 6   Three-phase, full-wave diode bridge rectifier with supply.

6.3.1   Mode I Operation ($0 \leq \mu \leq 60°$)

Operation that results in an overlap angle μ in the range ($0 \leq \mu \leq 60°$) is usually referred to as *mode I operation*. Much of the analysis of Sec. 4.4 is relevant here and overlap angle μ of Eq. (4.37) is again relevant.

$$\cos\mu = 1 - \frac{2\omega L_s I_{av}}{\sqrt{3}E_m} \tag{6.23}$$

For the present case of full-wave bridge operation, however, the average current in Eqs. (6.23) is twice the value for half-wave bridge operation in Eq. (4.40).

Combining Eqs. (6.16), (6.17), and (6.23) gives

$$\cos\mu = 1 - \frac{6}{\pi}\frac{\omega L_s}{R}\frac{E_{av}}{E_{av_o}} \tag{6.24}$$

which compares with Eq. (4.38) for half-wave operation. Waveforms are given in Fig. 6.8 for mode I operation where $\mu = 30°$. It is found that

$$E_{av} = \frac{3\sqrt{3}E_m}{2\pi}(1+\cos\mu) = \frac{E_{av_o}}{2}(1+\cos\mu) \tag{6.25}$$

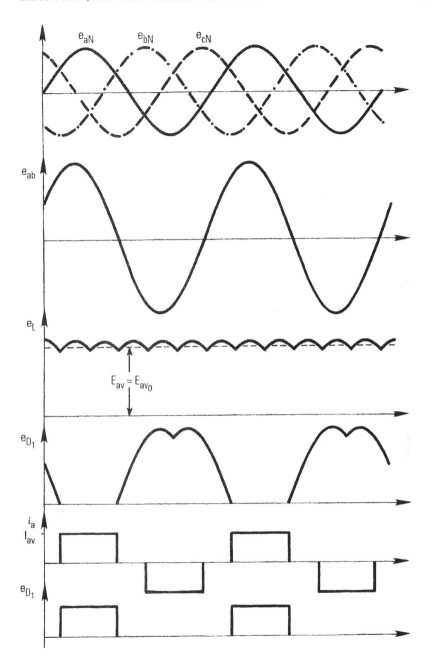

Fig. 7 Waveforms for full-wave diode bridge rectifier with highly inductive load ideal supply; $\mu = 0°$.

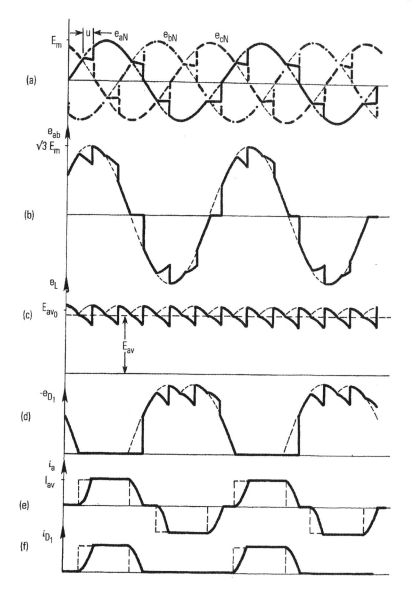

FIG. 8 Waveforms for full-wave diode bridge rectifier with highly inductive load: mode I, $\mu = 30°$, ideal supply.

which is identical in form to Eq. (4.33) for half-wave operation, except that E_{av_o} is now defined by the full-wave expression. With $\mu = 25°$, it is found that $E_{av}/E_{av_o} = 0.953$. This reduction of E_{av} on load compares with $E_{av}/E_{av_o} = 0.976$ for corresponding half-wave operation.

Combining Eqs. (6.24) and (6.25) enables one to calculate the overlap angle μ in terms of impedance parameters.

$$\cos\mu = \frac{1-(3/\pi)\,\omega L_s\,/\,R}{1+(3/\pi)\,\omega L_s\,/\,R} \tag{6.26}$$

Equation (6.26) differs from the corresponding expression (4.39) for half wave operation by a factor of 2 in the impedance ratios. Reduction of the average load voltage due to supply reactance can also be illustrated by combining Eqs. (6.23) and (6.25), eliminating μ,

$$E_{av} = \frac{3\sqrt{3}E_m}{\pi} - \frac{3\omega L_s I_{av}}{\pi}$$

or

$$E_{av} = E_{av_o} - \frac{3\omega L_s I_{av}}{\pi} \tag{6.27}$$

The second term of Eq. (6.27) represents the effective voltage drop across the supply inductor in each current loop. It is seen to be twice the corresponding value for half-wave operation in Example 4.7. The reduction of the load average voltage is therefore greater for a given supply inductances than with a half-wave rectifier.

Reduction of the load average voltage can also be calculated from the area of the phase voltage wave "missing due to overlap," as, for example, in Fig. 6.8. Such a calculation of the missing voltseconds involves integration of the instantaneous phase voltage since $i = 1/L \int e\,dt$. The definite integral $\int_0^T e\,dt$ is the voltage–time area supported by inductance L in the period T and represents some change of current level.

The average load current I_{av} is still related to the average load voltage by the relation

$$I_{av} = \frac{E_{av}}{R} = \frac{E_{av_o}}{R}(1+\cos\mu) \tag{6.28}$$

If the load resistance R in the circuit of Fig. 6.6 is constant, then I_{av} will remain constant, and μ will have a fixed value. Overlap angle μ will increase only if I_{av} increases, by reduction of the load resistor R, as shown in Fig. 6.9. It is seen in Fig. 6.7–Fig. 6.9 that the average load voltage E_{av} reduces as I_{av} increases. It can therefore be deduced that reduction of load resistance R causes I_{av} to increase and E_{av} to decrease but not in simple proportionality because of changes of the waveform and the value of the ripple voltage content. It can also be seen in Fig. 6.7–Fig. 6.9 that the conduction angles of individual diode elements also increase with μ, attaining the value 180° at $\mu = 60°$ and maintaining this value for $\mu > 60°$.

A useful "figure of merit" for practical rectifier circuits is the dimensionless ratio $\delta = \omega L_s\, I_{av}/E_{av}$. From Eqs. (6.26) it is found that for a full-wave diode bridge,

$$\delta = \frac{\omega L_s I_{av}}{E_{av}} = \frac{\omega L_s}{R} = \frac{\pi}{3}\frac{1-\cos\mu}{1+\cos\mu} \tag{6.29}$$

The ideal value for δ is zero, but δ increases as μ increases. At the limit of mode I operation, with $\mu = 60°$, δ is found to have the value $\pi/9 = 0.35$. The terminal voltage waveforms e_{ab} of Fig. 6.7–Fig. 6.12are seen to be symmetrical, compared with the corresponding waveforms of Figs. 4.9–4.13 for a half-wave bridge (which are asymmetrical). For phase voltage $e_{aN}\,(\omega t)$ (Fig. 6.8), the instantaneous value is represented by the relation,

$$e_{aN}(\omega t) = e_{AN}\bigg|_{0°,30°+\mu,150°+\mu,210°+\mu,330°+\mu}^{30°,150°,210°,330°,360°}$$
$$+\frac{1}{2}(e_{AN}+e_{CN})\bigg|_{30°,210°}^{30°+\mu,210°+\mu} + \frac{1}{2}(e_{AN}+e_{BN})\bigg|_{150°,330°}^{150°+\mu,330°+\mu} \tag{6.30}$$

where e_{AN}, e_{BN}, and e_{CN}, are defined by Eq. (4.26)–(4.28). The corresponding rms value E_{aN} of the supply-phase voltage is found from

$$E_{aN} = \sqrt{\frac{1}{2\pi}\int e_{aN}^2\, d\omega t} \tag{6.31}$$

The supply-phase current $i_a\,(\omega t)$ is seen, from Fig. 6.8, to have identical positive and negative alternations. The positive half cycle of $i_a\,(\omega t)$ is identical in form to that of Fig. 4.10e for the half-wave case, except that the maximum I_{av} is twice as great. Because of symmetry the rms value I_a of waveform $i_a\,(\omega t)$ in Fig. 6.8 is $\sqrt{2}$ times as great as the value due to the positive half wave alone.

Now with identical loads and identical supply voltages and source impedances, the use of full-wave and half-wave diode bridges would result in different values of overlap angle. The rms value of the supply line current for the full-wave bridge will be roughly $2\sqrt{2}$ times the value for the half-wave bridge, with

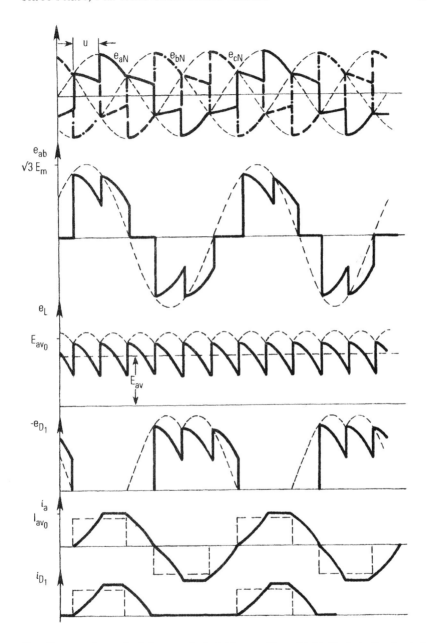

Fig. 9 Waveforms for full-wave diode bridge rectifier with highly inductive load: limit of mode I, $\mu = 60°$, ideal supply.

the same load resistor. Instantaneous current $i_a(\omega t)$ in Fig. 6.8 may be represented by the expression

$$i_a(\omega t) = i_a\Big|_{30°}^{30°+\mu} + i_a\Big|_{150°}^{150°+\mu} + i_a\Big|_{210°}^{210°+\mu} + i_a\Big|_{330°}^{330°+\mu} + I_{av}\Big|_{30°+\mu}^{150°} - I_{av}\Big|_{210°+\mu}^{330°} \quad (6.32)$$

The first of the six terms in Eqs. (6.32) is evaluated in Example 4.8. The rms value I_a of $i_a(\omega t)$ is found from

$$I_a = \sqrt{\frac{1}{2\pi} \int i_a^2(\omega t)\, d\omega t} \quad (6.33)$$

Since the load current is twice the value for the full-wave bridge as for the half-wave bridge, the corresponding load power dissipation $I_{av}^2 R$ is four times the value. It can be seen from Fig. 6.8 that the rms supply-phase voltage E_{aN} is lower than the corresponding value (Fig. 4.10a) for half-wave operation. The bridge power factor has a maximum value $3/\pi = 0.955$ when $\mu = 0$ and decreases as μ increases.

The input power to the bridge may be written (see the discussion in Example 6.8) as

$$P_{in} = 3EI_1 \cos \psi_1 \quad (6.34)$$

In Fig. 6.6 one may choose to calculate the input power from supply terminals ABC or bridge terminals abc. The current is the same in both cases. At the supply terminals one has the advantage of sinusoidal voltages but note that phase angle ψ_{A_1} between E_{A_1} and I_{a_1} is different from phase angle ψa_1 between E_{a_1} and I_{a_1}. The power P is the same at both points because no power is lost in the inductors L_s.

6.3.2   Mode II Operation ($\mu = 60°$)

Consider operation of the bridge when the load current is such that $\mu = 60°$ (Fig. 6.9). At the point $\omega t = 90°$, phase c has just finished conducting current through diode D_5 and handed over to phase a. But phase c is required to immediately conduct in the opposite direction from diode D_2. If the commutation or overlap angle had exceeded $60°$, phase c would have been required to conduct in two opposite directions simultaneously. What actually happens is that overlap between phases c and a, via diodes D_5 and D_1, continues past the instant $\omega t = 90°$. Since phase c is still conducting and is positive with respect to phase b, commutation between phases b and c cannot occur until current i_b in D_6 falls to zero. The additional interval of time for commutation to occur after $\mu = 60°$ is called the *inherent delay angle* α'. In Fig. 6.10 a value $\alpha' = 15°$, with $\mu = 60°$, is shown. As the load current increases further the value of α' also increases to

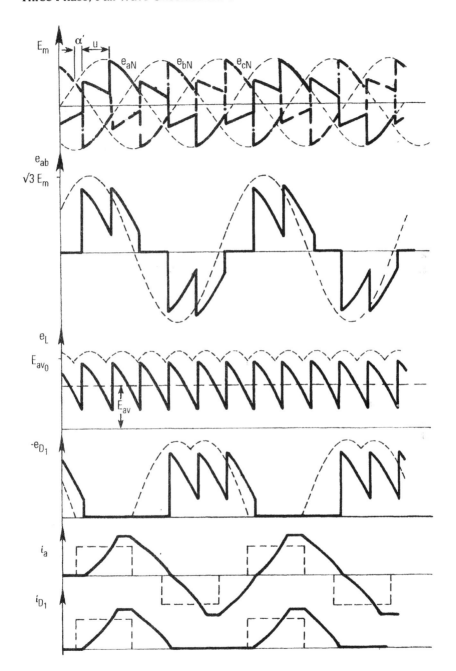

Fig. 10   Waveforms for full-wave diode bridge rectifier with highly inductive load: mode II, $\mu = 60°$, ideal supply.

a maximum value of 30° while μ remains constant at 60°. This form of bridge action is known as *mode II operation*. In the mode II interval, while $0 < \alpha' < 30°$, phase voltage $e_{aN}(\omega t)$ at the bridge terminals is represented by the equation

$$e_{aN}(\omega t) = E_m \sin \omega t \begin{vmatrix} 150°+\alpha', 330°+\alpha' \\ \\ 30°+\alpha'+\mu, 210°+\alpha'+\mu \end{vmatrix}$$

$$+ \frac{1}{2}(e_{AN} + e_{BN}) \begin{vmatrix} 30°+\alpha', 150°+\alpha'+\mu, 360° \\ \\ 0°, 150°+\alpha', 330°+\alpha' \end{vmatrix}$$

$$+ \frac{1}{2}(e_{AN} + e_{CN}) \begin{vmatrix} 30°+\alpha'+\mu, 210°+\alpha'+\mu \\ \\ 30°+\alpha', 150°+\alpha'+\mu \end{vmatrix} \tag{6.35}$$

It can be shown that mode II operation satisfies the relations

$$\frac{E_{av}}{E_{av_o}} = \frac{\sqrt{3}}{2} \sin\left(\frac{\pi}{3} - \alpha'\right) \tag{6.36}$$

$$\frac{I_{av}}{\hat{I}_{sc}} = \cos\left(\frac{\pi}{3} - \alpha'\right) \tag{6.37}$$

where, as before,

$$E_{av_o} = \frac{3\sqrt{3}}{\pi} E_m \tag{6.38}$$

and from Eq. (4.44),

$$\hat{I}_{sc} = \frac{\sqrt{3}E_m}{2\omega L_s} \tag{6.39}$$

Combining Eqs. (6.35) to (6.38) gives

$$\left(\frac{2}{\sqrt{3}}\right)^2 \left(\frac{E_{av}}{E_{av_o}}\right)^2 + \left(\frac{I_{av}}{\hat{I}_{sc}}\right)^2 = 1 \tag{6.40}$$

Waveforms of the rectifier for mode II operation, are given in Figs. 6.10 and 6.11.

6.3.3 Mode III Operation (60° ≤ μ ≤ 120°)

If, when $\alpha' = 30°$, the load current is further increased, commutation will begin, for example, between phases b and c while the commutation between phases c

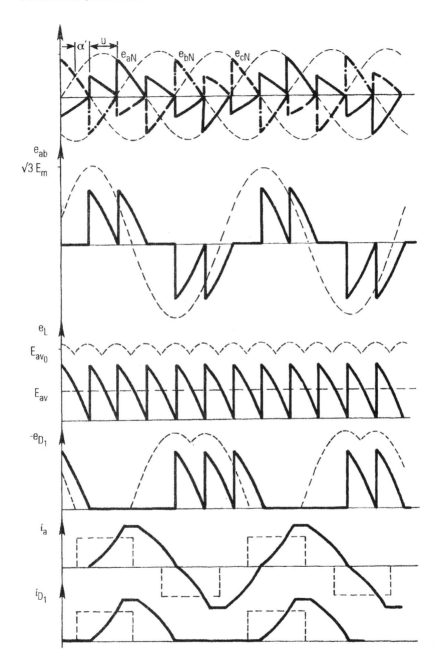

FIG. 11 Waveforms for full-wave diode bridge rectifier with highly inductive load: limit of mode II, $\mu = 60°$, $\alpha' = 30°$, ideal supply.

and a is still proceeding. During such intervals, a short circuit will exist between all three phases. For the remaining periods two phases are undergoing commutation so that mode III operation consists of alternate two-phase and three-phase short circuits. Current waveforms for the two-phase short-circuit condition are shown in Fig. 6.12. During mode III operation, the inherent delay angle α' is fixed at 30° while μ increases to a limiting value of 120° at which point there is a full terminal (i.e., three-phase) short circuit of terminals a, b and c. At that condition the average output voltage is zero, and the bridge currents become sinusoidal. For a full terminal short circuit the voltage regulation characteristic of the bridge is described by the relation

$$\frac{E_{av}}{E_{av_o}} = \sqrt{3} - \frac{3}{2}\frac{I_{av}}{\hat{I}_{sc}}$$

(6.41)

For the complete range of diode bridge operation, with highly inductive load, the load voltage–current characteristic is given in Fig. 6.13. This characteristic depicts the performance described by Eq. (4.44), (6.27), (6.40), and (6.41) for the three respective modes of operation.

6.3.4 Worked Examples

Example 6.9   A three-phase, full-wave diode bridge is required to supply load current of 50 A to a highly inductive load from a three-phase supply of 415 V, 50 Hz. Each supply line contains an effective series inductor $L_s = 1.3$ mH. Calculate the required value of load resistance, the average load voltage and the overlap angle.

From Eq. (6.27) the voltage reduction due to supply line reactance is indicated by the second term of the equation

$$E_{av} = \frac{3\sqrt{3}E_m}{\pi} - \frac{3\omega L_s I_{av}}{\pi}$$

The specified voltage of 415 V can be assumed to be the rms line-to-line voltage of the three-phase supply. Therefore,

$$E_{av} = \frac{3}{\pi} \times \sqrt{2} \times 415 - \frac{3}{\pi} \times \frac{2\pi \times 50 \times 1.3 \times 50}{1000}$$

$$= 560.45 - 19.5$$

$$= 541 \text{ V}$$

With a load current of 50 A,

$$R = \frac{E_{av}}{I_{av}} = \frac{541}{50} = 10.82 \text{ } \Omega$$

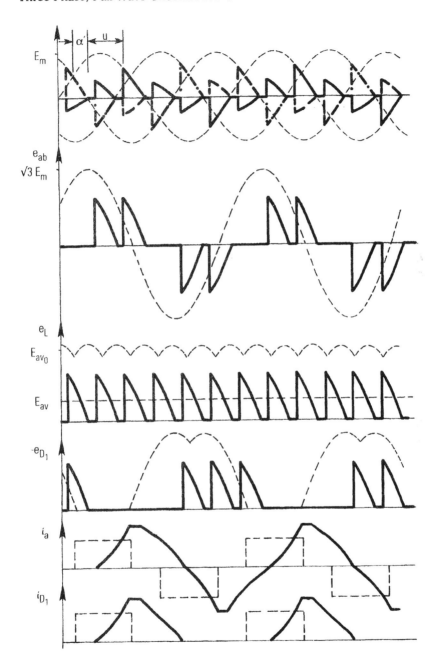

Fig. 12 Waveforms for full-wave diode bridge rectifier with highly inductive load: mode III, $\mu = 60°$, $\alpha' = 30°$, ideal supply.

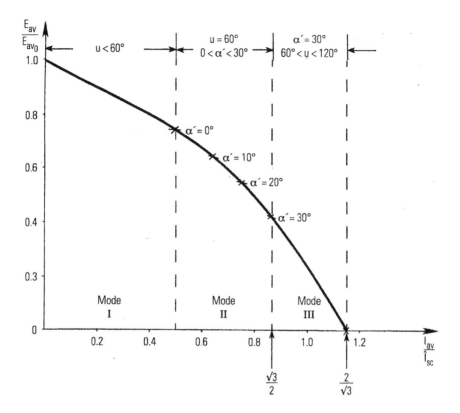

FIG. 13 Voltage regulation characteristic for full-wave diode bridge rectifier with highly inductive load.

Now

$$E_{av_o} = \frac{3\sqrt{3}E_m}{\pi}$$
$$= 560.45 \text{ V}$$

From Eqs. (6.25),

$$\cos\mu = 2\frac{E_{av}}{E_{av_o}} - 1 = 2\times\frac{541}{560.45} - 1 = 0.931$$

Therefore, overlap angle μ is

$$\mu = \cos^{-1}(0.931) = 21.41°$$

Example 6.10   A three-phase, full-wave diode bridge rectifier supplies a load resistor R in the presence of a highly inductive, series load filter. Each three-phase supply line contains a series inductance L_s. Calculate the overlap angle and the reduction of average load voltage due to overlap for the cases (1) $\omega L_s/R = 0.05$ and (2) $\omega L_s/R = 0.2$.

The case $\omega L_s/R = 0.05$ represents a fairly good voltage source, whereas the case $\omega L_s/R = 0.2$ represents a poor supply. From Eq. (6.26),

$$\mu = \begin{cases} 24.7° & \text{at } \dfrac{\omega L_s}{R} = 0.05 \\[2mm] 47.2° & \text{at } \dfrac{\omega L_s}{R} = 0.2 \end{cases}$$

Substituting values of μ into Eq. (6.25) gives

$$\frac{E_{av}}{E_{av_o}} = \begin{cases} 0.954 & \text{at } \mu = 24.7° \\ 0.84 & \text{at } \mu = 47.2° \end{cases}$$

A voltage regulation of 4.6% (when $\mu = 24.7°$) may be acceptable in many applications. A voltage regulation of 16% (when $\mu = 47.2°$) is much greater than would normally be acceptable in (say) an ac distribution system.

Example 6.11   A three-phase, full-wave diode bridge with a highly inductive load is fed from a three-phase supply of balanced voltages. The combination of load resistance R, supply line inductance L_s, and load current I_{av} is such that the bridge operates in mode I. Show that the normalized load voltage–load current characteristic in mode I is defined by the linear relationship

$$\frac{E_{av}}{E_{av_o}} + \frac{I_{av}}{2\hat{I}_{sc}} = 1$$

From Eqs. (6.27),

$$E_{av} = E_{av_o} - \frac{3\omega L_s I_{av}}{\pi}$$

where

$$E_{av_o} = \frac{3\sqrt{3}}{\pi} E_m$$

From Eqs. (6.39) the peak value of the short-circuit current (which is not reached in mode I operation) is

$$\hat{I}_{sc} = \frac{\sqrt{3}E_m}{2\omega L_s}$$

Substituting $\hat{I}_{sc}$ into the above equation, eliminating ωL_s, gives

$$E_{av} = E_{av_o} - \frac{E_{av_o} I_{av}}{2\hat{I}_{sc}}$$

Therefore,

$$\frac{E_{av}}{E_{av_o}} + \frac{I_{av}}{2\hat{I}_{sc}} = 1$$

Q.E.D.

PROBLEMS

Three-Phase, Full-Wave Diode Bridge with Resistive Load and Ideal Supply

6.1 A three-phase, star-connected, sinusoidal voltage supply has a peak voltage E_m per phase. This supplies power to a load resistor through a three-phase, full-wave, diode bridge rectifier. Sketch the circuit arrangement and give the waveform of a phase current in correct time phase to the corresponding supply-phase voltage. Sketch the waveform of the load current for a supply cycle and derive an expression for its average value in terms of E_m and R. Explain what you would expect to be the lowest order of harmonic ripple frequency in the load current.

6.2 Sketch the circuit of a full-wave, six-diode bridge rectifier to supply a resistive load from a balanced three-phase, sinusoidal supply. Sketch also the per-phase supply current and the load current and calculate the average values of these.

6.3 For the full-wave, three-phase bridge of Problem 6.2 derive expressions for the rms supply current, the load power, and hence the power factor for operation with resistive load.

6.4 A full-wave, three-phase diode bridge rectifier supplies power to a resistive load. With an ideal, three-phase power supply, the waveform of the supply line current is shown in Fig. 6.3. Use Fourier analysis to calculate the magnitude of the fundamental component of this waveform. What is the time average value of this current waveform over a supply cycle?

6.5 A set of balanced, sinusoidal, three-phase voltages is applied to a three-phase, full-wave diode bridge with a resistive load $R = 50\ \Omega$. If the supply voltages are 415 V, 50 Hz, calculate the load power dissipation and the rms load current.

6.6 For the three-phase bridge of Problem 6.4 calculate the rms value of the supply current and hence the necessary current rating of the bridge diodes.

6.7 For a three-phase, full-wave, diode bridge with resistive load R, show that the input power per phase is given by

$$P = \frac{E_m^2}{4\pi R}(2\pi + 3\sqrt{3})$$

where E_m is the peak value of the phase voltage.

6.8 For a three-phase, full-wave diode bridge rectifier with resistive load R and ideal supply, show that for all values of R and supply voltage, the power factor of operation is given by

$$PF = \sqrt{\frac{2\pi + 3\sqrt{3}}{4\pi}} = 0.956$$

6.9 Explain why the connection of equal capacitors C across the input terminals of a three-phase, full-wave diode bridge rectifier with resistive load would cause the operating power factor of the combined load to reduce (i.e., become worse).

6.10 Show that the average load voltage of a p pulse, uncontrolled, three-phase rectifier with peak load voltage E_m is given by

$$E_{av} = \frac{p}{\pi}E_m \sin\frac{\pi}{p}$$

6.11 In a three-phase p pulse rectifier the rms value of the nth load voltage harmonic is denoted by E_{L_n}. The peak value of the nth load voltage harmonic is therefore given by $\sqrt{2}\,E_{L_n}$. Show that these values of load voltage harmonic are related to the average load voltage E_{av_o} by the relation

$$\frac{\sqrt{2}E_{L_n}}{E_{av_o}} = \frac{2}{n^2 - 1}$$

Note that the above relation is only true for $n = p, 2p, 3p$, etc.

6.12 For a three-phase, full-wave diode bridge rectifier show that the peak value of the lowest order ac load voltage harmonic is 5.7% of the average load voltage.

Three-Phase, Full-Wave Diode Bridge Rectifier with Highly Inductive Load and Ideal Supply

6.13   A three-phase, full-wave diode bridge rectifier contains a load resistor R in series with a high inductance filter. Sketch consistent waveforms of the voltage and current per phase on the supply side, assuming ideal smoothing.

6.14   For the three-phase bridge of Problem 6.13 show that the rms value I_a of a supply current is related to the smooth load current I_{av} by relation

$$I_a = \sqrt{\frac{2}{3}} I_{av}$$

6.15   Show that the input power per phase of the three-phase bridge circuit of Problem 6.13 is given by

$$P = \frac{9}{\pi^2} \frac{E_m}{R}$$

where E_m is the peak value of the supply phase voltage.

6.16   A three-phase, full-wave diode bridge rectifier with highly inductive load cannot be power-factor corrected at all by the use of terminal capacitance. Use the Fourier components of the supply line current to explain why this is so.

6.17   A three-phase, full-wave diode bridge rectifier with highly inductive load requires a load current of 100 A from a 240-V, 50-Hz, three-phase supply. Calculate the required voltage and current rating of the bridge diodes.

6.18   Show that for a full-wave, three-phase diode bridge rectifier with highly inductive load, the power factor, irrespective of voltage and current values, is given by $PF = 3/\pi = 0.955$.

6.19   For a three-phase, full-wave diode bridge rectifier with highly inductive load, calculate the magnitude and phase angle of the fundamental component of the supply current per phase. Sketch the supply voltage and current per phase and sketch, on the same diagram, the fundamental current component.

6.20   A three-phase, full-wave diode bridge rectifier with highly inductive load carries a load current I_{av}. Show that the rms current rating I_D of each diode is given by

$$I_D = \frac{I_{av}}{\sqrt{3}}$$

6.21 For a full-wave, three-phase, diode, bridge rectifier with highly inductive load, obtain the Fourier series for the supply current waveform. Use this to deduce the magnitudes of the fundamental and principal higher harmonic components.

Three-Phase, Full-Wave Diode Bridge Rectifier with Highly Inductive Load in the Presence of Supply Inductance

6.22 A three-phase supply 240 V, 50 Hz provides power to a three-phase, full-wave uncontrolled bridge. The highly inductive load requires a constant current of 30 A. Each supply line contains an effective series inductance of 1 mH. Calculate the average load voltage and the overlap angle.

6.23 For the three-phase bridge of Problem 6.22 calculate the reduction of the average load voltage due to overlap for the cases (a) $\omega L_s/R = 0.03$ and (b) $\omega L_s/R = 0.25$.

6.24 A three-phase, full-wave diode bridge transfers power from a balanced three-phase supply to a highly inductive load. The supply lines each contain series reactance ωL_s such that operation occurs at overlap angle $\mu = 30°$. Sketch consistent diagrams for the per-phase voltage and current at the bridge terminals and calculate the per-unit reduction of average load voltage.

6.25 For the three-phase bridge circuit of Problem 6.24 sketch the variation of ratio $\omega L_s/R$ versus overlap angle μ for $0 < \mu < 60°$.

6.26 A three-phase, full-wave uncontrolled bridge circuit supplies power to a variable load resistor R in series with a load inductor of constant, high value. The supply lines contain series inductance L_s. Use appropriate equations to show that the load characteristic E_{av}/E_{av_o} versus $I_{av}/\hat{I}_{av}$ is represented by Fig. 6.13.

7

Three-Phase, Full-Wave Controlled Bridge Rectifier Circuits with Passive Load Impedance

Two half-wave, three-pulse controlled rectifiers of the type shown in Fig. 7.1 can be combined in the formation of Fig. 7.1a. If the switches in the top half of the bridge ($Th_1 Th_3 Th_5$) are gated and fired in antiphase with the switches in the bottom half ($Th_4 Th_6 Th_2$) and the neutral wire N is omitted, the two half-wave bridge effects are added. The resultant load voltage e_L (ωt) is then the sum of the two individual half bridges. The average load voltage E_{av} is then twice the average value of either of the two half-wave bridges acting individually.

Figure 7.1b shows the basic form of a three-phase, full-wave bridge rectifier circuit. This is the same form as the uncontrolled bridge Fig. 6.1, and the numbering notation is the same. Although the semiconductor switches in Fig. 7.1 are shown as silicon controlled rectifiers, they could equally well be any of the other three-terminal, gate-controlled switches of Table 1.1.

With a supply of zero impedance the three supply-phase voltages for the circuit of Fig. 7.1 retain balanced sinusoidal form for any load condition. These voltages are defined by the equations

$$e_{aN} = E_m \sin \omega t \tag{7.1}$$

$$e_{bN} = E_m \sin(\omega t - 120°) \tag{7.2}$$

$$e_{cN} = E_m \sin(\omega t - 240°) \tag{7.3}$$

The corresponding line-to-line voltages at the supply point are

216

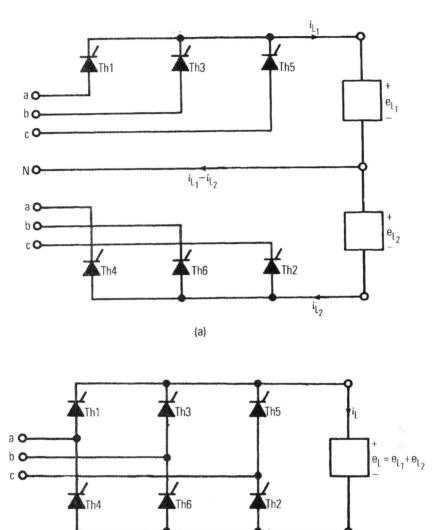

Fig. 1 Three-phase, full-wave controlled bridge rectifier circuit: (a) depicted as two half-wave bridges with neutral connection and (b) conventional formation, without neutral.

$$e_{ab} = e_{aN} + e_{Nb} = e_{aN} - e_{bN} = \sqrt{3}E_m \sin(\omega t + 30°) \qquad (7.4)$$

$$e_{bc} = \sqrt{3}E_m \sin(\omega t - 90°) \qquad (7.5)$$

$$e_{ca} = \sqrt{3}E_m \sin(\omega t - 210°) \qquad (7.6)$$

Waveforms of the supply voltages are given in Figs. 6.2 and 7.2.

The device numbering notation shown in the bridge rectifier circuit of Fig. 7.1 is standard for the three-phase, full-wave controlled bridge in both its rectifier and inverter modes of operation. To provide a current path from the supply side to the load side requires the simultaneous conduction of at least two appropriate switches. When one element of the upper group of switches and one of the lower group conducts, the corresponding line-to-line voltage is applied directly to the

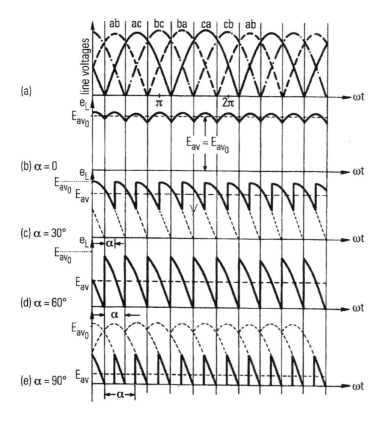

Fig. 2 Voltage waveforms of the three-phase, full-wave controlled bridge rectifier with resistive load and ideal supply: (a) supply line voltages, (b) load voltage ($\alpha = 0°$), (c) load voltage ($\alpha = 30°$), (d) load voltage ($\alpha = 60°$), and (e) load voltage ($\alpha = 90°$).

load. In Fig. 7.1 the switches are depicted as silicon controlled rectifier types of thyristor. For this reason the terminology *Th* is used in their description. If, for example, the switches Th_1 and Th_6 conduct simultaneously, then line voltage e_{ab} is applied across the load. There are some switch combinations that are not permissible. If, for example, the switches in any leg conduct simultaneously from both the top half and the bottom half of the bridge, then this would represent a short circuit on the ac supply. To provide load current of the maximum possible continuity and smoothness appropriate bridge switches must conduct in pairs sequentially, for conduction intervals up to 120° or $\pi/3$ radius of the supply voltage. The average load voltage and current are controlled by the firing-angle of the bridge thyristors, each measured from the crossover point of its respective phase voltages.

7.1 RESISTIVE LOAD AND IDEAL SUPPLY

When the thyristor firing-angle α is 0°, the bridge operates like a diode rectifier circuit with the waveforms given in Fig. 4.2. The corresponding conduction sequence of the circuit devices is given in Fig. 6.2a, in which the upper tier represents the upper half of the bridge. Supply line current i_a (ωt) for the first cycle (Fig. 6.2f) is made up of four separate components contributed by the four separate circuits shown in Fig. 7.3. An alternative representation of the device conduction pattern is shown in Fig. 6.2b in which the upper and lower tiers do not represent the upper and lower halves of the bridge, but this pattern shows that the thyristor switches conduct in chronological order. The circuit operation possesses two different modes, depending on the value of the firing angle. In the range $0 < \alpha < \pi/3$ the load voltage and current are continuous (Fig. 7.2c and d), and an oncoming thyristor will instantly commutate an off-going thyristor. In the range $\pi/3 < \alpha < 2\pi/3$, the load current becomes discontinuous because an off-going thyristor extinguishes before the corresponding on-coming thyristor is fired. For resistive loads with negligible supply reactance, both the load current and the supply current are always made up of parts of sinusoids, patterned from the line voltages. For all firing angles the sequence order of thyristor conduction in the circuit of Fig. 7.1 is always that shown in Fig. 6.2a. However, the onset of conduction is delayed, after the phase voltage crossover at $\omega t = 30°$, until the appropriate forward biassed thyristors are gated and fired.

Consider operation at $\alpha = 30°$, for example. In Fig. 7.2 forward-bias voltage occurs on thyristors Th_1 and Th_6 at $\omega t = 30°$. If the firing-angle is set at $\alpha = 30°$, conduction via Th_1 and Th_6 (Fig.7.3a) does not begin until $\omega t = \alpha + 30° = 60°$ and then continues for 60°. At $\omega t = \alpha + 90° = 120°$, the dominant line voltage is e_{ac}, thyristor Th_6 is reverse biassed, and conduction continues via the newly fired thyristor Th_2 (Fig. 7.3b) for a further 60°. At $\omega t = 180°$, thyristor Th_1 is commutated off by the switching in of Th_3 and line current i_a (Fig. 7.4c)

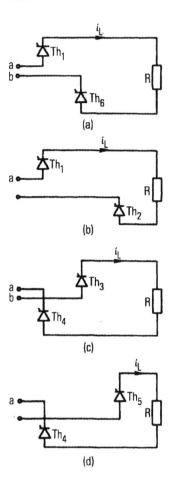

FIG. 3 Equivalent circuits of conduction with resistive load and ideal supply: (a) $30° \leq \omega t \leq 90°$ (Th_1, Th_6 on), (b) $90° \leq \omega t \leq 150°$ (Th_1, Th_2 on), (c) $210° \leq \omega t \leq 270°$ (Th_3, Th_4 on), and (d) $270° \leq \omega t \leq 330°$ (Th_5, Th_4 on).

becomes zero so that the load current path is provided by Th_2 and Th_3 for a further 60° interval. When $\omega t = 240°$, the dominant line voltage is e_{ba}. The firing of Th_4 transfers the load current from Th_2, (Fig. 7.3c), and supply current resumes in phase a in the opposite direction. After a further 60°, at $\omega t = 300°$, line voltage e_{ca} is dominant (Fig. 7.4a), and the switching in of Th_5 causes the commutation of Th_3. Thyristors Th_4Th_5 then provide the load current path which is fed from phase c to phase a, as shown in Fig. 7.3d.

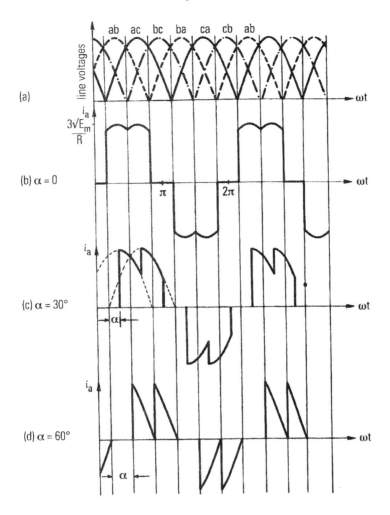

Fig. 4 Waveforms of the three-phase, full-wave controlled bridge rectifier with resistive load and ideal supply: (a) supply line voltages, (b) supply line current ($\alpha = 0°$), (c) supply line current ($\alpha = 30°$), and (d) supply line current ($\alpha = 60°$).

7.1.1 Load-Side Quantities

The sequence of thyristor firing creates the load voltage (and current) waveforms shown in Fig. 7.2. In mode I operation, where $0 \leq \alpha \leq 60°$, the average voltage can be obtained by taking any $60°$ interval of $e_L (\omega t)$.

For $\alpha + 30° \leq \omega t \leq \alpha + 90°$,

$$e_L(\omega t) = \sqrt{3}E_m \sin(\omega t + 30°) \Big|_{\alpha+30°}^{\alpha+90°} \qquad (7.7)$$

The average value of Eq. (7.7) in terms of peak phase voltage E_m is

$$E_{av} = \frac{3}{\pi}\int_{\alpha+30°}^{\alpha+90°} \sqrt{3}E_m \sin(\omega t + 30°)\, d\omega t$$

$$= \frac{3\sqrt{3}}{\pi}E_m \cos\alpha$$

$$= E_{av_0}\cos\alpha \qquad (7.8)$$

where

$$E_{av_0} = \frac{3\sqrt{3}}{\pi}E_m = 1.654E_m \qquad (7.9)$$

The average current I_{av} is, therefore, a function of of α

$$I_{av} = \frac{E_{av}}{R} = \frac{E_{av_0}}{R}\cos\alpha \qquad (7.10)$$

With resistive load the instantaneous load voltage is always positive. When the anode voltage of a thyristor goes negative extinction occurs. At a firing angle $\alpha > 60°$, the load voltage and current therefore become discontinuous, as shown in Fig. 7.2. This represents a different mode of operation from $\alpha < 60°$ and the load voltage is then described by the following equation:

For $60° \leq \alpha \leq 120°$,

$$e_L(\omega t) = \sqrt{3}E_m \sin(\omega t + 30°) \Big|_{\alpha+30°}^{150°} \qquad (7.11)$$

The average value of Eq. (7.11) is given by

$$E_{av} = \frac{3\sqrt{3}}{\pi}E_m \left[1 + \cos(\alpha + 60°)\right] \qquad (7.12)$$

When $\alpha = 60°$, Eq. (7.8) and (7.12) give identical results. At $\alpha = 120°$, the average load voltage becomes zero.

The waveforms of Fig. 7.2 show that the load voltage waveform has a repetition rate six times that of the phase voltage. This means that the lowest ripple frequency is six times fundamental frequency. If a Fourier analysis is performed on the load voltage waveform the two lowest order harmonics are the dc level (i.e., the average value) followed by the sixth harmonic.

Load power dissipation can be found from the rms load current. The rms or effective load current I_L is defined as

$$I_L = \sqrt{\frac{1}{2\pi} \int_0^{2\pi} i_L^2(\omega t)\, d\omega t} \tag{7.13}$$

where $i_L = e_L/R$ from Eq. (7.7) or (7.11).

Comparing waveforms of the supply and load currents at a given firing angle, one would anticipate that $I_L > I_a$ because $i_L(\omega t)$ has a greater area under the curve than does $i_a(\omega t)$ and therefore $i_L^2(\omega t)$ is likely to be greater than $i_a^2(\omega t)$. The substitution of Eq. (7.7) or (7.11) respectively into Eq. (7.13) gives

$$I_L\Big|_{0 \le \alpha \le 60^\circ} = \frac{\sqrt{3}E_m}{2R} \sqrt{\frac{2\pi + 3\sqrt{3}\,\cos 2\alpha}{\pi}} \tag{7.14}$$

$$I_L\Big|_{60^\circ \le \alpha \le 120^\circ} = \frac{\sqrt{3}E_m}{2R} \sqrt{\frac{4\pi - 6\alpha - 3\sin(2\alpha - 60^\circ)}{\pi}} \tag{7.15}$$

Power dissipation in the bridge circuit of Fig. 7.1 is assumed to occur entirely in the load resistor. This may be obtained from the rms (not the average) load current.

$$P_L = I_L^2 R \tag{7.16}$$

Combining Eqs. (7.14) and (7.15) with Eq. (7.16)

$$P_L\Big|_{0 \le \alpha \le 60^\circ} = \frac{3E_m^2}{4\pi R}(2\pi + 3\sqrt{3}\cos 2\alpha) \tag{7.17}$$

$$P_L\Big|_{60^\circ \le \alpha \le 120^\circ} = \frac{3E_m^2}{4\pi R}\Big[4\pi - 6\alpha - 3\sin(2\alpha - 60^\circ)\Big] \tag{7.18}$$

The load side properties of the bridge are summarised in Table 7.1.

7.1.2 Supply-Side Quantities

Waveforms of the currents on the supply side of the bridge are shown in Fig. 7.4 for mode I operation. The instantaneous supply currents for the two modes of operation are defined by the following: For $0 \le \alpha \le 60^\circ$,

$$i_a(\omega t) = \frac{\sqrt{3}E_m}{R}\sin(\omega t + 30^\circ)\Big|_{\alpha + 30^\circ,\dots}^{\alpha + 90^\circ,\dots} + \frac{\sqrt{3}E_m}{R}\sin(\omega t - 30^\circ)\Big|_{\alpha + 90^\circ,\dots}^{\alpha + 150^\circ,\dots} \tag{7.19}$$

TABLE 7.1   Three-Phase, Full-Wave Controlled Bridge Rectifier with Ideal Supply: Load-Side Properties

	Resistive load	Highly inductive load
Instantaneous load voltage	$0 \leq \alpha \leq 60°$, $\sqrt{3}E_m \sin(\omega t + 30°)\Big|_{\alpha+30°}^{\alpha+90°}$ $60° \leq \alpha \leq 120°$, $\sqrt{3}E_m \sin(\omega t + 30°)\Big|_{\alpha+30°}^{150°}$	$E_{av_0} \cos \alpha$
Average load voltage	$0 \leq \alpha \leq 60°$, $E_{av_0} \cos \alpha$ $60° \leq \alpha \leq 120°$, $E_{av_0}\left[1 + \cos(\alpha + 60°)\right]$	$E_{av_0} \cos \alpha$
RMS load current	$0 \leq \alpha \leq 60°$, $\dfrac{\sqrt{3}E_m}{2R}\sqrt{\dfrac{2\pi + 3\sqrt{3}\cos 2\alpha}{\pi}}$ $60° \leq \alpha \leq 120°$, $\dfrac{\sqrt{3}E_m}{2R}\sqrt{\dfrac{4\pi - 6\alpha - 3\,\sin(2\alpha - 60°)}{\pi}}$	$\dfrac{E_{av_0}}{R}\cos \alpha$
Load power	$0 \leq \alpha \leq 60°$, $\dfrac{\sqrt{3}E_m^2}{4R}(2\pi + 3\sqrt{3}\cos 2\alpha)$ $60° \leq \alpha \leq 120°$, $\dfrac{\sqrt{3}E_m^2}{4R}\left[4\pi - 6\alpha - 3\,\sin(2\alpha - 60°)\right]$	$\dfrac{E_{av_0}^2}{R}\cos^2 \alpha$

For $60° \leq \alpha \leq 120°$,

$$i_a(\omega t) = \frac{\sqrt{3}E_m}{R}\sin(\omega t + 30°)\Big|_{\alpha+30°}^{150°} + \frac{\sqrt{3}E_m}{R}\sin(\omega t - 30°)\Big|_{\alpha+90°}^{210°} \qquad (7.20)$$

The rms values of the supply line currents may be obtained via the defining integral Eq. (7.13). Substituting Eqs. (7.19) and (7.20) into the form of Eq. (7.13) gives

$$I_a \Big|_{0 \le \alpha \le 60^\circ} = \frac{E_m}{\sqrt{2R}} \sqrt{\frac{2\pi + 3\sqrt{3}\cos 2\alpha}{\pi}} \tag{7.21}$$

$$I_a \Big|_{60^\circ \le \alpha \le 120^\circ} = \frac{E_m}{\sqrt{2R}} \sqrt{\frac{4\pi - 6\alpha - 3\sin(2\alpha - 60^\circ)}{\pi}} \tag{7.22}$$

At $\alpha = 60^\circ$, Eq. (7.21) and (7.22) are found to be identical.

Comparison of the rms supply and load currents gives, for both modes of operation,

$$I_L = \sqrt{\frac{3}{2}} I_a \tag{7.23}$$

The relationship of Eq. (7.23) is found to be identical to that obtained for an uncontrolled, full-wave bridge with resistive load (Table 6.1).

7.1.3 Operating Power Factor

With perfect switches the power dissipated at the load must be equal to the power at the supply point. This provides a method of calculating the operating power factor

$$PF = \frac{P_L}{3E_a I_a} \tag{7.24}$$

Substituting for P_L, Eq. (7.17) or (7.18), and for I_a, Eq. (7.21) or (7.22), into Eq. (7.24), noting that $E_a = E_m/\sqrt{2}$

$$PF \Big|_{0 \le \alpha \le 60^\circ} = \sqrt{\frac{2\pi + 3\sqrt{3}\cos 2\alpha}{4\pi}} \tag{7.25}$$

$$PF \Big|_{60^\circ \le \alpha \le 120^\circ} = \sqrt{\frac{4\pi - 6\alpha - 3\sin(2\alpha - 60^\circ)}{4\pi}} \tag{7.26}$$

When $\alpha = 0$, Eq. (7.25) has the value

$$PF \Big|_{\alpha = 0} = \sqrt{\frac{2\pi + 3\sqrt{3}}{4\pi}} = 0.956 \tag{7.27}$$

which agrees with Eq. (6.11) for the uncontrolled (diode) bridge. The power factor of the three-phase bridge rectifier circuit, as for any circuit, linear or nonlinear, with sinusoidal supply voltages, can be represented as the product of a distortion factor and a displacement factor The *distortion factor* is largely related to load impedance nonlinearity; in this case, the switching action of the thyristors. The *displacement factor* is the cosine of the phase angle between the fundamental components of the supply voltage and current. This angle is partly due to the load impedance phase angle but mainly due here to the delay angle of the current introduced by the thyristors. Both the current displacement factor and the current distortion factor are functions of the fundamental component of the supply current. This is calculated in terms of the Fourier coefficients a_1 and b_1, quoted from the Appendix for the order $n = 1$.

$$a_1 = \frac{1}{\pi} \int_0^{2\pi} i(\omega t) \cos \omega t \; d\omega t$$

$$b_1 = \frac{1}{\pi} \int_0^{2\pi} i(\omega t) \sin \omega t \; d\omega t$$

Expressions for the coefficients a_1 and b_1 are given in Table 7.2.

The current displacement factor and current distortion factor are given by

$$\text{Displacement factor of supply current} = \cos \psi_1 = \cos \left(\tan^{-1} \frac{a_1}{b_1} \right) \qquad (7.28)$$

$$\text{Distortion factor of supply current} = \frac{I_{a_1}}{I_a} = \frac{I_1}{I} \qquad (7.29)$$

where

$$I_{a_1} = \frac{c_1}{\sqrt{2}} = \frac{\sqrt{a_1^2 + b_1^2}}{\sqrt{2}} \qquad (7.30)$$

Expressions for the current displacement factor and current distortion factor, for both modes of operation, are also given in Table 7.2. The product of these is seen to satisfy the defining relation

$$PF = (\text{displacement factor})(\text{distortion factor}) \qquad (7.31)$$

7.1.4 Shunt Capacitor Compensation

Some degree of power factor correction can be obtained by connecting equal lossless capacitors C across the supply terminals (Fig. 7.5). The bridge voltages and currents are unchanged and so is the circuit power dissipation. The capacitor

TABLE 7.2 Three-Phase, Full-Wave Controlled Bridge Rectifier with Ideal Supply: Supply-Side Properties

	Resistive Load		Highly inductive load
	Mode I ($0 \leq \alpha \leq 60°$)	Mode II ($0 \leq \alpha \leq 120°$)	
Fourier coefficients of supply current a_1	$-\dfrac{3\sqrt{3}E_m}{2\pi R}\sin 2\alpha$	$-\dfrac{3E_m}{2\pi R}\left[1+\cos(2\alpha-60°)\right]$	$-\dfrac{9}{\pi^2}\dfrac{E_m}{R}\sin 2\alpha$
b_1	$\dfrac{E_m}{2\pi R}\left(2\pi+3\sqrt{3}\cos 2\alpha\right)$	$\dfrac{E_m}{2\pi R}\left[4\pi-6\alpha-3\sin(2\alpha-60°)\right]$	$\dfrac{9}{\pi^2}\dfrac{E_m}{R}\left(1+\cos 2\alpha\right)$
Fundamental rms supply current I_1	$\dfrac{1}{\sqrt{2}}\sqrt{a_1^2+b_1^2}$	$\dfrac{1}{\sqrt{2}}\sqrt{a_1^2+b_1^2}$	$\dfrac{1}{\sqrt{2}}\sqrt{a_1^2+b_1^2}$
RMS current I	$\dfrac{E_m}{\sqrt{2}R}\sqrt{\dfrac{2\pi+3\sqrt{3}\cos 2\alpha}{\pi}}$	$\dfrac{E_m}{\sqrt{2}R}\sqrt{\dfrac{4\pi-6\alpha-3\sin(2\alpha-60°)}{\pi}}$	$\dfrac{3\sqrt{2}}{\pi}\dfrac{E_m}{R}\cos\alpha$
Current displacement factor (cos ψ_1)	$\dfrac{2\pi+3\sqrt{3}\cos 2\alpha}{\sqrt{27+4\pi^2+12\sqrt{3}\pi\cos 2\alpha}}$	$\dfrac{4\pi-6\alpha-3\sin(2\alpha-60°)}{\sqrt{9\left[1+\cos(2\alpha-60°)\right]^2+\left[4\pi-6\alpha-3\sin(2\alpha-60°)\right]^2}}$	$\cos\alpha$
Current distortion factor (I_1/I)	$\dfrac{27+4\pi^2+12\sqrt{3}\pi\cos 2\alpha}{4\pi(2\pi+3\sqrt{3}\cos 2\alpha)}$	$-\dfrac{3\sqrt{3}E_m}{2\pi R}\sin 2\alpha$	$\dfrac{3}{\pi}$
Power factor	$\sqrt{\dfrac{2\pi+3\sqrt{3}\cos 2\alpha}{4\pi}}$	$\sqrt{\dfrac{4\pi-6\alpha-3\sin(2\alpha-60°)}{4\pi}}$	$\dfrac{3}{\pi}\cos\alpha$

current i_c (ωt) is a continuous function unaffected by thyristor switching. In phase a, for example, the instantaneous capacitor current i_{ca} (ωt) is given by

$$i_{ca}(\omega t) = \frac{E_m}{X_c}\sin(\omega t + 90°) = \frac{E_m}{X_c}\cos \omega t \tag{7.32}$$

where $X_c = 1/2\pi fC$.

The corresponding instantaneous supply current i_{s_a} (ωt) in phase a is now

$$i_{s_a}(\omega t) = i_a(\omega t) + i_{c_a}(\omega t) \tag{7.33}$$

Therefore,

$$\left. i_{s_a} \right|_{0 \le \alpha \le 60°} = \left. \frac{E_m}{X_c}\cos \omega t + \frac{\sqrt{3}E_m}{R}\sin(\omega t + 30°) \right|_{\alpha+30°}^{\alpha+90°}$$

$$+ \left. \frac{\sqrt{3}E_m}{R}\sin(\omega t - 30°) \right|_{\alpha+90°}^{210°} \tag{7.34}$$

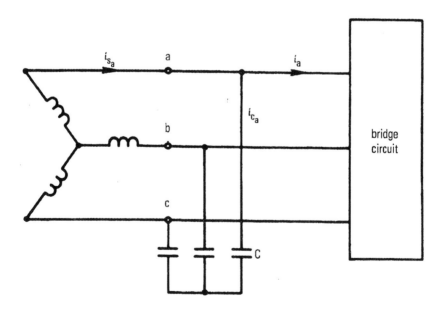

FIG. 5   Three phase-bridge circuit with supply side capacitors.

$$i_{s_a}\bigg|_{60°\leq\alpha\leq120°} = \frac{E_m}{X_c}\cos\omega t + \frac{\sqrt{3}E_m}{R}\sin(\omega t + 30°)\bigg|_{\alpha+30°}^{150°}$$

$$+\frac{\sqrt{3}E_m}{R}\sin(\omega t - 30°)\bigg|_{\alpha+90°}^{210°} \tag{7.35}$$

The substitution of Eqs. (7.34) and (7.35), respectively, into Eq. (7.13) gives modified expressions for the rms supply current,

$$I_{s_a}\bigg|_{0\leq\alpha\leq60°} = \frac{E_m}{\sqrt{2}R}\sqrt{\frac{R^2}{X_c^2} + \frac{2\pi + 3\sqrt{3}\cos2\alpha}{\pi} - \frac{3\sqrt{3}R}{\pi X_c}\sin2\alpha} \tag{7.36}$$

$$I_{s_a}\bigg|_{60°\leq\alpha\leq120°} = \frac{E_m}{\sqrt{2}R}\sqrt{\frac{R^2}{X_c^2} + \frac{4\pi - 6\alpha - 3\sin(2\alpha - 60°)}{\pi} - \frac{3R}{\pi X_c}\left[\sin(2\alpha + 30°) + 1\right]}$$

$$\tag{7.37}$$

Since the system voltages and power are unchanged by the presence of the capacitor the power factor will be improved if the ms supply current with the capacitor is reduced below the level of the bridge rms current (which is the supply rms current in the absence of the capacitor).

If the compensated power factor is denoted by PF_c, the ratio of compensated to uncompensated power factor is found to be

$$\frac{PF_c}{PF}\bigg|_{0\leq\alpha\leq60°} = \sqrt{\frac{(1/\pi)(2\pi + 3\sqrt{3}\cos2\alpha)}{R^2/X_c^2 + (1/\pi)(2\pi + 3\sqrt{3}\cos2\alpha) - (3\sqrt{3}R/\pi X_c)\sin2\alpha}} \tag{7.38}$$

$$\frac{PF_c}{PF}\bigg|_{60°\leq\alpha\leq120°} = \sqrt{\frac{(1/\pi)\left[4\pi - 6\alpha - 3\sin(2\alpha - 60°)\right]}{R^2/X_c^2 + (1/\pi)\left[4\pi - 6\alpha - 3\sin(2\alpha - 60°)\right] - (3R/\pi X_c)\left[\sin(2\alpha + 30°) + 1\right]}} \tag{7.39}$$

The ratio PF_c/PF will be greater than unity, indicating that power factor improvement has occurred, when the following inequalities are true:
For $0 \leq \alpha \leq 60°$,

$$\frac{R}{X_c}\left(\frac{R}{X_c} - \frac{3\sqrt{3}}{\pi}\sin2\alpha\right) < 0 \tag{7.40}$$

For $60° \leq \alpha \leq 120°$,

$$\frac{R}{X_c}\left\{\frac{R}{X_c} - \frac{3}{\pi}\left[\sin(2\alpha + 30°) + 1\right]\right\} < 0 \tag{7.41}$$

For the limiting values of firing angle α, being zero in Eq. (7.40) and 120° in Eq. (7.41) it is found that R/X_c would need to be negative to cause power factor improvement. In other words, when $\alpha = 0$, the use of capacitance does not give improvement but actually makes the power factor worse.

The use of supply-point capacitance aims to reduce the displacement angle ψ_{s_1} to zero so that displacement factor $\cos \psi_{s_1} = 1.0$, which is its highest realizable value. From Eq. (7.28) it is seen that $\psi_{s_1} = 0$ when $a_{s_1} = 0$. If Eqs. (7.34) and (7.35) are substituted into Eq. (A.9) in the Appendix, it is found that

$$a_{s_1}\Big|_{0\leq\alpha\leq60°} = \frac{E_m}{X_c} - \frac{3\sqrt{3}E_m}{2\pi R}\sin 2\alpha \tag{7.42}$$

$$a_{s_1}\Big|_{60°\leq\alpha\leq120°} = \frac{E_m}{X_c} - \frac{3E_m}{2\pi R}\left[1 + \cos(2\alpha - 60°)\right] \tag{7.43}$$

Unity displacement factor and maximum power factor compensation are therefore obtained by separately setting Eq. (7.42) and (7.43) to zero:

For $0 \leq \alpha \leq 60°$,

$$\frac{R}{X_c} - \frac{3\sqrt{3}}{2\pi}\sin 2\alpha = 0 \tag{7.44}$$

For $60° \leq \alpha \leq 120°$,

$$\frac{R}{X_c} - \frac{3}{2\pi}\left[1 + \cos(2\alpha - 60°)\right] = 0 \qquad \text{for } 60° \leq \alpha \leq 120° \tag{7.45}$$

When the conditions of (7.44),(7.45) are realized the power factor has attained its maximum possible value due to capacitor compensation. For $0 \leq \alpha \leq 60°$,

$$PF_c\Big|_{\cos\psi_1=1} = \frac{2\pi + 3\sqrt{3}\cos 2\alpha}{\sqrt{4\pi(2\pi + 3\sqrt{3}\cos 2\alpha) - 27\sin^2 2\alpha}} \tag{7.46}$$

For $60° \leq \alpha \leq 120°$,

$$PF_c\Big|_{\cos\psi_1=1} = \frac{4\pi - 6\alpha - 3\sin(2\alpha - 60°)}{\sqrt{4\pi\left[4\pi - 6\alpha - 3\sin(2\alpha - 60°)\right] - 9\left[1 + \cos(2\alpha - 60°)\right]^2}} \tag{7.47}$$

The degree of power factor improvement realizable by capacitor compensation is zero at $\alpha = 0$ and is small for small firing angles. For firing angles in the mid range, $30° \leq \alpha \leq 60°$, significant improvement is possible.

Note that the criteria of Eqs. (7.44) and (7.45) are not the same as the criteria of Eqs. (7.40) and (7.41) because they do not refer to the same constraint.

7.1.5 Worked Examples

Example 7.1 A three-phase, full-wave controlled bridge rectifier has a resistive load, $R = 100\ \Omega$. The three-phase supply 415 V, 50 Hz may be considered ideal. Calculate the average load voltage and the power dissipation at (1) $\alpha = 45°$ and (2) $\alpha = 90°$.

At $\alpha = 45°$ from Eq. (7.8),

$$E_{av} = \frac{3\sqrt{3}}{\pi} E_m \cos \alpha$$

where E_m is the peak value of the phase voltage. Assuming that 415 V represents the rms value of the line voltage, then E_m has the value

$$E_m = 415 \times \frac{\sqrt{2}}{\sqrt{3}}$$

Therefore,

$$E_{av_0} = \frac{3\sqrt{3}}{\pi} \times 415 \times \frac{\sqrt{2}}{\sqrt{3}} \times \frac{1}{\sqrt{2}} = 396.3 \text{ V}$$

The power is given by Eq. (7.17),

$$P_L = \frac{3E_m^2}{4\pi R}(2\pi + 3\sqrt{3}\cos 2\alpha)$$

At $\alpha = 45° = \pi/4$,

$$P_L = \frac{3}{4\pi} \times \frac{415^2}{100} \times \frac{2}{3} \times 2\pi = 1722 \text{ W}$$

At $\alpha = 90° = \pi/2$, from Eq. (7.12),

$$E_{av} = \frac{3\sqrt{3}}{\pi} E_m \left[1 + \cos(2\alpha - 60°)\right]$$

$$= \frac{3\sqrt{3}}{\pi} \times 415 \times \frac{\sqrt{2}}{\sqrt{3}}\left(1 - \frac{\sqrt{3}}{2}\right) = 75.1 \text{ V}$$

The power is now given by Eq. (7.18)

$$P_L = \frac{3E_m^2}{4\pi R}\left[4\pi - 6\alpha - 3\,\sin(2\alpha - 60°)\right]$$

$$= \frac{3}{4\pi} \times \frac{415^2}{100} \times \frac{2}{3}\left(4\pi - 3\pi - \frac{3\sqrt{3}}{2}\right)$$

$$= \frac{415^2}{200\pi}(\pi - 2.589) = 149 \text{ W}$$

Example 7.2   For a three-phase, full-wave controlled bridge rectifier with resistive load and ideal supply, obtain a value for the load current ripple factor, when $\alpha = 60°$, compared with uncontrolled operation.

The rms values of the load current in the two modes of operation are given by Eq. (7.14) and (7.15). The average values are given in Eqs. (7.8), (7.10), and (7.12). Taking the ratio I_L/I_{av} it is found that for $0 \le \alpha \le 60°$,

$$\frac{I_L}{I_{av}} = \frac{(\sqrt{3}E_m/2R)\sqrt{(2\pi + 3\sqrt{3}\cos 2\alpha)/\pi}}{(3\sqrt{3}E_m/\pi R)\cos \alpha}$$

$$= \frac{\pi}{6}\sqrt{\frac{2\pi + 3\sqrt{3}\cos 2\alpha}{\pi \cdot \cos^2 \alpha}} \tag{Ex. 7.2a}$$

and for $60° \le \alpha \le 120°$,

$$\frac{I_L}{I_{av}} = \frac{(\sqrt{3}E_m/2R)\sqrt{[4\pi - 6\alpha - 3\sin(2\alpha - 60°)]/\pi}}{(3\sqrt{3}E_m/\pi R)\left[1 + \cos(\alpha + 60°)\right]}$$

$$= \frac{\pi}{6}\sqrt{\frac{4\pi - 6\alpha - 3\sin(2\alpha - 60°)}{\pi\left[1 + \cos(\alpha + 60°)\right]^2}} \tag{Ex. 7.2b}$$

From each of the relations (Ex. 7.2a) and (Ex. 7.2b) at $\alpha = 60°$, it is found that

$$\frac{I_L}{I_{av}} = 1.134$$

The ripple factor is, from Eq. (2.11),

$$RF = \sqrt{\left(\frac{I_L}{I_{av}}\right)^2 - 1} = 0.535$$

From ratio (Ex. 7.2a) above, at $\alpha = 0$,

$$\frac{I_L}{I_{av}} = \frac{\pi}{6}\sqrt{\frac{2\pi+3\sqrt{3}}{\pi}} = \frac{\pi}{6}(1.9115) = 1.001$$

This latter result confirms the values given in Eqs. (6.4) and (6.7). The ripple factor at $\alpha = 0$ is, therefore, zero, but it becomes significant as the firing angle is retarded.

Example 7.3   Calculate the operating power factor for the three-phase, full-wave, bridge rectifier of Example 7.1 at (1) $\alpha = 45°$ and (2) $\alpha = 90°$. If the maximum possible compensation by capacitance correction is realized, calculate the new values of power factor and the values of capacitance required.

At $\alpha = 45°$, from Eq. (7.25),

$$PF = \sqrt{\frac{2\pi+3\sqrt{3}\cos 90°}{4\pi}}$$

$$= \sqrt{\frac{1}{2}} = 0.707$$

At $\alpha = 90°$, from Eq. (7.26),

$$PF = \sqrt{\frac{4\pi-3\pi-3\sin 120°}{4\pi}}$$

$$= \sqrt{\frac{\pi-\dfrac{3\sqrt{3}}{2}}{4\pi}} = 0.21$$

If the maximum realizable compensation is achieved the power factor is then given by Eqs. (7.46) and (7.47).

At $\alpha = 45°$, from Eq. (7.46),

$$PF_c = \frac{2\pi+3\sqrt{3}\cos 90°}{\sqrt{4\pi(2\pi+3\sqrt{3}\cos 90°)-27\sin^2 90°}}$$

$$= \frac{2\pi}{\sqrt{8\pi^2-27}} = \frac{2\pi}{7.21} = 0.87$$

At $\alpha = 90°$, from Eq. (7.47),

$$PF_c = \frac{4\pi - 3\pi - 3\sin 120°}{\sqrt{4\pi(4\pi - 3\pi - 3\sin 120°) - 9(1 + \cos\ 120°)^2}}$$

$$= \frac{\pi - 3\sqrt{3}/2}{\sqrt{4\pi\left(\pi - \dfrac{3\sqrt{3}}{2}\right) - 9\left(1 - \dfrac{1}{2}\right)^2}}$$

$$= \frac{0.544}{\sqrt{6.83 - 2.25}} = 0.254$$

The criteria for zero displacement factor are given in Eqs. (7.44) and (7.45). At $\alpha = 45°$, from Eq. (7.44),

$$\frac{1}{X_c} = 2\pi f C = \frac{3\sqrt{3}}{2\pi R}\ \sin 2\alpha$$

$$C = \frac{1}{2\pi 50}\frac{3\sqrt{3}}{2\pi 100} = 52.6\ \mu F$$

At $\alpha = 90°$, from Eq. (7.45),

$$\frac{1}{X_c} = \frac{3}{2\pi R}\Big[1 + \cos(2\alpha - 60°)\Big]$$

$$C = \frac{1}{2\pi 50}\frac{3}{2\pi 100}\left(1 - \frac{1}{2}\right)$$

$$= 15.2\mu F$$

7.2 HIGHLY INDUCTIVE LOAD AND IDEAL SUPPLY

7.2.1 Load-Side Quantities

The three-phase, full-wave, controlled bridge rectifier is most commonly used in applications where the load impedance is highly inductive. Load inductance is often introduced in the form of a large inductor in series with the load resistor (Fig. 7.6). If the load-side inductance smooths the load current to make it, very nearly, a pure direct current as shown in Fig. 7.7b, then

$$i_L(\omega t) = I_{av} = I_L = I_m \tag{7.48}$$

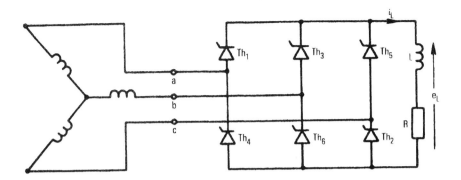

FıG. 6 Three-phase, full-wave controlled bridge rectifier circuit with series R-L load.

With a smooth load current there is zero average voltage on the smoothing inductor and the average load voltage falls entirely on the load resistor so that Eq. (7.10) remains true. The patterns of the load current and supply currents are shown in Fig. 7.7 for firing angles up to $\alpha = 60°$. Unlike the case with resistive load, the load current is continuous for all values of a in the control range, and only one mode of operation occurs. The average voltage, for all firing-angles, is identical to that derived in Eq. (7.7) with the corresponding average current in Eq. (7.10).

It is seen from Eq. (7.10) that the average load current becomes zero at $\alpha = 90°$. The controlled range with highly inductive load is therefore smaller than with resistive load, as shown in Fig. 7.8. With a smooth load current there is no ripple component at all, and the current ripple factor has the ideal value of zero.

For $\alpha \leq 60°$, the instantaneous load voltage, with highly inductive load, is the same as for resistive load. At $\alpha = 75°$, $e_L(\omega t)$ contains a small negative component for part of the cycle. When α 90°, the instantaneous load voltage has positive segments identical to those in Fig. 7.3e, but these are balanced by corresponding negative segments to give an average value of zero. Although the load current ripple factor is zero, the load voltage ripple factor is determined by the ratio E_L/E_{av}. From Eqs. (7.8) and (7.14), with $\alpha \leq 60°$,

$$\frac{E_L}{E_{av}} = \frac{\pi}{6} \sqrt{\frac{2\pi + 3\sqrt{3}\cos 2\alpha}{\pi \cos^2 \alpha}} \tag{7.49}$$

The load power dissipation is proportional to the square of the load rms current, and therefore, substituting Eqs. (7.10) into Eq. (7.16),

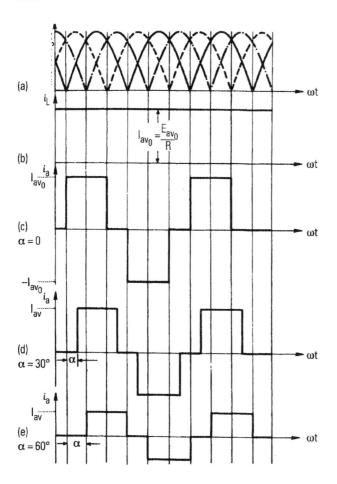

FIG. 7 Waveforms of the three-phase, full-wave controlled bridge rectifier circuit with highly inductive load and ideal supply: (a) supply line voltages, (b) load current ($\alpha = 0°$), (c) supply line current i_a ($\alpha = 0°$), (d) supply line current i_a ($\alpha = 30°$), and (e) supply line current i_a ($\alpha = 60°$).

$$P_L = I_L^2 R = I_{av}^2 R$$

$$= \frac{E_{av_0}^2}{R} \cos^2 \alpha = \frac{27 E_m^2}{\pi^2 R} \cos^2 \alpha \qquad (7.50)$$

The load-side properties are summarized in Table 7.1.

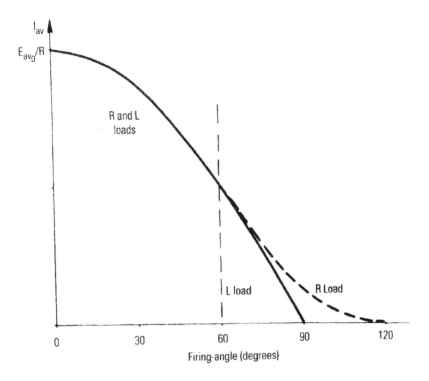

FIG. 8 Average load current versus SCR firing angle for the three-phase, full-wave controlled bridge rectifier circuit with ideal supply.

7.2.2 Supply-Side Quantities

The supply current $i_a(\omega t)$ shown in Fig. 7.7 is defined by the equation

$$i_a(\omega t) = \frac{E_{av_0}}{R}\cos\alpha\bigg|_{\alpha+30°}^{\alpha+150°} - \frac{E_{av_0}}{R}\cos\alpha\bigg|_{\alpha+210°}^{\alpha+330°} \qquad (7.51)$$

Since the rms values of the negative and positive parts of the wave are identical, the rms supply current I_a is given by

$$I_a = \sqrt{\frac{1}{\pi}\int_{\alpha+30°}^{\alpha+150°}\left(\frac{E_{av_0}}{R}\cos\alpha\right)^2 d\omega t}$$

$$= \frac{E_{av_0}}{R}\cos\alpha\sqrt{\frac{1}{\pi}\left(\omega t\right)_{\alpha+30°}^{\alpha+150°}}$$

$$= \sqrt{\frac{2}{3}} \frac{E_{av_0}}{R} \cos \alpha$$

$$= \frac{3\sqrt{2} E_m \cos \alpha}{\pi R}$$

$$= \sqrt{\frac{2}{3}} I_{av} \qquad (7.52)$$

The value in Eq. (7.52) is found to be $\sqrt{2}$ times the corresponding value for a half-wave rectifier, given in Eq. (5.39), and is identical to Eq. (7.23) for the case of resistive load.

The operating power factor of the bridge can be obtained by substituting Eqs. (7.50) and (7.51) into Eq. (7.24), noting that $E_a = E_m/\sqrt{2}$.

$$PF = \frac{P_L}{3E_a I_a}$$

$$= \frac{(27 E_m^2 / \pi^2 R) \cos^2 \alpha}{3(E_m / \sqrt{2})(3\sqrt{2} E_m \cos \alpha) / \pi R}$$

$$= \frac{3}{\pi} \cos \alpha \qquad (7.53)$$

The power factor also is found to be $\sqrt{2}$ times the value for a three-phase, half-wave controlled bridge rectifier and has the well-known value of $3/\pi$, or 0.955, for $\alpha = 0$ (or diode bridge) operation. A Fourier analysis of the supply current $i_a(\omega t)$ shows that the coefficients a_1 and b_1 below are valid for the fundamental (supply frequency) component

$$a_1 = \frac{1}{\pi} \int_0^{2\pi} i_a(\omega t) \cos \omega t \; d\omega t$$

$$= \frac{2}{\pi} \frac{E_{av_0}}{R} \cos \alpha (\sin \omega t) \Big|_{\alpha + 30°}^{\alpha + 150°}$$

$$= -\frac{2\sqrt{3}}{\pi} \frac{E_{av_0}}{R} \sin \alpha \cos \alpha$$

$$= -\frac{9}{\pi^2} \frac{E_m}{R} \sin 2\alpha \qquad (7.54)$$

$$b_1 = \frac{1}{\pi} \int_0^{2\pi} i_a(\omega t) \sin \omega t \, d\omega t$$

$$= \frac{2}{\pi} \frac{E_{av_0}}{R} \cos \alpha \left(-\cos \omega t\right) \Bigg|_{\alpha + 30°}^{\alpha + 150°}$$

$$= \frac{2\sqrt{3}}{\pi} \frac{E_{av_0}}{R} \cos^2 \alpha$$

$$= \frac{9}{\pi^2} \frac{E_m}{R} (1 + \cos 2\alpha) \tag{7.55}$$

Equations (7.54) and (7.55) can be used to obtain a very important relationship

$$\frac{a_1}{b_1} = \frac{-\sin 2\alpha}{1 + \cos 2\alpha} = -\tan \alpha = \tan \psi_1 \tag{7.56}$$

From Eq. (7.56) it can be seen that the displacement angle ψ_1 of the input current is equal to the firing angle (the negative sign representing delayed firing):

$$\alpha = \psi_1 \tag{7.57}$$

The displacement factor $\cos \psi_1$ is therefore equal to the cosine of the delayed firing angle

$$\cos \psi_1 = \cos \alpha \tag{7.58}$$

The relationship of Eq. (7.58) is true for both half-wave and full-wave bridges with highly inductive load. It is not true for bridges with purely resistive loading. The distortion factor of the input current is obtained by combining Eqs. (7.29), (7.30), (7.52), (7.54), and (7.55).

$$\begin{aligned}
\text{Distortion factor of the supply currents} &= \frac{I_{a_1}}{I_a} \\[2mm]
&= \frac{(1/\sqrt{2})(9/\pi^2)(E_m/R)\sqrt{\sin^2 2\alpha + (1 + \cos 2\alpha)^2}}{(3\sqrt{2}E_m/\pi R)\cos \alpha} \\[2mm]
&= \frac{3}{\pi}
\end{aligned} \tag{7.59}$$

The product of the displacement factor Eq. (7.58) and distortion factor Eq. (7.59) is seen to give the power factor Eq. (7.53). Some of the supply-side properties of the inductively loaded bridge are included in Table 7.2.

For any balanced three-phase load with sinusoidal supply voltage, the real or active power P is given by

$$P = 3EI_1 \cos \psi_1 \tag{7.60}$$

where I_1 the rms value of the fundamental component of the supply current and $\cos\psi_1$ is the displacement factor (not the power factor). Substituting Eqs. (7.52) and (7.58) into Eq. (7.60) gives

$$P = \frac{27}{\pi^2} \frac{E_m^2}{R} \cos^2 \alpha \tag{7.61}$$

which is seen to be equal to the power P_L dissipated in the load resistor, Eq. (7.50).

7.2.3 Shunt Capacitor Compensation

If equal capacitors C are connected in star at the supply point (Fig. 7.5), the instantaneous supply current is given by

$$i_{s_a}(\omega t) = \frac{E_m}{X_c} \cos \omega t + \frac{3\sqrt{3}E_m}{\pi R} \cos\alpha \bigg|_{\alpha+30^\circ}^{\alpha+150^\circ} - \frac{3\sqrt{3}E_m}{\pi R} \cos\alpha \bigg|_{\alpha+210^\circ}^{\alpha+330^\circ} \tag{7.62}$$

The substitution of Eq. (7.62) into Eq. (7.13) gives an expression for the rms supply current

$$I_{s_a} = \sqrt{\frac{1}{\pi} \int_0^\pi \left[\left(\frac{E_m}{X_c}\cos\omega t\right) + \left(\frac{3\sqrt{3}E_m}{\pi R}\cos\alpha\right) \bigg|_{\alpha+30^\circ}^{\alpha+150^\circ} \right]^2 d\omega t}$$

$$= \frac{E_m}{\sqrt{2}} \sqrt{\frac{R^2}{X_c^2} - \frac{18}{\pi^2}\frac{R}{X_c}\sin 2\alpha + \frac{36}{\pi^2}\cos^2\alpha} \tag{7.63}$$

When the capacitance is absent, X_c becomes infinitely large and Eq. (7.63) reduces to Eq. (7.52). The power flow and the terminal voltage are unaffected by the connection of the capacitors. The compensated power factor is given by combining Eqs. (7.61) and (7.63)

$$PF_c = \frac{P}{3E_a I_{s_a}}$$

$$= \frac{(9/\pi^2)\cos^2\alpha}{\sqrt{R^2/4X_c^2 - (9/2\pi^2)(R/X_c)\sin 2\alpha + (9/\pi^2)\cos^2\alpha}} \tag{7.64}$$

The ratio of the compensated power factor to the uncompensated power factor is given by the ratio of the load current to the supply current

$$\frac{PF_c}{PF} = \frac{I_a}{I_{s_a}} = \sqrt{\frac{(18/\pi^2)\cos^2\alpha}{R^2/2X_c^2 - (9/\pi^2)(R/X_c)\sin 2\alpha + (18/\pi^2)\cos^2\alpha}} \quad (7.65)$$

The power factor is therefore improved when $PF_c/PF > 1$ which occurs when

$$\frac{R}{2X_c}\left(\frac{R}{X_c} - \frac{18}{\pi^2}\sin 2\alpha\right) < 0 \tag{7.66}$$

Examination of the inequality (7.66) shows that power factor improvement occurs when $0 < C < (18\sin\alpha)/\omega\pi^2 R$

Fourier coefficient a_{s_1} for the fundamental component of the compensated supply current is given by

$$a_{s_1} = \frac{1}{\pi}\int_0^{2\pi} i_{s_a}(\omega t)\cos\omega t\, d\omega t$$

$$= \frac{2}{\pi}\int\left[\left(\frac{E_m}{X_c}\cos^2\omega t\right)_0^\pi + \left(\frac{3\sqrt{3}E_m}{\pi R}\cos\alpha\cos\omega t\right)_{\alpha+30°}^{\alpha+150°}\right] d\omega t$$

$$= \frac{E_m}{X_c} - \frac{9}{\pi^2}\frac{E_m}{R}\sin 2\alpha \tag{7.67}$$

When $C = 0$, Eq. (7.67) reduces to Eq. (7.54).

To obtain the maximum value of the displacement factor, coefficient a_{s_1} must be zero. The condition for maximum realizable capacitor compensator is therefore, from Eq. (7.67),

$$C = \frac{9}{\pi^2}\frac{1}{\omega R}\sin 2\alpha \tag{7.68}$$

Setting Eq. (7.67) to zero and substituting into Eq. (7.64) gives the maximum power factor achievable by terminal capacitor compensation.

$$PF_{C_{max}} = \frac{\frac{3}{\pi}\cos\alpha}{\sqrt{1 - \frac{9}{\pi^2}\sin^2\alpha}} \tag{7.69}$$

For any nonzero value of α, it is seen that the uncompensated power factor $(3/\pi)\cos\alpha$ is improved due to optimal capacitor compensation, as illustrated in Fig. 7.9. Over most of the firing-angle range, the possible degree of power factor improvement is substantial. A disadvantage of power factor compensation by the

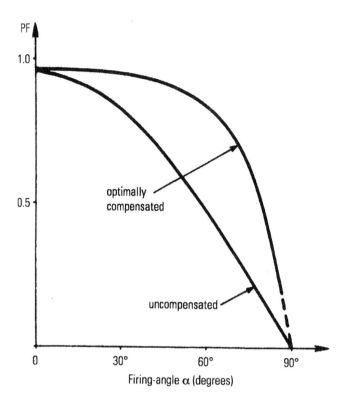

Fig. 9   Power factor versus firing angle for the three-phase, full-wave controlled bridge rectifier with highly inductive load and ideal supply.

use of capacitors is that for fixed-load resistance, the value of the optimal capacitor varies with firing angle.

7.2.4   Worked Examples

Example 7.4   A three-phase, full-wave, controlled bridge rectifier contains six ideal thyristor switches and is fed from an ideal three-phase voltage source of 240 V, 50 Hz. The load resistor $R = 10\ \Omega$ is connected in series with a large smoothing inductor. Calculate the average load voltage and the power dissipation at (1) $\alpha = 30°$ and (2) $\alpha = 60°$.

If 240 V. represents the rms value of the line voltage, then the peak phase voltage E_m is given by

$$E_m = \frac{\sqrt{2}}{\sqrt{3}} 240$$

From Eq. (7.8)

$$E_{av} = \frac{3\sqrt{3}}{\pi} \times \frac{\sqrt{2}}{\sqrt{3}} \times 240 \cos\alpha$$

$$= 324 \cos\alpha$$

At $\alpha = 30°$, $E_{av} = 280.6$V
At $\alpha = 60°$, $E_{av} = 162$V
The power dissipation is given by Eq. (7.50),

$$P = I_{av}^2 R = \frac{E_{av}^2}{R}$$

At $\alpha = 30°$, $P = 7.863$ kW. At $\alpha = 60°$, $P = 2.625$ kW.

Example 7.5 For the three-phase bridge of Example 7.4 calculate the displacement factor, the distortion factor, and the power factor at (1) $\alpha = 30°$ and (2) $\alpha = 60°$.

From Eq. (7.58) it is seen that the displacement factor is given by

Displacement factor $= \cos\psi_1 = \cos\alpha$

At $\alpha = 30°$, $\cos\alpha = 0.866 = \sqrt{3/2}$
At $\alpha = 60°$, $\cos\alpha = 0.5$

Because the wave shape of the supply current is not affected by the firing-angle of the bridge thyristors (although the magnitude is affected) the supply current distortion factor is constant. From Eq. (7.59),

Distortion factor $= \frac{3}{\pi} = 0.955$

For loads with sinusoidal supply voltage the power factor, seen from the supply point, is the product of the displacement factor and the distortion factor:

$$PF = \frac{3}{\pi} \cos\alpha$$

At $\alpha = 30°$, $PF = 0.827$
At $\alpha = 60°$, $PF = 0.478$

Example 7.6 For the three-phase bridge rectifier of Example 7.4 calculate the required voltage and current ratings of the bridge thyristors

In a three-phase, full-wave, thyristor bridge the maximum voltage on an individual switch is the peak value of the line voltage

$$E_{max} = \sqrt{2}E_{line}$$

where E_{line} is the rms value of the line voltage. Therefore,

$$E_{max} = \sqrt{2} \times 240 = 339.4 \text{ V}$$

(Note that E_{max} is $\sqrt{3}$ times the peak value E_m of the phase voltage.)

From Eq. (7.52) the rms value of the supply current is

$$I_a = \sqrt{\frac{2}{3}}I_{av} = \frac{3\sqrt{2}E_m}{\pi R}\cos\alpha$$

but each thyristor conducts only one (positive) pulse of current every supply voltage cycle. In Fig. 7.6, for example, thyristor Th_1 conducts only the positive pulses of current i_a (ωt) shown in Fig. 7.7. Therefore,

$$I_{Th_1} = \sqrt{\frac{1}{2\pi}\int_{\alpha+30°}^{\alpha+150°} i_a^2(\omega t)\, d\omega t}$$

The defining expression for I_{Th} above is seen to have the value $1/\sqrt{2}$ that of I_a in Eq. (7.52),

$$I_{Th_1} = \frac{1}{\sqrt{2}I_a}$$

$$= \frac{3E_m}{\pi R} \quad \text{at } \alpha = 0°$$

$$= \frac{3}{\pi} \times \frac{\sqrt{2}}{\sqrt{3}} \times 240 \times \frac{1}{10} = 18.7 \text{ A}$$

Example 7.7 The three-phase, full-wave bridge rectifier of Example 7.4 is to have its power factor compensated by the connection of equal, star-connected capacitors at the supply point. Calculate the maximum value of capacitance that will result in power factor improvement and the optimum capacitance that will give the maximum realizable power factor improvement at (1) $\alpha = 30°$ and (2) $\alpha = 60°$. In each ease compare the compensated power factor with the corresponding uncompensated value.

The criterion for power factor improvement is defined by Eq. (7.66), which shows that

$$C_{max} = \frac{18\sin 2\alpha}{\omega \pi^2 R}$$

At $\alpha = 30°$,

$$C_{max} = 503 \ \mu F$$

At $\alpha = 60°$,

$$C_{max} = 503 \ \mu F$$

The optimum value of capacitance that will cause unity displacement factor and maximum power factor is given in Eq. (7.68)

$$C_{opt} = \frac{9}{\omega \pi^2 R} \sin \alpha$$

Note that $C_{opt} = C_{max}/2$.
  For both firing angles,

$$C_{opt} = 251.5 \ \mu F$$

In the presence of optimum capacitance the power factor is obtained from Eq. (7.69)

$$PF_{c_{max}} = \frac{3 \cos \alpha}{\sqrt{\pi^2 - 9 \sin^2 \alpha}}$$

At $\alpha = 30°$, $PF_c = 0.941$, which compares with the uncompensated value $PF = 0.827$ (Example 7.4). At $\alpha = 60°$, $PF_c = 0.85$, which compares with the uncompensated value $PF = 0.478$ (Example 7.4).

7.3 HIGHLY INDUCTIVE LOAD IN THE PRESENCE OF SUPPLY IMPEDANCE

Let the three-phase, full-wave bridge rectifier circuit now contain a series inductance L_s in each supply line. Because of the nonsinusoidal currents drawn from the supply, the voltages at the bridge input terminals abc (Fig. 7.10) are not sinusoidal but are given by

$$e_{aN} = e_{AN} - L_s \frac{di_a}{dt} \tag{7.70}$$

$$e_{bN} = e_{BN} - L_s \frac{di_b}{dt} \tag{7.71}$$

$$e_{cN} = e_{CN} - L_s \frac{di_c}{dt} \tag{7.72}$$

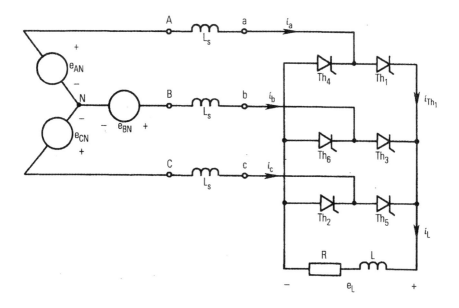

FIG. 10   Three-phase, full-wave controlled bridge rectifier circuit including supply inductance.

where e_{AN}, e_{BN}, and e_{CN} are now defined by the form of Eqs. (7.1) to (7.3), respectively.

The onset of ignition through any particular rectifier is delayed due both to the firing angle α, as described in the previous sections, and also due to overlap created by the supply inductance. For normal bridge operation at full load the overlap angle μ is typically 20° to 25° and is usually less than 60°. Operation of the bridge can be identified in several different modes.

7.3.1   Mode I Operation ($0 \leq \mu \leq 60°$)

A common mode of operation is where two thyristors conduct for most of the cycle, except in the commutation or overlap intervals when a third thyristor also conducts. Referring to Fig. 7.10, the sequence of conduction is 12, 123, 23, 234, 34, 345, 45, 456, 56, 561, 61, 612. The resulting waveforms are given in Fig. 7.11.

Consider operation at the instant $\omega t = \pi$ of the cycle. Thyristors Th_1 and Th_2 have been conducting (Fig. 7.11a and c), so that

At $\omega t = \pi = \alpha + 150°$,

chain

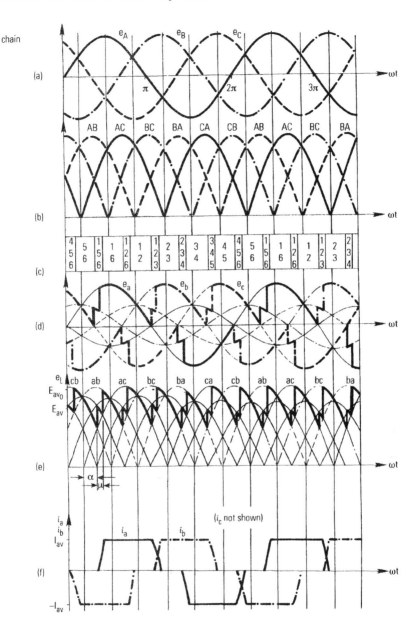

FIG. 11   Waveforms of the three-phase, full-wave controlled bridge rectifier with highly inductive load, in the presence of supply inductance: (a) supply phase voltages, (b) supply line voltages, (c) thyristor firing pattern, (d) supply voltages ($\alpha = 30°$, $\mu \approx 10°$), (e) load voltage ($\alpha = 30°$, $\mu \approx 10°$), and (f) supply currents ($\alpha = 30°$, $\mu \approx 10°$).

$$i_1 = i_a = I_{av} \tag{7.73}$$

$$i_2 = i_c = -i_a = -I_{av} \tag{7.74}$$

$$i_b = 0 \tag{7.75}$$

$$e_L = e_{ac} = e_{aN} - e_{cN} = \sqrt{3}E_m \cos(\omega t - 120°) \tag{7.76}$$

At $\omega t = \pi$, thyristor Th_3 is gated. Since e_b is positive, the thyristor switches on connecting its cathode to the positive load terminal (Fig. 7.12b). Since points a and b are now joined, they have the same potential with respect to neutral N, which is the average of the corresponding open circuit voltages, $(e_{aN} + e_{bN})/2$ $= (E_m/2)\sin(\omega t - 60°)$. But, simultaneously, the negative point of the load (point c) has a potential with respect to N of $E_m \sin(\omega t - 240°)$. Therefore, during the overlap period

$$e_L(\omega t) = \frac{E_m}{2}\sin(\omega t - 60°) - E_m \sin(\omega t - 240°)$$

$$= \frac{3}{2}E_m \sin(\omega t - 60°) \tag{7.77}$$

This is seen to be three times the value of $(e_{aN} + e_{bN})/2$ and is in time phase with it.

At $\omega t = \pi + \mu = \alpha + 150° + \mu$, thyristor Th_1 is extinguished by natural commutation after its current i_a has fallen to zero. Current i_b, which started to flow at $\omega t = \pi$ when Th_3 was fired, reaches the value I_{av} at $\omega t = \pi + \mu$ and takes over the load current relinquished by Th_1. Current then flows from b to c and voltage e_{bc} is impressed upon the load. During the overlap interval $\alpha + 150°$ $\le \omega t \le \alpha + 150° + \mu$ (Fig. 7.12b), the net voltage, proceeding clockwise around loop BNAB, is

$$V_{BNAB} = e_{BN} - e_{AN} + L_s\frac{di_a}{dt} - L_s\frac{di_b}{dt} = 0 \tag{7.78}$$

But

$$e_{BN} - e_{AN} = -\sqrt{3}E_m \cos(\omega t - 60°) \tag{7.79}$$

and

$$i_a + i_b = I_{av} = -i_c \tag{7.80}$$

Substituting Eqs. (7.79) and (7.80) into Eq. (7.78), noting that $dI_{av}/dt = 0$, gives

$$-\sqrt{3}E_m \cos(\omega t - 60°) = 2L_s\frac{di_b}{dt} = 2\omega L_s\frac{di_b}{d\omega t} \tag{7.81}$$

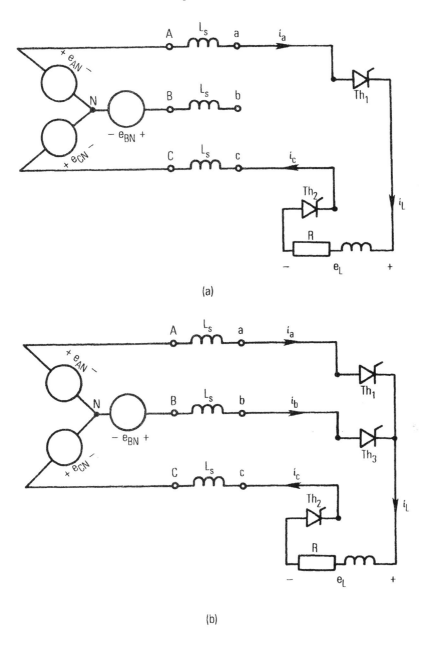

(a)

(b)

FIG. 12 Equivalent circuits of conduction for the three-phase, full-wave controlled bridge rectifier, including supply inductance: (a) $\omega t = \alpha + 150°$ and (b) $\omega t = \alpha + 150° + \mu$.

The time variation of a supply line current or a thyristor switch current during overlap can be deduced by the use of a running upper limit in the definite integration of Eq. (7.81).

For $\alpha + 150° \le \omega t \le \alpha + 150° + \mu$,

$$\int_0^{i_b(\omega t)} di_b = \frac{-\sqrt{3}E_m}{2\omega L_s} \int_{\alpha+150°}^{\omega t} \cos(\omega t - 60°)\, d\omega t \tag{7.82}$$

from which

$$i_b(\omega t) = \hat{I}_{sc}[\cos\alpha - \sin(\omega t - 60°)] = I_{av} - i_a \tag{7.83}$$

where $\hat{I}_{sc} = \sqrt{3}E_m/2\omega L_s$ is the peak value of the circulating current in the short-circuited section ANBN of Fig. 7.12b. The term $\hat{I}_{sc}$ was previously used in the analysis of the uncontrolled rectifier in Chapter 6.

Waveforms of $i_b(\omega t)$ and $i_a(\omega t)$ are given in Fig. 7.13. It is clearly illustrated that the portions of $i_a(\omega t)$, $i_b(\omega t)$ during overlap are parts of sine waves. The definition of $i_a(\omega t)$ for a complete period (not given here) would involve four terms similar in style to Eq. (7.83) plus a term in I_{av}.

7.3.1.1 Load-Side Quantities

Integrating both sides of Eq. (7.81),

$$\int di_b = i_b = \frac{-\sqrt{3}E_m}{2\omega L_s} \int \cos(\omega t - 60°)\, d\omega t$$

$$= \frac{-\sqrt{3}E_m}{2\omega L_s} \sin(\omega t - 60°) + K \tag{7.84}$$

where K is a constant of integration. At $\omega t = \alpha + \mu + 150°$, $i_b = I_{av}$ so that

$$K = I_{av} + \frac{\sqrt{3}E_m}{2\omega L_s} \sin(\alpha + \mu + 90°)$$

$$= I_{av} + \frac{\sqrt{3}E_m}{2\omega L_s} \cos(\alpha + \mu) \tag{7.85}$$

At $\omega t = \alpha + 150°$, $i_b = 0$ in Fig. 7.11 so that

$$K = \frac{\sqrt{3}E_m}{2\omega L_s} \sin(\alpha + 90°)$$

$$= \frac{\sqrt{3}E_m}{2\omega L_s} \cos\alpha \tag{7.86}$$

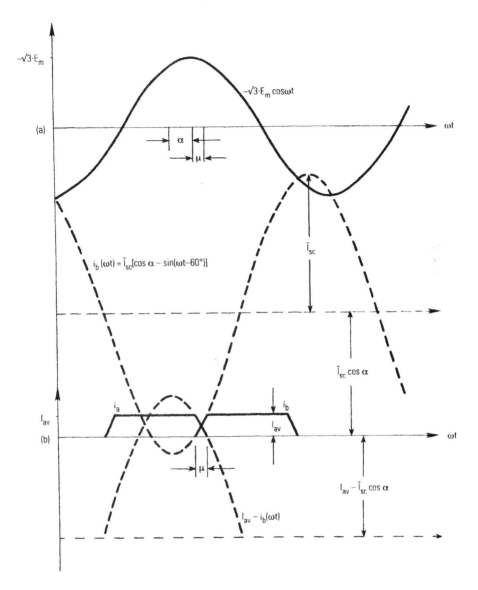

FIG. 13 Waveforms of the three-phase, full-wave controlled bridge rectifier with highly inductive load, in the presence of supply inductance $\alpha = 30°$, $\mu \approx 15°$: (a) generator line voltage e_{BC} ωt and (b) supply line currents i_A and i_B (ωt).

Eliminating K between Eq. (7.85) and (7.86) gives an expression for the average load current in the presence of supply inductance L_s per phase

$$I_{av} = \frac{\sqrt{3}E_m}{2\omega L_s}\left[\cos\alpha - \cos(\alpha+\mu)\right]$$

(7.87)

Expression (7.87) is identical to the corresponding expression (5.70) for half-wave operation The average load voltage can be found from the $e_L(\omega t)$ characteristic of Fig. 7.11e. Consider the 60° section defined by $150° \le \omega t \le 210°$.

$$E_{av} = \frac{6}{2\pi}\left[\int_{150°}^{150°+\alpha} e_{ac}\,d\omega t + \int_{150°+\alpha}^{150°+\alpha+\mu}\frac{3}{2}(e_{aN}+e_{bN})\,d\omega t + \int_{150°+\alpha+\mu}^{210°} e_{bc}\,d\omega t\right]$$

(7.88)

Elucidation of Eq. (7.88) is found to give

$$E_{av} = \frac{3\sqrt{3}E_m}{2\pi}\left[\cos\alpha + \cos(\alpha+\mu)\right]$$

$$= \frac{E_{av_0}}{2}\left[\cos\alpha + \cos(\alpha+\mu)\right]$$

(7.89)

The average value E_{av} in Eq. (7.89) is seen to be twice the value obtained in Eq. (5.61) for a half-wave bridge. The variation of E_{av} with α is demonstrated in Fig. 7.14.

If the term $\cos(\alpha+\mu)$ is eliminated between Eqs. (7.85) and (7.87), it is found that

$$E_{av} = E_{av_0}\cos\alpha - \frac{3\omega L_s}{\pi}I_{av}$$

(7.90)

The first term of Eq. (7.90) is seen to represent the average load voltage with ideal supply, consistent with Eq. (7.8). The second term of Eq. (7.90) represents the reduction of average load voltage due to voltage drop in the supply line inductances and is seen to be consistent with Eq. (6.27) for the diode bridge. Since, as always, $I_{av} = E_{av}/R$, Eq. (7.90) can be rearranged to show that

$$E_{av} = \frac{E_{av_0}\cos\alpha}{1+3\omega L_s/\pi R}$$

(7.91)

The power dissipated in the load P_L is conveniently expressed in terms of average load quantities (since the current is smooth) and $I_{rms} = I_{av}$

$$P_L = E_{av}I_{av} = \frac{E_{av}^2}{R} = I_{av}^2 R$$

(7.92)

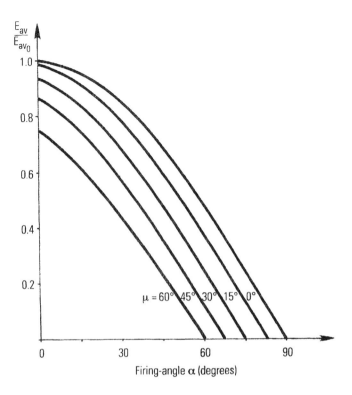

FIG. 14   Effect of supply reactance on the normalized average load voltage versus firing angle of a three-phase, full-wave controlled bridge rectifier with highly inductive load.

Substituting Eqs. (7.89) and (7.90) into Eq. (7.92) gives

$$P_L = \frac{9E_m^2}{4\pi\omega L_s}\left[\cos^2\alpha - \cos^2(\alpha+\mu)\right]$$

(7.93)

7.3.1.2 Supply-Side Quantities

All of the load power passes into the bridge from the supply. For a balanced three-phase load with sinusoidal voltage of peak value E_m per phase and periodic nonsinusoidal current with a fundamental component of rms value I_1, the input power maybe given by

$$P_{in} = \frac{3E_m}{\sqrt{2}}I_1\cos\psi_1$$

(7.94)

where $\cos \psi_1$ is the current displacement factor. Neglecting any power loss in the conducting thyristors, $P_L = P_{in}$. Equating Eqs. (7.92) and (7.9) gives

$$I_1 \cos \psi_1 = \frac{\sqrt{6}}{\pi} I_{av} \frac{\cos \alpha + \cos(\alpha + \mu)}{2}$$

$$= I_1(0) \frac{\cos \alpha + \cos(\alpha + \mu)}{2} \qquad (7.95)$$

The term $(\sqrt{6}/\pi) I_{av}$ in Eq. (7.95) is the rms fundamental supply current with zero overlap, $I_1(0)$.

It is reasonable to assume that overlap makes only a small difference in the value of I_1. In fact, for $\mu \leq 30°$, there is only 1.1% difference. Therefore, if

$$I_1 \cong I_1(0) = \frac{\sqrt{6}}{\pi} I_{av} \qquad (7.96)$$

then, very nearly,

$$\cos \psi_1 = \frac{1}{2}\left[\cos \alpha + \cos(\alpha + \mu)\right] \qquad (7.97)$$

Combining Eqs. (7.96), (7.97), and (7.98) gives

$$\cos \psi_1 \cong \frac{E_{av}}{E_{av_0}} = \cos \alpha - \frac{3\omega L_s}{\pi} \frac{I_{av}}{E_{av_0}} \qquad (7.98)$$

With an ideal supply $L_s = 0$ and Eq. (7.98) reduces to Eq. (7.58).

The rms value of the input current can be obtained from the defining integral

$$I_a = \sqrt{\frac{1}{2\pi} \int_0^{2\pi} i_a^2(\omega t) \, d\omega t} \qquad (7.99)$$

Each cycle of $i_a(\omega t)$ contains six parts and the necessary mathematics to solve Eq. (7.99) is very lengthy (see Ref. 4). It is found, after much manipulation, that

$$I_a = \sqrt{\frac{2}{3}} I_{av} \sqrt{1 - f(\alpha, \mu)} \qquad (7.100)$$

where

$$f(\alpha, \mu) = \frac{3}{2\pi} \frac{\sin \mu \left[2 + \cos(2\alpha + \mu)\right] - \mu \left[1 + 2 \cos \alpha \cos(\alpha + \mu)\right]}{\left[\cos \alpha - \cos(\alpha + \mu)\right]^2} \qquad (7.101)$$

The term $\sqrt{2/3}\, I_{av}$ in Eq. (7.100) is seen, from Eq. (7.52), to be the rms supply current with zero overlap. It should be noted that the two numerator parts of Eq. (7.101) are very similar in value and great care is required in a numerical solution. It is found that for $\mu \leq 30°$ the effect of overlap reduces the rms supply current by less than 5% for all firing angles. One can therefore make the approximation

$$I_a \cong \sqrt{\frac{2}{3}} I_{av} \tag{7.102}$$

For small values of μ the power factor is given by combining Eq. (7.95) with Eq. (7.102)

$$PF = \frac{I_1}{I_a} \cos \psi_1$$

$$= \frac{3}{2\pi} \left[\cos \alpha + \cos(\alpha + \mu) \right] \tag{7.103}$$

When $\mu = 0$, Eq. (7.103) reduces to Eq. (7.53). For any finite value of μ the power factor is reduced compared with operation from an ideal supply.

Terminal capacitance can be used to obtain some measure of power factor improvement by reducing the displacement angle to zero. For overlap conditions such that $\mu \leq 30°$, the equations of this section provide the basis for approximate calculations. It is significant to note, however, that an important effect of supply inductance is to render the bridge terminal voltages nonsinusoidal. Equation (7.24), which defines the power factor, remains universally valid, but the relationship of Eq. (7.31) no longer has any validity.

7.3.2 Mode II Operation ($\mu = 60°$)

In mode II operation three rectifier thyristors conduct simultaneously, which only occurs during overload or with-short circuited load terminals. The average current through the load is then found to be

$$I_{av} = \frac{E_m}{2\omega L_s} \left[\cos(\alpha - 30°) - \cos(\alpha + \mu + 30°) \right] \tag{7.104}$$

The corresponding average load voltage is found to be

$$E_{av} = \frac{\sqrt{3}}{2} E_{av_0} \left[\cos(\alpha - 30°) - \cos(\alpha + \mu + 30°) \right]$$

$$= \frac{9}{2\pi} \frac{E_m}{\omega L_s} \left[\cos(\alpha - 30°) - \cos(\alpha + \mu + 30°) \right] \tag{7.105}$$

7.3.3 Mode III Operation 60° ≤ μ ≤ 120°

With mode III operation there is alternate conduction between three and four rectifier thyristors. Three thyristor operation is equivalent to a line-to-line short circuit on the supply, while four thyristor working constitutes a three-phase short circuit. Equations (7.104) and (7.105) still apply. If the term $\cos(\alpha + \mu + 30°)$ is eliminated between these, it is found that

$$\frac{E_{av}}{E_{av_0}} + \frac{3}{2}\frac{I_{av}}{\hat{I}_{sc}} = \sqrt{3}\cos(\alpha - 30°)$$

(7.106)

7.3.4 Worked Examples

Example 7.8   A fully-controlled, three-phase bridge rectifier supplies power to a load via a large, series filter inductor. The supply lines contain series inductance so as to cause overlap. Sketch the variation of the average load voltage with firing angle α for a range of overlap angle μ.

The average voltage is given by Eq. (7.89)

$$E_{av} = \frac{E_{av_0}}{2}\left[\cos\alpha + \cos(\alpha + \mu)\right]$$

The range of firing angles is $0 \le \alpha \le 90°$, and the realistic range of μ is $0 \le \mu \le 60°$. Values of the ratio E_{av}/E_{av_0} are given in Fig. 7.14.

Example 7.9   A three-phase, full-wave controlled bridge rectifier with highly inductive load operates from a 440-V, 50-Hz supply. The load current is maintained constant at 25 A. If the load average voltage is 400 V when the firing angle is 30°, calculate the load resistance, the source inductance, and the overlap angle.

The peak phase voltage E_m is

$$E_m = 440\sqrt{\frac{2}{3}} = 359.3 \text{ V}$$

From Eq. (7.90),

$$E_{av} = \frac{3\sqrt{3}}{\pi} \times 440\sqrt{\frac{2}{3}} \times \frac{\sqrt{3}}{2} - \frac{3 \times 2\pi 50 L_s}{\pi} \times 25 = 400 \text{ V}$$

which gives

$$L_s = \frac{514.6 - 400}{7500} = 15.3 \text{ mH/phase}$$

The load resistance is given in terms of the (presumed average values of the) load voltage and current

$$R = \frac{E_{av}}{I_{av}} = \frac{400}{25} = 16 \ \Omega$$

From Eq. (7.89),

$$\cos(\alpha + \mu) = \frac{2E_{av}}{E_{av_0}} - \cos\alpha$$

Now

$$E_{av_0} = \frac{3\sqrt{3}E_m}{\pi} = 594 \ \text{V}$$

Therefore,

$$\cos(\alpha + \mu) = \frac{2(400)}{594} - \frac{\sqrt{3}}{2} = 0.48$$

$$\alpha + \mu = 61.3°$$

$$\mu = 31.3°$$

Example 7.10   For the three-phase bridge rectifier of Example 7.9 calculate the reduction of the power transferred and the change of power factor due to overlap compared with operation from an ideal supply.

The average load current and voltage are both reduced due to overlap. From Eq. (7.92),

$$P_L = E_{av}I_{av} = \frac{E_{av}^2}{R}$$

and from Eq. (7.89)

$$E_{av} = \frac{E_{av_0}}{2}\left[\cos\alpha + \cos(\alpha + \mu)\right]$$

With $\alpha = 30°$, it was shown in Example 7.9 that $\mu = 31.3°$ when $E_{av} = 400$ V. Therefore,

$$P_L = \frac{400^2}{16} = 10 \ \text{kW}$$

This compares with operation from an ideal supply where, from Eq. (7.50),

$$P_{L_{\mu=0}} = \frac{E_{av_0}^2}{R}\cos^2\alpha = \frac{594^2}{16}\times\frac{3}{4} = 16.54 \text{ kW}$$

The large overlap angle has therefore reduced the power transferred to 60% of its "ideal" value.

An accurate calculation of power factor is not possible within the present work for an overlap angle $\mu = 31.3°$. An approximate value, from Eq. (7.103), is

$$PF \cong \frac{3}{2\pi}(\cos 30° + \cos 61.3°)$$

$$= 0.642$$

Example 7.11   A three-phase, full-wave controlled bridge rectifier with highly inductive load is operated with a thyristor firing angle $\alpha = 30°$. The supply line inductance is such that overlap occurs to an extent where $\mu = 15°$. Sketch the waveforms of the voltage across a bridge thyristor and the associated current.

When a thyristor switch is conducting current the voltage drop across, it can be presumed to be zero. While a thyristor is in extinction, the line-to-line voltage occurs across it, except during overlap, so that the peak value of a thyristor voltage is $\sqrt{3}E_m$. During the overlap intervals the peak value of the load (and thyristor) voltages is $(3/2)E_m$, as developed in Eq. (7.77). At the end of the first supply voltage cycle $\omega t = \pi$, thyristor Th_1 is conducting current i_a and the voltage e_{Th_1} is zero (Figs. 7.11 and 7.15). At $\omega t = \pi$, thyristor Th_3 is fired and Th_3 then conducts current i_b simultaneously. At $\omega t = \pi + \mu$ the overlap finishes with $i_a = 0$ and Th_1 switching off. The voltage e_{Th_1} then jumps to the appropriate value of e_{ab} $(\pi + \mu)$ and follows e_{ab} (ωt) until $\omega t = 240°$ and Th_4 is switched in. During the overlap of Th_2 and Th_4, $240° < \omega t < 240° + \mu$, the voltage across Th_1 is $-3e_{bN}/2$. When Th_2 is commutated off at $\omega t = 240° + \mu$, a negative current i_a (ωt) is flowing through Th_4, and the voltage across Th_1 jumps back to e_{ab} $(240° + \mu)$. At $\omega t = 300°$, thyristor Th_5 is switched in and overlaps with Th_3. The anode of Th_1 is held at point a, but the overlap of Th_3 and Th_5 causes e_{Th_1} to jump to $3e_{aN}/2$ during overlap. When Th_3 switches off at $\omega t = 300° + \mu$, current i_b falls to zero and voltage e_{Th_1} $(\omega t + \mu)$ follows e_{ac} and so on. The overall waveform for e_{Th_1} is given in Fig. 7.15 with the associated current i_{Th_1} also shown.

7.4 SUMMARY OF THE EFFECTS OF SUPPLY REACTANCE ON FULL-WAVE BRIDGE RECTIFIER OPERATION (WITH HIGHLY INDUCTIVE LOAD)

The presence of supply line inductance inhibits the process of current commutation from one thyristor switch to the next. Instead of the instantaneous current

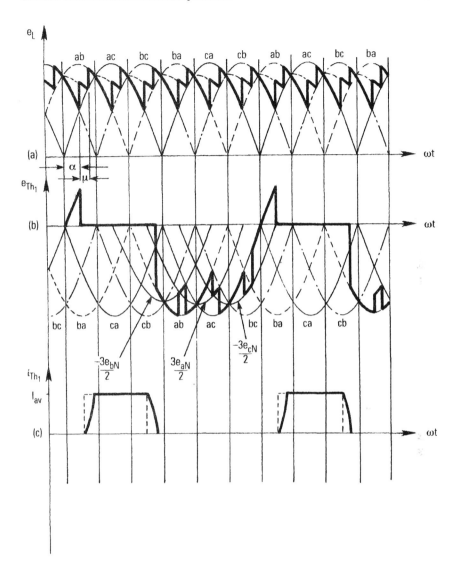

FIG. 15 Waveforms of the three-phase, full-wave controlled bridge rectifier with highly inductive load in the presence of supply inductance $\alpha = 30°$, $\mu \approx 15°$; (a) load voltage, (b) thyristor voltage e_{Th1} (ωt), and (c) thyristor Th_1 current.

transfer that occurs between thyristors when the supply is ideal a finite time is required to accomplish a complete transfer of current. The duration of the current transfer time, usually called the overlap period, depends on the magnitude of the supply voltage, the load resistor, the supply current level, the thyristor firing angle, and the source inductance.

1. The average load voltage E_{av} is reduced, Eq. (7.89).
2. The waveform of the load voltage $e_L (\omega t)$, (Fig. 7.11e) is modified compared with corresponding "ideal supply" operation (Fig. 7.3c). The additional piece "missing" from the waveform in Fig. 7.11e, for example, can be used to calculate the reduction of the average load voltage.
3. Because the waveform $e_L (\omega t)$ is changed, its harmonic properties as well as its average value are changed. The basic ripple frequency (i.e., six times supply frequency) is unchanged, and therefore the order of the harmonics of $e_L (\omega t)$ remains $6n$, where $n = 1, 2, 3, \cdots$ Therefore, the magnitudes of the harmonics of $e_L (\omega t)$ are affected by overlap.
4. The average load current I_{av} is affected by the reduction of average load voltage because $I_{av} = E_{av}/R$.
5. Load power dissipation is reduced due to overlap, Eq. (7.92), since $P_L = E_{av}I_{av}$.
6. The waveform of the supply line currents are modified. Current i_a (ωt) in Fig. 7.7e, for example, is modified to the waveform of Fig. 7.11f. Its conduction angle is extended from $120°$ to $120° + \mu$ in each half cycle. Because of the change of waveform, the magnitudes of its harmonic components are modified.
7. The rms value of the supply current is reduced, Eq. (7.100).
8. The rms value of the fundamental component of the supply current is reduced.
9. The modified shape of the supply current causes the current displacement angle ψ_1 to increase and therefore the displacement factor $\cos\psi_1$ to decrease.
10. The power factor of operation is reduced due to overlap, Eq. (7.103), at small values of μ.
11. The waveform of the bridge terminal voltage is no longer sinusoidal but contains "notches" during the overlap periods. This reduces the rms supply voltage and also the rms value of the fundamental component of the supply voltage.
12. The power factor relationship

     $PF = $ (displacement factor)(distortion factor)

     is no longer valid because of the nonsinusoidal terminal voltage.
13. The notching of the supply voltage can give rise to the spurious firing

of silicon controlled rectifiers by forward breakover and also to interference with electronic circuits connected at the same or adjacent supply points.

PROBLEMS

Three-Phase, Full-Wave, Controlled Bridge with Resistive Load and Ideal Supply

7.1 A three-phase, full-wave bridge rectifier of six ideal thyristor switches is connected to a resistive load. The ideal three-phase supply provides balanced sinusoidal voltages at the input terminals. Show that the average load voltage E_{av} is given by Eqs. (7.8) and (7.12) in the two respective modes of operation. Sketch E_{av} versus firing angle α over the full operating range.

7.2 A three-phase, full-wave bridge rectifier containing six ideal thyristors supplies a resistive load $R = 100 \ \Omega$. The ideal supply 240 V, 50 Hz provides balanced sinusoidal voltages. Calculate the average load current and power dissipation at (a) $\alpha = 30°$, (b) $\alpha = 60°$, and (c) $\alpha = 90°$.

7.3 For the three-phase bridge circuit of Problem 7.2 deduce and sketch the voltage waveform across a thyristor at $\alpha = 30°$.

7.4 For the three-phase bridge circuit of Problem 7.1 show that the rms values of the supply current are given by Eqs. (7.21) and (7.22).

7.5 For a three-phase, full-wave bridge circuit with resistive load, show that for both modes of operation, the rms supply current I_a is related to the rms load current I_L by the relation Eq. (7.23).

7.6 Expressions for the fundamental component of the supply current into a three-phase, full-wave controlled bridge rectifier supplying a resistive load are given in Table 7.2. Calculate the rms values of this fundamental component with a supply of 240 V, 50 Hz and a load resistor $R = 100 \ \Omega$ at (a) $\alpha = 30°$, (b) $\alpha = 60°$, and (c) $\alpha = 90°$.

7.7 The power input to a three-phase, full-wave, controlled bridge rectifier is given by the relation $P = 3EI_{a_1} \cos \psi_1$, where E is the rms phase voltage, I_{a_1} is the rms value of the fundamental component of the supply current, and $\cos \psi_1$ is the current displacement factor (not the power factor!). Calculate P for the bridge circuit of Problem 7.6 and check that the values obtained agree with the power dissipation calculated on the load side.

7.8 Show that the Fourier coefficients a_1 and b_1 of the fundamental component of the supply line current for a full-wave controlled bridge with resistive load R, at firing angle α, are given by

$$a_1 = -\frac{3\sqrt{3}E_m}{2\pi R}\sin 2\alpha \quad b_1 = \frac{E_m}{2\pi R}(2\pi + 3\sqrt{3}\cos 2\alpha)$$

7.9 Derive expressions for the current displacement factor $\cos \psi_1$ and the current distortion factor I_1/I for a three-phase, full-wave controlled bridge rectifier with resistive load. Show that the respective products of these are consistent with the expressions (7.25) and (7.26) for the power factor.

7.10 Calculate and sketch the variation of the power of a three-phase, full-wave bridge rectifier with resistive load over the operating range of thyristor firing angles.

7.11 Use the information of Problem 7.7 to derive expressions for the reactive voltamperes Q into a three-phase, full-wave bridge rectifier with resistive load, where $Q = 3EI_{a_1} \sin \psi_1$. Does a knowledge of real power P and reactive voltamperes Q account for all the apparent voltamperes S (= $3EI_a$) at the bridge terminals?

7.12 Three equal capacitors C are connected in star across the terminals of a full-wave, three-phase bridge rectifier with resistive load. If $X_c = R$, sketch waveforms of a capacitor current, a bridge input current, and the corresponding supply current at $\alpha = 30°$. Does the waveform of the supply current seem to represent an improvement compared with the uncompensated bridge?

7.13 For the three-phase bridge circuit of Problem 7.2 what will be the minimum value of supply point capacitance per phase that will cause power factor improvement at (a) $\alpha = 30°$, (b) $\alpha = 60°$ and (c) $\alpha = 90°$?

7.14 For the three-phase bridge circuit of Problem 7.2. what must be the respective values of the compensating capacitors to give the highest realizable power factor (by capacitor correction) at the three values of firing angle?

7.15 For the three-phase bridge circuit of Problem 7.2 calculate the operating power factor at each value of firing angle. If optimum compensation is now achieved by the use of the appropriate values of supply point capacitance, calculate the new values of power factor.

Three-Phase, Full-Wave Controlled Bridge Rectifier with Highly Inductive Load and Ideal Supply

7.16 A three-phase, full-wave controlled bridge rectifier contains six ideal thyristors and is fed from an ideal, three-phase supply of balanced sinusoidal voltages. The load consists of a resistor R in series with a large filter inductor. Show that, for all values of thyristor firing angle α, the average load voltage is given by Eq. (7.8). Sketch E_{av} versus α and compare the result with that obtained for purely resistive load.

7.17 For the three-phase, inductively loaded bridge of Problem 7.16 calculate the Fourier coefficients a_1 and b_1 of the fundamental component of the supply current. Use these to show that the current displacement angle ψ_1 $[\tan^{-1}(a_1/b_1)]$ is equal to the thyristor firing angle α.

7.18 A three-phase, full-wave controlled bridge rectifier is supplied from an ideal three-phase voltage source of 415 V, 50 Hz. The load consists of resistor $R = 100 \ \Omega$ in series with a very large filter inductor. Calculate the load power dissipation at (a) $\alpha = 30°$ and (b) $\alpha = 60°$, and compare the values with those that would be obtained in the absence of the load filter inductor.

7.19 Show that for the inductively loaded bridge of Problem 7.16 the distortion factor of the supply current is independent of thyristor firing angle.

7.20 Show that the waveform of the supply current into a controlled bridge rectifier with highly inductive load is given by

$$i(\omega t) = \frac{2\sqrt{3}I_{av}}{\pi}\left[\sin(\omega t - \alpha) - \frac{1}{5}\sin 5(\omega t - \alpha) - \frac{1}{7}\sin 7(\omega t - \alpha)^- ...\right]$$

where I_{av} is the average load current.

7.21 For the three-phase bridge rectifier of Problem 7.16 show that the power input is equal to the load power dissipation.

7.22 Derive an expression for the load voltage ripple factor (RF) for a three-phase inductively loaded bridge rectifier and show that this depends only on the thyristor firing angle. Obtain a value for the case $\alpha = 0$, and thereby show that the RF is zero within reasonable bounds of calculation.

7.23 For the inductively loaded bridge rectifier of Problem 7.16 show that the rms supply current is given by

$$I = \frac{3\sqrt{2}E_m}{\pi R}\cos\alpha$$

Calculate this value for the cases (a) $\alpha = 30°$ and (b) $\alpha = 60°$.

7.24 For the inductively loaded bridge of Problem 7.18 calculate the rms current and peak reverse voltage ratings required of the bridge thyristors.

7.25 Show that the average load voltage of a three-phase, full-wave controlled bridge circuit with highly inductive load can be obtained by evaluating the integral

$$E_{av} = \frac{6}{2\pi} \int_{\alpha-30°}^{\alpha+30°} E_m \cos \omega t \, d\omega t$$

Sketch the waveform of the instantaneous load voltage e_L (ωt) for $\alpha = 75°$, and show that it satisfies the above relationship.

7.26   A three-phase, full-wave, thyristor bridge is fed from an ideal three-phase supply and transfers power to a load resistor R. A series inductor on the load side gives current smoothing that may be considered ideal. Derive an expression for the rms value of the fundamental component of the supply current. Use this expression to show that the reactive voltamperes Q entering the bridge is given by

$$Q = \frac{27E_m^2}{4\pi^2 R} \sin 2\alpha$$

7.27   For the three-phase, bridge rectifier of Problem 7.18 calculate the power factor. If equal capacitors C are now connected in star at the supply calculate the new power factor when $X_c = R$. What is the minimum value of firing angle at which compensation to the degree $X_c = R$ renders a power factor improvement?

7.28   For the bridge rectifier circuit of Problem 7.16 derive an expression for the terminal capacitance that will give maximum power factor improvement.

7.29   The bridge rectifier circuit of Problem 7.18 is compensated by the use of equal capacitors C connected in star at the supply terminals. Calculate the values of capacitance that will give unity displacement factor at (a) $\alpha = 30°$ and (b) $\alpha = 60°$. In each case calculate the degree of power factor improvement compared with uncompensated operation.

7.30   For the bridge circuit of Problem 7.28 sketch, on squared paper, consistent waveforms of the bridge line current, the capacitor current and the supply line current. Does the waveform of the supply current appear less distorted than the rectangular pulse waveform of the bridge current?

7.31   A three-phase, full-wave, bridge rectifier circuit, Fig. 7.6, supplies power to load resistor R in the presence of a large load filter inductor. Equal capacitors are connected at the supply terminals to give power factor improvement by reducing the current displacement angle ψ_1 to zero at the fixed thyristor firing-angle α. Derive a general expression for the supply current distortion factor in the presence of supply capacitance. For the case when C has its optimal value so that the displacement factor is increased to unity is the distortion factor also increased?

Three-Phase, Full-Wave Controlled Bridge Rectifier with Highly Inductive Load in the Presence of Supply Inductance

7.32   A full-wave controlled bridge rectifier circuit transfers power to a load resistor R in series with a large filter inductor. The three-phase supply contains a series inductance L_s in each supply line and has sinusoidal open-circuit voltages where E_m is the peak phase voltage. Show that at thyristor firing angle α the average load current is given by

$$I_{av} = \frac{\sqrt{3}E_m}{2\omega L_s}\left[\cos\alpha - \cos(\alpha+\mu)\right]$$

where μ is the overlap angle.

7.33   For the full-wave bridge of Problem 7.32, use Eq. (7.89), or otherwise, to show that the average load voltage is given by

$$E_{av} = \frac{3\sqrt{3}E_m}{2\pi}\left[\cos\alpha + \cos(\alpha+\mu)\right]$$

7.34   A three-phase, full-wave controlled bridge rectifier with highly inductive load operates from a 240-V, 50-Hz supply. The load current is required to remain constant at 15 A. At firing angle $\alpha = 15°$, the load voltage is found to be 200 V. Calculate the source inductance and the overlap angle.

7.35   For the three-phase rectifier of Problem 7.32 show that overlap angle μ may be obtained from

$$\frac{\cos(\alpha+\mu)}{\cos\alpha} = \frac{1 - 3\omega L_s/\pi R}{1 + 3\omega L_s/\pi R}$$

7.36   For the three-phase bridge rectifier of Problem 7.34 calculate the reduction of the power transferred due to overlap compared with operation from an ideal supply.

7.37   A three-phase bridge rectifier with highly inductive load operates at a firing angle $\alpha = 30°$ and results in an overlap angle $\mu = 15°$. Calculate the per-unit reduction of rms supply current compared with operation from an ideal supply.

7.38   A resistor $R = 20\Omega$ is supplied from a three-phase controlled bridge rectifier containing a large series filter on the load side. The supply is 240 V, 50 Hz, and each supply line contains a series inductance $L_s = 10$ mH.

Calculate the approximate power factor of operation for (a) $\alpha = 0$, (b) $\alpha = 30°$, and (c) $\alpha = 60°$.

7.39 A three-phase, full-wave controlled bridge rectifier supplies a highly inductive load. Show that in the overlap intervals caused by supply-line inductance, the load voltage is 1.5 times the relevant phase voltage.

7.40 For a three-phase, full-wave bridge rectifier with highly inductive load, it was shown in Eq. (7.58) that the current displacement $\cos \psi_1$ is related to the thyristor firing angle α by a relationship $\cos \psi_1 = \cos \alpha$ when the supply is ideal. Show that in the presence of significant supply inductance this relationship is no longer valid but that

$$\cos \psi_1 \cong \frac{1}{2}\left[\cos \alpha + \cos(\alpha + \mu)\right]$$

8

Rectifier Power Factor and Pulse-Width Modultion Controlled Rectifier Circuits

8.1 POWER FACTOR AND SUPPLY CURRENT DISTORTION IN THREE-PHASE, PHASE-CONTROLLED BRIDGE RECTIFIERS

The three-phase controlled rectifier is a much-used circuit device with many applications. For both passive impedance and active (motor) loads it is common to use a dc output filter inductor to smooth the load current. In addition, a shunt-connected filter capacitor C smooths the output to a largely ripple-free adjustable direct voltage V_{dc} (Fig. 8.1).

Operation of the three-phase bridge rectifier using conventional phase-delay (i.e., phase-angle) control was extensively described in Chapter 7. For all forms of load the method of phase-angle control has two serious drawbacks: (1) low lagging power factor and (2) supply current distortion. Each of these features is briefly discussed below.

8.1.1 Power Factor of Phase-controlled Bridge Rectifiers

The supply point power factor of a three-phase bridge rectifier with conventional phase-retardation control reduces greatly as the switch firing angle is retarded to accomplish load voltage reduction. This is shown in Fig. 7.9, which is a graphical

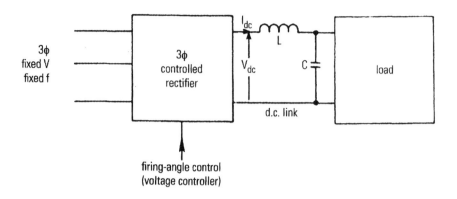

FIG. 1   Three-phase controlled rectifier with output filter.

version of Eq. (7.53). Because the current has the same waveform for all firing angles with inductive and motor loads (Fig. 7.7) the distortion factor I_1/I has the constant value $3/\pi$, or 0.955. But the displacement factor $\cos \psi_1$ is equal to the cosine of the delayed firing angle $\cos \alpha$, as developed in Sec. 7.2.2. This results in progressive reduction of the displacement factor (and hence power factor) as firing angle α increases.

The power factor of a three-phase, phase-controlled rectifier is reproduced here from Eq. (7.53).

$$\text{Power factor} = \frac{3}{\pi} \cos \alpha \qquad (8.1)$$

Power factor improvement is a major item of industrial practice to reduce electricity supply costs. From Eq. (8.1) it is seen that this can be approached in terms of distortion factor improvement, which involves changing the supply current waveform or in terms of reducing the input current displacement angle ψ_1 (i.e., firing angle α).

8.1.2   Supply Current Distortion of Phase-Controlled Bridge Rectifiers

Waveforms of the supply currents for three-phase bridge rectifier operation are given in Fig. 7.4 for resistive loads and in Fig. 7.7 for inductive and motor loads. These mathematically odd functions do not contain even order harmonic components. Also, because of three-phase symmetry, neither do they contain the odd triple values $n = 3, 9, 15, 21$, etc. The dominant harmonic components, in descending order of magnitude are the components $n = 5, 7, 11, 13, 17, 19$, etc.

Power supply utilities are operated on the basic feature that the voltages and currents generated and transmitted are sinusoidal. Any departure from the ideal sinusoidal waveforms causes deterioration of the supply system performance, mostly in the form of increased losses. In addition, the operation of loads that draw nonsinusoidal currents cause consequent nonsinusoidal voltage drops across the series impedances of transformers and transmission lines. The result is that the system voltages at the point of load coupling may become distorted. This affects not only the customer with the distorting load, such as a three-phase rectifier, but all other customers connected to the transmission system at that point.

Because the connection of distorting loads is now so widespread, the United States and the European Union both have guidelines governing the maximum amount of distortion that is acceptable. These guidelines detail the levels of rms harmonic phase currents, of all harmonic numbers up to $n = 19$, that are acceptable at the various voltage levels of the system. In the United Kingdom for example, at the standard three-phase distribution level of 415 V, the maximum permitted harmonic phase current levels are 56 A at $n = 5$ and 40 A at $n = 7$, for each consumer.

8.2 METHODS OF POWER FACTOR IMPROVEMENT

Significant improvement (i.e., increase) of multiphase rectifier power factor can be realized in several different ways.

1. Supply side capacitors or/and filter circuits
2. Tap changing transformers
3. Multiple rectifier connections with sequential operation
4. Modifying the rectifier firing strategy to achieve pulse-width modulation control

Items 1, 2, 3 and above are mentioned only briefly here since they are covered in detail in other books (see for example, References 1, 2, 3, 6, 17, 19, 23).

8.2.1 Supply-Side Capacitors

Great improvement in the displacement factor can be achieved by the use of supply-side compensating capacitors (Fig. 7.5). For any value of rectifier firing angle there is an optimum value of capacitance that will give the maximum realizable value of displacement factor (and hence power factor).

In order to realize the optimum improvement shown in Fig. 7.9, it is necessary to change the value of the capacitors at each different firing angle. This is

not practicable in engineering terms and would be very expensive if a wide range
of operation was desired (i.e., if a wide range of load voltage variation is needed)

It is common with rectifiers of large rating to use a range of supply-side,
shunt-connected filters, each tuned to a particular harmonic frequency. The operation of the rectifier itself is unchanged but the filters act to prevent rectifier
generated harmonic currents from circulating in the power supply. The result is
that the rectifier plus filters, seen from the supply system, operates at increased
power factor and reduced electricity tariffs.

8.2.2   Transformer Tap Changing

Some types of industrial load, such as battery charging and electrolysis, require
only slow changes and involve relatively small variations of voltage of the order
$\pm 10\%$. A supply of this kind can be provided by the use of an adjustable ratio
transformer. For on-load or off-load operation variations of output voltage can
be obtained by transformer tap changing using solid-state switches. In high-power
applications silicon controlled rectifiers are usually incorporated.

The principle is illustrated in Fig. 8.2. Any output voltage between v_2 and
v_1 can be obtained by smooth adjustment of the switch firing-angles. The output
voltage waveform (Fig. 8.2b) is much less distorted than those of conventional
three-phase rectifier operation, such as Fig. 7.2. If waveforms such as those of
Fig. 8.2b are used to supply a three-phase diode rectifier, like that of Fig. 6.1, a
high value of displacement factor can be maintained over a wide range of output
voltage.

8.2.3   Sequential Operation of Multiple Rectifiers

The half-wave rectifiers of Chapters 4 and 5 and the full-wave rectifiers of Chapters 6 and 7 can all be extended by the use of more than three phases.

In Fig. 7.1 it was shown that the full-wave, six-pulse bridge is equivalent
to two three-pulse bridges. Two six-pulse bridges, such as that of Fig. 7.1b, can
be added in series to produce what is effectively a 12-pulse bridge (Fig. 8.3).
The use of separate wye and delta secondary windings from a common primary
winding creates output phase voltages with a 30° phase displacement, which
effectively increases the pulse number. Many different transformer arrangements
are available. By dividing transformer phase windings into two sections with
different voltage phase angles or/and by the use of interphase reactor windings,
it is possible to produce 24 pulse supplies.

All of the many multiple rectifier connections, however, use phase-angle
delay as the basis of their switching. The result is still to produce reduced, lagging
power factor and supply current distortion.

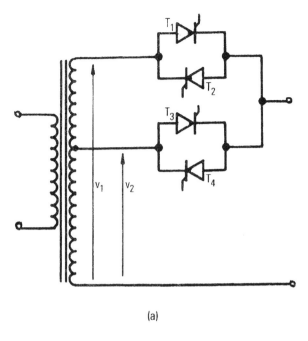

(a)

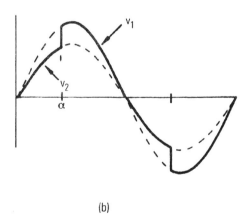

(b)

FIG. 2 Single-phase transformer tap changer $\alpha = 60°$.

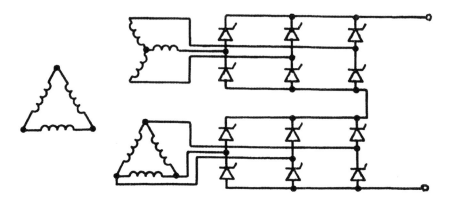

FIG. 3 Two six-pulse bridges added to produce a 12-pulse output.

8.2.4 Modification of Rectifier Firing Strategy

Waveforms obtained by the retarded triggering of controlled switches are typified by Figs. 3.2, 3.5, 3.6, and 3.11 for single-phase operation and Figs. 7.4, 7.7, and 7.11 for three-phase operation. The reduced lagging power factor is, in all cases, due to delayed triggering of the rectifier switches.

An alternative to delayed triggering is to notch the waveform symmetrically in the manner shown, for example, in Fig. 8.4. Each half sine wave is notched to give (in this case) five equal width conduction bands with respectively equal nonconduction notches between them. Compared with the symmetrical phase-delay waveforms of Fig. 3.6 or Figs. 14.3 and 14.4, the waveform of Fig. 8.4 is seen to possess a very important feature—its fundamental component is symmetrically in phase with the waveform itself. For Fig. 8.4 the displacement angle ψ_1 is zero, and the displacement factor $\cos \psi_1$ has its maximum possible value of unity. The waveform of Fig. 8.4 does not represent unity power factor because it has a distortion factor due to the notched wave shape. But it is reasonable to expect that a current waveform such as that of Fig. 8.4, combined with sinusoidal supply voltage, would represent high power factor operation.

A wide range of notched waveforms can be devised, with even or uneven conduction bands. It is sometimes found to be advantageous to use pulse waveforms in which the pulse widths are not uniform but are obtained by modulation techniques. The most versatile and useful approach is found to be in terms of pulse-width modulation (PWM), which is described extensively in the following section.

Notched or pulse waveforms both imply a significant engineering challenge concerning switch commutation. All of the rectifier circuits described in Chapters

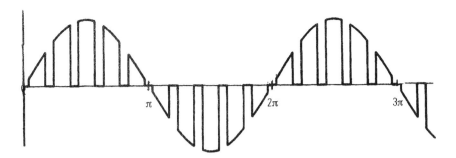

FIG. 4 Sine wave with symmetrical uniform notching.

1 to 7 above are actuated by self- or natural commutation-conducting switches are extinguished by natural cycling of the supply voltages. For a waveform such as Fig. 8.4 conduction may be initiated and terminated at several arbitrary intervals in each half cycle. The necessity to switch off the conducting element requires forced commutation by gating or by auxiliary commutation circuits. This is a more complicated and much bigger design challenge than a natural commutation circuit and is much more expensive to implement.

8.3 PROPERTIES OF PULSE WAVEFORMS

8.3.1 Single-Pulse Modulation

The basic pulse waveform consists of a fixed duration single pulse in each half wave. More flexible forms of control would permit variation of this single pulse by (1) fixing the leading edge but varying the trailing edge, (2) fixing the trailing edge but varying the leading edge or (3) varying the pulse width while keeping the pulses symmetrical about $\pi/2$, $3\pi/2$, etc. Figure 8.5a shows a single-pulse waveform of pulse width δ symmetrical about $\pi/2$ and $3\pi/2$. This waveform has the Fourier series

$$\upsilon(\omega t) = \frac{4V}{\pi}\left(\sin\frac{\delta}{2}\sin\ \omega t - \frac{1}{3}\sin\frac{3\delta}{2}\sin 3\omega t + \frac{1}{5}\ \sin\ \frac{5\delta}{2}\ \sin\ 5\omega t \cdots \right) \quad (8.2)$$

Pulse width δ has a maximum value of π radians at which the fundamental term in Eq. (8.2) is a maximum. An individual harmonic of order n may be eliminated by making $\delta = 2\pi/n$, but this is likely also to reduce the value of the fundamental component. The rms value of the single-pulse waveform of Fig. 8.5a is found to be

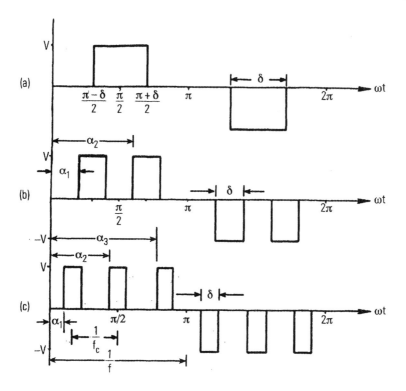

Fig. 5 Pulse voltages waveforms: (a) single-pulse modulation ($N = 1$), (b) two-pulse modulation, ($N = 2$), and (c) three-pulse modulation ($N = 3$).

$$V_{rms} = V \sqrt{\frac{\delta}{\pi}} \tag{8.3}$$

The nth-order harmonic in Eq. (8.2) is seen to have a peak value

$$V_n = \frac{4V}{n\pi} \sin \frac{n\delta}{2} \tag{8.4}$$

A Fourier analysis of Fig. 8.5(a) shows that $a_n = 0$, which makes the phase displacement angle zero. This confirms the visual impression that since the pulses are symmetrical within their half waves, then there is no phase displacement of the voltage harmonics.

The distortion factor of the single-pulse waveform is

$$\text{Distortion factor} = \frac{V_1/\sqrt{2}}{V_{rms}} = \frac{2\sqrt{2}}{\sqrt{\pi\delta}} \sin \frac{\delta}{2} \qquad (8.5)$$

which has a maximum value of 0.9 when $\delta = \pi$. This is consistent with the data of Table 10.2.

8.3.2 Fourier Properties of a Pulse Train

When a wave consists of N identical pulses per half cycle, symmetrically spaced, the coefficients of the nth Fourier harmonic are found to be given by

$$c_n = \sum_{m=1}^{N} \frac{4V}{n\pi} \sin \frac{n\delta}{2} \sin n\left(\alpha_m + \frac{\delta}{2}\right) \qquad (8.6)$$

When $N = 1$, as in Fig. 8.5a, then $\alpha_1 = (\pi - \delta)/\pi$ and Eq. (8.6) reduces to Eq. (8.4). Examples are given in Fig. 8.5 for $N = 2$ and $N = 3$. In all cases coefficient $\alpha_1 = 0$, which confirms the visual impression that the fundamental component of the pulse wave is symmetrical with the pulse train itself. The number of pulses N and the pulse spacing greatly affect the magnitudes and orders of the higher harmonic components. For example, with $N = 3$,

$$c_n = \frac{4V}{n\pi} \sin \frac{n\delta}{2} \left[\sin n\left(\alpha_1 + \frac{\delta}{2}\right) + \sin n\left(\alpha_2 + \frac{\delta}{2}\right) + \sin n\left(\alpha_3 + \frac{\delta}{2}\right) \right] \qquad (8.7)$$

The greater is the number of pulses N, the higher will be the value of c_n for a fixed value of δ. Certain harmonic orders can be suppressed or totally eliminated by the appropriate choice of N, α, and δ.

In the pulse trains of Fig. 8.5 the duty cycle is $N\delta/\pi$, which represents the ratio of conduction time to total period time and is also the mean height of the pulse train

$$\text{Duty cycle} = \frac{\text{conduction time}}{\text{total period time}} = \frac{N\delta}{\pi} \qquad (8.8)$$

With a high duty cycle a high value of fundamental component is realizable (Fig. 8.6a). As the duty cycle reduces with the same number of pulses N, caused by reduction of the pulse width δ, the fundamental component is also progressively reduced (Fig. 8.6b).

The rms values of the pulse waveforms of Fig. 8.6 are given by

$$V_{rms} = V\sqrt{\frac{N\delta}{\pi}} \qquad (8.9)$$

When $N = 1$, Eq. (8.9) reduces to Eq. (8.3).

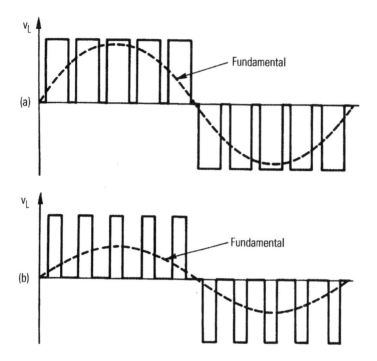

FIG. 6 Effect of duty cycle on the fundamental component amplitude: (a) high duty cycle and (b) low duty cycle.

8.3.3 Production of a Uniformly Notched Pulse Train by Amplitude Modulation

A train of evenly spaced identical pulses can be produced by amplitude modulating a single-sided triangular carrier wave v_c ($\omega_c t$) by a single-sided, square-wave modulating signal v_m (ωt) of lower frequency, as shown in Fig. 8.7. The ratio of the modulating signal peak amplitude V to the carrier signal peak amplitude V_c is a basic parameter of all modulated waveforms and is called the modulation ratio or modulation index M.

$$\text{Modulation ratio } = M = \frac{V}{V_c} \tag{8.10}$$

The pulse height V of the resulting modulated signal v_o (ωt) (Fig. 8.7) can be adjusted in the range $0 \leq V \leq V_c$, and the pulse width varied in the range $0 \leq \delta \leq \pi$. The width of the equal pulses is related to the signal voltages by a relation

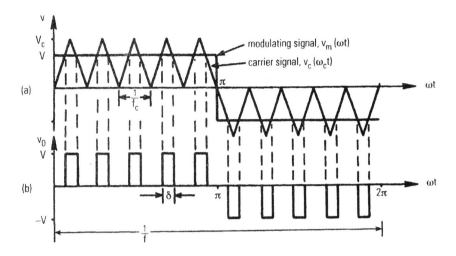

Fig. 7 Multiple-pulse voltage waveforms: (a) carrier signal and modulating signal and (b) output (modulated) signal.

$$\frac{\delta}{N} = (1 - \frac{V}{V_c}) \frac{\pi}{N}$$

(8.11)

or

$$\frac{\delta}{\pi} = 1 - \frac{V}{V_c} = 1 - M$$

(8.12)

Ratio δ/π in Eq. (8.11) is seen to represent the duty cycle per number of pulses.

The rms value of the multipulse waveform given by Eq. (8.9) can be combined with Eq. (8.12) to give

$$\frac{V_{rms}}{V} = \sqrt{\frac{N\delta}{\pi}} = \sqrt{N(1 - \frac{V}{V_c})} = \sqrt{N(1 - M)}$$

(8.13)

It is a characteristic of all modulated waves that the fundamental frequency component of the output (modulated) wave is equal to the frequency of the modulating wave. Frequency variation of the output signal is therefore obtained by frequency adjustment of the modulating (control) signal.

Let the carrier pulse frequency be f_c and the overall cycle frequency $\omega/2\pi$ be f, where $f_c > f$, as illustrated in Figs. 8.5 and 8.7. When the number of equal, symmetrical pulses per half cycle is N, then

$$N = \frac{f_c}{2f} = \frac{\omega_c}{2\omega} = \text{integer}$$

(8.14)

A further basic property of modulated waveforms is the ratio between the carrier frequency and the modulating frequency, known as the *frequency ratio p.*

$$\text{Frequency ratio } = \frac{f_c}{f} = p$$

(8.15)

When p is an integer this is defined as an example of synchronous modulation.

If p is an odd integer, then the modulated waveform contains half-wave symmetry (i.e., the positive and negative half cycles are antisymmetrical), and there are no even-order harmonics.

8.4   PROPERTIES OF SINUSOIDAL PULSE-WIDTH MODULATION WAVEFORMS

8.4.1   Pulse-Width Modulation

An alternative to the uniformly notched waveforms of Figs. 8.6 and 8.7 is to use a wave pattern produced by pulse-width modulation (PWM). A feature of a PWM waveform (Fig. 8.8b) is that the on periods are not uniform but are greatest at the desired peak of the fundamental output component wave. In this way low-order higher harmonics, such as $n = 3, 5, 7 \cdots$, may be greatly reduced compared with an evenly notched waveform having the same fundamental component value (Fig. 8.8a).

8.4.2   Single-Phase Sinusoidal Modulation with Natural Sampling

The principle of single-phase sinusoidal PWM is illustrated in Fig. 8.9. A sinusoidal modulating signal $v_m (\omega t) = V_m \sin \omega t$ is applied to a single-sided triangular carrier signal $v_c (\omega t)$ of maximum height V_c. The natural intersections of $v_m (\omega t)$ and $v_c (\omega_c t)$ determine both the onset and duration of the modulated pulses so that the pulse pattern is described as being due to natural sampling. The circuitry actuating the turn-on and turn-off of the converter switches also is controlled by sensing these intersections.

In naturally sampled, sinusoidal PWM the duration of the output voltage pulses is proportional to the instantaneous value of the sinusoidal modulating waveform at the center of the pulse. The pulse of greatest width is coincident with the peak of the modulating wave (Fig. 8.9). The pulse area and the fundamental component of the output voltage are proportional to the corresponding magnitude V of the modulating sinewave. In Fig. 8.9, reducing V proportionately reduces

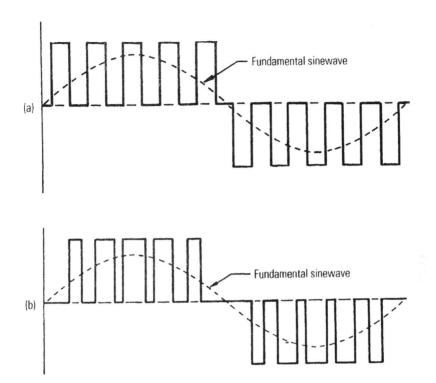

FIG. 8   Comparison of pulse patterns: (a) uniformly notched pulse train and (b) PWM wave.

the modulation index M and the peak value of the fundamental output (i.e., modulated) component.

There are several other different pulse-width modulation techniques used in power electronics applications in electrical engineering. For example, a method widely used in the variable-frequency inverter control of ac motors is described extensively in Chapter 11.

8.4.3   Harmonic Elimination in PWM Waveforms

A double-sided PWM waveform with several arbitrary switching angles is shown in Fig. 8.10. Switchings occur at angles defined as $\alpha_1, \alpha_2, \cdots, \alpha_n$ over the repetitive period of 2π radians. Because the waveform contains both half-wave and quarter-wave symmetry, a complete cycle can be fully defined by the switching angles for only a quarter cycle of the waveform.

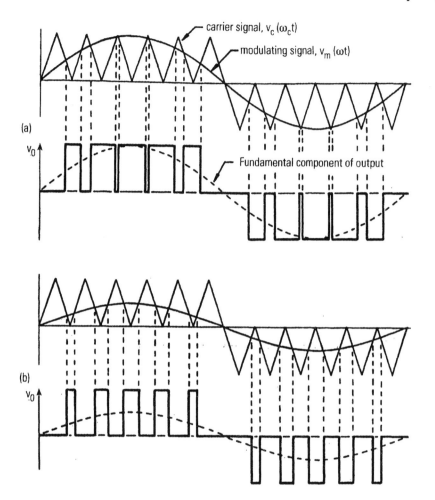

FIG. 9 Principle of sinusoidal modulation: (a) $M = V/V_c = 1.0$ and (b) $M = V/V_c = 0.5$.

The switching angles in Fig. 8.10 can be calculated in order that the PWM waveform possesses a fundamental component of a desired magnitude while, simultaneously, optimizing a certain performance criterion. For example, the criterion might be to eliminate certain selected harmonics, such as the fifth and/or the seventh, from the waveform. Alternatively, the criterion might be to minimize the total harmonic content and thereby maximize the distortion factor.

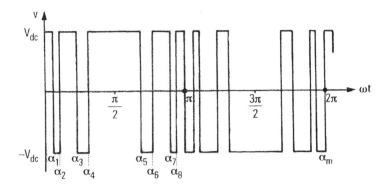

FIG. 10 PWM voltage waveform with eight arbitrary switchings per half cycle.

From the Fourier equations of the Appendix it can be inferred, for the quarter-cycle range, that coefficient $\alpha_n = 0$ and that b_n is given by

$$b_n = \frac{4}{\pi}\sin\int_0^{\pi/2} v(\omega t)\sin n\omega t \, d\omega t \tag{8.16}$$

For the example in Fig. 8.10 containing two notches (four switchings) per quarter cycle, the waveform is defined by

$$v(\omega t) = V_{dc}\begin{vmatrix}\alpha_1,\alpha_3,\pi/2 \\ 0,\alpha_2,\alpha_4\end{vmatrix} - V_{dc}\begin{vmatrix}\alpha_2,\alpha_4 \\ \alpha_1,\alpha_3\end{vmatrix} \tag{8.17}$$

Combining Eqs. (8.16) and (8.17) gives

$$b_n = \frac{4V_{dc}}{\pi}(1 - 2\cos n\alpha_1 + 2\cos n\alpha_2 - 2\cos n\alpha_3 + 2\cos n\alpha_4) \tag{8.18}$$

The pattern of Eq. (8.18) can be extended to accommodate any desired number of notches or switchings per quarter wave. Each switching angle in the quarter wave represents an unknown to be determined.

A generalized from of Eq. (8.18) is given by

$$V_n = \frac{4V_{dc}}{n\pi}\left[1 + 2\sum_{i=1}^{m} (-1)^i \cos n\alpha_i\right] \tag{8.19}$$

where m is the number of switchings per quarter cycle. The solution of Eq. (8.19) requires m independent, simultaneous equations; the particular case of Fig. 8.10 and Eq. (8.18), for example, has $m = 4$. This means that with two notches per quarter wave it is possible to limit or eliminate four harmonics, one of which may be the fundamental component. In balanced three-phase systems the triplen harmonics are suppressed naturally. It may therefore be logical to suppress the 5-, 7-, 11- and 13-order harmonics, which results in the following equations:

$$b_5 = \frac{4}{5}\frac{V_{dc}}{\pi}\left(1 - 2\cos\ 5\alpha_1 + 2\cos\ 5\alpha_2 - 2\cos\ 5\alpha_3 + 2\cos\ 5\alpha_4\right) = 0 \tag{8.20}$$

$$b_7 = \frac{4V_{dc}}{7\pi}\left(1 - 2\cos\ 7\alpha_1 + 2\cos\ 7\alpha_2 - 2\cos\ 7\alpha_3 + 2\cos\ 7\alpha_4\right) = 0 \tag{8.21}$$

$$b_{11} = \frac{4V_{dc}}{11\pi}\left(1 - 2\cos\ 11\alpha_1 + 2\cos\ 11\alpha_2 - 2\cos\ 11\alpha_3 + 2\cos\ 11\alpha_4\right) = 0 \tag{8.22}$$

$$b_{13} = \frac{4V_{dc}}{13\pi}\left(1 - 2\cos\ 13\alpha_1 + 2\cos\ 13\alpha_2 - 2\cos\ 13\alpha_3 + 2\cos\ 13\alpha_4\right) = 0 \tag{8.23}$$

Solution of the four simultaneous equations [Eqs. (8.20)–(8.23)] gives the results $\alpha_1 = 10.55°$, $\alpha_2 = 16.09°$, $\alpha_3 = 30.91°$, and $\alpha_4 = 32.87°$. Increase of the number of notches per quarter cycle increases the number of harmonics that may be suppressed, but has the concurrent effects of reducing the fundamental component and increasing the switching losses.

In general, the set of simultaneous, nonlinear equations describing particular performance criteria need to be solved or optimized using numerical methods. Precomputed values of switching angle may be stored in a ROM-based lookup table from which they are accessed by a microprocessor in order to generate the necessary switching pulses. It would not be possible to solve numerically the set of equations in real time, as would be needed in a motor control application. The larger the number of notchings per quarter cycle, the more refined becomes the waveform. This may entail solving a large set of nonlinear equations for which a solution is not always practicable. Furthermore, these equations need to be solved repetitively, once for each desired level of output.

8.5 THREE-PHASE BRIDGE RECTIFIER IN PWM MODE

The general arrangement of a three-phase, full-wave bridge rectifier is shown in Fig. 8.11. For PWM operation the switching elements must be capable of reverse voltage blocking, such as a MOSFET or a power transistor in series with a fast recovery diode. The reverse connected diodes across the switches facilitate regen-

erative action when there is an appropriate dc source and also participate in the
rectifier action.

Power enters the bridge from the three-phase sinusoidal supply. Reference
modulating sine waves (Fig. 8.12) map the supply voltages and intersect a double-
sided triangular carrier wave. Natural sampling PWM is used to produce phase
voltages (relative to a hypothetical mid-load point) V_A, V_B, and V_C at the bridge
entry terminals. It can be seen in Fig. 8.12 that the fundamental harmonic components of the phase voltages will be in phase with the waveforms themselves.

The profile of the dc output voltage follows the form of the six-pulse waveforms of Fig. 6.2c and Fig. 7.3c. With notched or PWM operation, however, the
profile will be built up using the middle 60° sections of the line-to-line voltages
represented by (say) waveform V_{AB} of Fig. 8.12. Note that the d.c. output voltage
level is controlled by adjustment of the modulation ratio M. The ripple frequency
on the dc voltage is determined by the frequency ratio p. Since the modulating
voltages are of supply frequency the ripple frequency is, in effect, determined
by the selected carrier frequency. Capacitor C in Fig. 8.11 acts to filter the ripple
component of voltage leaving a very smooth output voltage.

It is possible to devise switch firing strategies to reduce or eliminate particular harmonics. Because the voltage waveforms have sinusoidal profiles, rather
than the fixed levels of Figs. 8.7–8.10, however, the harmonic elimination method
described in Sec. 8.4.3 does not apply.

Because the supply voltages at points a, b, and c in Fig. 8.11 are sinusoidal
but the bridge terminal voltages $V_A V_B$, and V_C are either evenly notched or pulse-

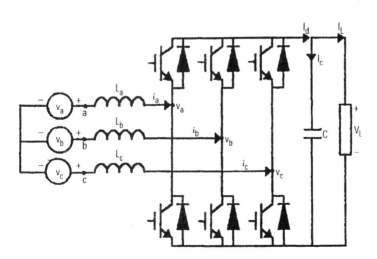

FIG. 11   Power circuit diagram of a three-phase PWM bridge rectifier.

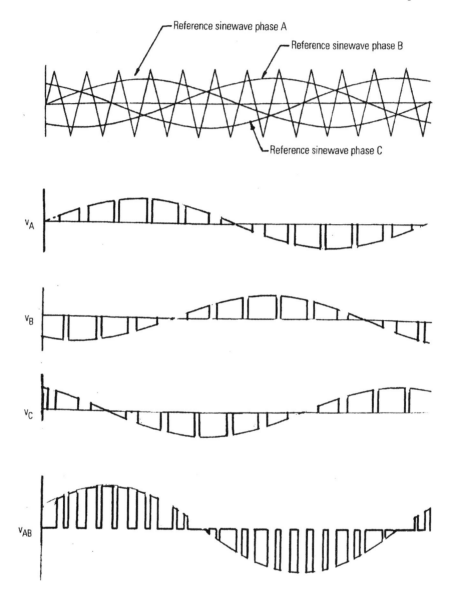

Fig. 12 PWM waveforms of a three-phase rectifier $M = 0.65$, $p = 12$. (Adapted from Ref. 11.)

width modulated, the difference voltages lie across the line reactors L_a, L_b, L_c. The action is very similar, in principle, to a line-commutated rectifier–inverter, as described in Chapter 9.

If high-frequency harmonics are disregarded the bridge operation, for the fundamental frequency component, may be approximated by the equivalent circuit of Fig. 8.13a. With a high value of frequency ratio, (say) $p = 27$, the input current waveforms are close to being sinusoidal. At some values of the dc output voltage and switching conditions these currents will be in time phase with the fundamental components of the PWM phase voltages, to result in unity displacement factor (not unity power factor). The fundamental frequency components can be represented by a phasor diagram (Fig. 8.13b). Some types of device known as *boost rectifiers* have a dc output voltage greater than the line-to-line voltages at the entry terminals. If $V_A > V_a$, the resulting input current I_a leads its phase voltage in time phase (Fig. 8.13c).

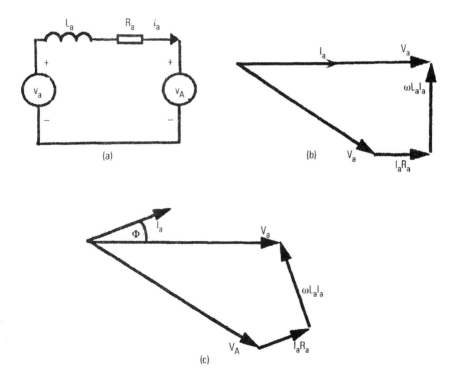

FIG. 13 Operation of a PWM rectifier: (a) equivalent circuit for fundamental components, (b) phase diagram for unity displacement factor (buck operation), and (c) leading displacement factor (boost operation).

The design and implementation of a PWM rectifier system would be complex and expensive. Each rectifier would probably need a unique base drive logic, depending on the application. This would need to be interfaced to the switches from a microcontroller and backup protection systems incorporated. Currently the available switches do not possess an adequate combination of power handling capability, switching speed and cost to make the PWM rectifier commercially attractive.

8.6 WORKED EXAMPLES

Example 8.1   A double-pulse notched voltage waveform of the type shown in Fig. 8.5b has a peak amplitude V. Calculate the fundamental component and compare with this the value obtained by the use of a single-pulse waveform of the same total area.

In the waveform $v(\omega t)$ in Fig. 8.5b, $N = 2$ and the instantaneous value is therefore

$$v(\omega t) = V \begin{vmatrix} \alpha_1+\delta, \alpha_2+\delta \\ \alpha_1, \alpha_2 \end{vmatrix} - V \begin{vmatrix} \pi+\alpha_1+\delta, \pi+\alpha_2+\delta \\ \pi+\alpha_1, \pi+\alpha_2 \end{vmatrix}$$

Since the waveform is anti-symmetrical about $\omega t = 0$ the fundamental component $v_1(\omega t)$ passes through the origin, $\psi_1 = 0$ and $\alpha_1 = 0$. Fourier coefficient b_1 is given by

$$b_1 = \frac{2}{\pi} \int_0^\pi v(\omega t) \sin \omega t \, d\omega t$$

$$= -\frac{2V}{\pi} (\cos \omega t) \begin{vmatrix} \alpha_1+\delta, \alpha_2+\delta \\ \alpha_1, \alpha_2 \end{vmatrix}$$

$$= \frac{2V}{\pi} \left[\cos\alpha_1 - \cos(\alpha_1 + \delta) + \cos\alpha_2 - \cos(\alpha_2 + \delta) \right]$$

$$= \frac{4V}{\pi} \sin\frac{\delta}{2} \left[\sin\left(\alpha_1 + \frac{\delta}{2}\right) + \sin\left(\alpha_2 + \frac{\delta}{2}\right) \right]$$

This equation is seen to be valid for the fundamental frequency $n = 1$, when $N = 2$, from Eq. (8.6). If the condition is one of symmetry with $\alpha_1 = \delta = \pi/5 = 36°$, then $\alpha_2 = 3\delta = 108°$ and

$$b_1 = \frac{4V}{\pi} \sin\frac{\delta}{2} \left(\sin\frac{3\delta}{2} + \sin\frac{7\delta}{2} \right)$$

$$= \frac{4V}{\pi} \sin 18° \left(\sin 54° + \sin 126° \right)$$

$$= \frac{4V}{\pi} \times 0.31 \times (0.809 + 0.809)$$

$$b_1 = \frac{4V}{\pi} \times 0.05 = \frac{2V}{\pi} = 0.637 \text{ V} \quad \text{(peak value)}$$

In comparison, for a single-pulse waveform of the same area, Eq. (8.4) gives

$$V_1 = \frac{4V}{\pi} \sin 36°$$

The use of two pulses per half cycle, with the same total area therefore results in reduction of the fundamental component, even though the rms value is unchanged. The distortion factor in therefore also reduced proportionately.

Example 8.2 A double-pulse, single-sided notched voltage waveform has the pattern of Fig. 8.5b. What restriction would require to be imposed on the design parameters α_1, α_2, and δ so that the third harmonic component was completely suppressed?

The Fourier coefficient b_n for the waveform of Fig. 8.5b was shown in the previous example to be

$$b_n = \frac{4V}{n\pi} \sin \frac{n\delta}{2} \left[\sin n \left(\alpha_1 + \frac{\delta}{2} \right) + \sin n \left(\alpha_2 + \frac{\delta}{2} \right) \right]$$

The may be expressed, alternatively, as

$$b_n = \frac{4V}{n\pi} \sin \frac{n\delta}{2} \left[2 \sin n \left(\frac{\alpha_1 + \alpha_2}{2} + \frac{\delta}{2} \right) \cos n \left(\frac{\alpha_2 - \alpha_1}{2} \right) \right]$$

To make $b_n = 0$ for the case $n = 3$

$$b_3 = \frac{4V}{3\pi} \sin \frac{3\delta}{2} \left[2 \sin 3 \left(\frac{\alpha_1 + \alpha_2}{2} + \frac{\delta}{2} \right) \cos \frac{3}{2} (\alpha_2 - \alpha_1) \right] = 0$$

Three options arise:

   1. $\sin(3\delta/2) = 0$.
   2. $\sin 3 (\alpha_1 + \alpha_2)/2 + \delta/2) = 0$.
   3. $\cos(3/2)(\alpha_1 - \alpha_3) = 0$.

This leads to the following restrictions:

   1. $\delta = 0$, $(2/3)\pi$, $(4/3)\pi$, etc.
   2. $3 [(\alpha_1 + \alpha_2)/2 + \delta/2] = 0$, π, 2π, $\cdots$, or $\alpha_1 + \alpha_2 = -\delta$, $2\pi/3 - \delta$, $\cdots$,
   3. $(3/2) (\alpha_1 - \alpha_2) = \pi/2$, $3\pi/2$, $5\pi/2$, $\cdots$, or $\alpha_1 - \alpha_2 = \pi/3$, π, $5\pi/3$, $\cdots$.

     Condition 1 is not admissible. From 2 and 3 it is seen that, for example,
   4. $\alpha_1 + \alpha_2 = -\delta$.

5. $\alpha_1 - \alpha_2 = \pi/3$.

Combining 4 and 5 gives

$$\alpha_1 = \frac{\pi}{6} - \frac{\delta}{2}$$

$$\alpha_2 = -\frac{\pi}{6} - \frac{\delta}{2}$$

Other combinations are possible showing that there is no unique solution. For example, if

$$\alpha_1 + \alpha_2 = \frac{2\pi}{3} - \delta$$

$$\alpha_1 - \alpha_2 = \pi$$

then

$$\alpha_1 = \frac{5\pi}{6} - \frac{\delta}{2} \quad \alpha_2 = -\frac{\pi}{6} - \frac{\delta}{2}$$

PROBLEMS

8.1 Show that the rms value of the single-pulse waveform of Fig. 8.5a is given by Eq. (8.3).

8.2 Calculate the values of the fundamental components of the pulse waveforms of Fig. 8.5a and b if $\delta = 108°$

8.3 The voltage waveform in Fig. 8.14 contains three single-sided pulses in each half cycle, spaced symmetrically with respect to $\pi/2$. Obtain an expression for the amplitude of the nth harmonic if $\alpha_1 = \pi/6$ and $\alpha_2 = \pi/3$, and compare this with the corresponding expression for a single-pulse waveform of the same area. What are the respective fundamental values?

8.4 For the waveform of Fig. 8.14, calculate the values of α_1 and α_2 that will permit the 3rd and 5th harmonic components to be eliminated.

8.5 For the voltage waveform of Fig. 8.10 show that the Fourier coefficient b_n, in terms of the switching angles α_1 and α_2, is given by

$$b_n = \frac{4V}{n\pi}(1 - \cos n\alpha_1 + \cos n\alpha_2)$$

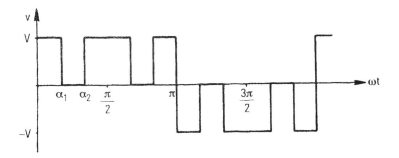

FIG. 14   Voltage waveform for Problem 8.3.

8.6   Define relationships for the switching angles that need to be satisfied if the 3rd and 5th harmonic components are to be eliminated from the waveform of Fig. 8.15. Calculate appropriate values of α_1 and α_2.

8.7   A single-sided triangular carrier wave of peak height V_c contains six pulses per half cycle and is modulated by a sine wave $v_m(\omega t) = V_m \sin \omega t$ synchronized to the origin of a triangular pulse. Sketch waveforms of the resultant modulated wave if (a) $V_m = 0.5V_c$, (b) $V_m = V_c$ and (c) $V_m = 1.5 \, V_c$. Which of these waveforms appears to contain the greatest fundamental (i.e, modulating frequency) value?

8.8   For the waveforms described in Problem 8.7 estimate, graphically, the values of ωt at which intersections occur between $v_c(\omega_c t)$ and $v_m(\omega t)$ when $V_m = V_c$. Use these to calculate values of the harmonics of the modulated wave up to $n = 21$ and thereby calculate the rms value.

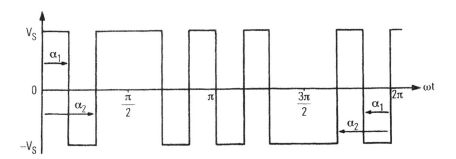

FIG. 15   PWM voltage waveform with two arbitrary switchings per quarter cycle.

9

Three-Phase, Naturally Commutated, Controlled Bridge Rectifier–Inverter

9.1 THEORY OF OPERATION

The process of transferring power from a dc supply to an ac load is known as *inversion*, and the necessary form of converter is usually known as an *inverter*. Inverters can be classed according to their manner of commutation, that is, by the process by which conducting switches are extinguished.

In a small number of applications the inverter feeds directly into an established three-phase voltage source (Fig. 9.1(a). The commutation function is then performed by natural cycling of the ac voltages, as in the rectifier circuits described in chapters 1-8. Naturally commutated inverters, described in Chapter 9, invariably use silicon controlled rectifiers as switching elements, and the output frequency is maintained constant by the three-phase busbars.

The three-phase bridge rectifier of Fig. 7.1b can be used as an inverter if the passive load is replaced by a dc supply with reversed polarity voltage, as shown in Fig. 9.1. The current direction on the dc side is unchanged—direct current enters the common anodes of Th_4, Th_6, Th_2 and leaves the common cathodes of Th_1, Th_3, Th_5. With inverter operation the battery acts as a power source, and current leaves the positive plate of the battery. This is unlike rectifier operation (Fig. 7.1b), where direct current enters the positive terminal of the load.

For inverter operation the voltage level, frequency and waveform on the ac side are set by the three-phase bus and cannot be changed. As with rectifier

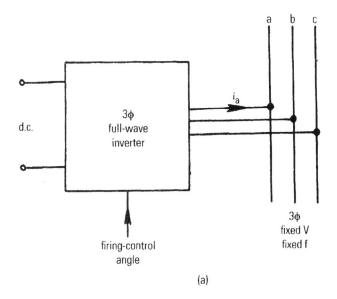

(a)

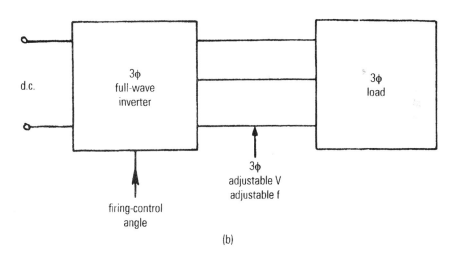

(b)

FIG. 1   Basic forms of inverters: (a) naturally commutated and (b) artificially commutated

operation, described in Sec. 7.1., the anode voltages of the switches undergo cyclic variation and are therefore switched off by natural commutation. There is no advantage to be gained here by the use of gate turn-off devices. But it is a requirement that the switches can be turned on at controlled switching angles. An inverter cannot therefore operate using uncontrolled (diode) switches.

In the operation of the inverter circuit of Fig. 9.2 certain restrictions must be imposed on the switching sequences of the switches. For example, both switches of any inverter arm, such as Th_1 and Th_4, cannot conduct simultaneously. Similarly, only one switch of the lower bridge (Th_2, Th_4, Th_6) and one switch of the upper bridge (Th_1, Th_3, Th_5) can conduct simultaneously. If sequential firing is applied to the six SCRs, the three phase currents are identical in form but mutually displaced in phase by 120°. The detailed operation of the circuit for rectifier operation, described in Sec. 7.1, is again relevant here, except that the range of firing angles is now $90° \leq \alpha \leq 180°$.

The waveform of the instantaneous voltage e_L (ωt) on the dc side of the bridge, with ideal ac supply, is shown in Fig. 9.3 for both rectifier and inverter operation. This waveform is derived from the line-to-line ac-side voltages. The level of the dc-side current I_{dc}, and hence the power transfer, is determined by the relation between V_{dc} and the magnitude and polarity of the equivalent average value of e_L (ωt).

In the circuit of Fig. 9.2 the source voltage V_{dc} is of such polarity as to forward bias the SCR switches, whereas during inversion, the instantaneous polarities of the ac-side voltages act to reverse bias them. When the dc-side current I_{dc} tends to decrease, the induced voltage of the filter inductor will assume a polarity that acts to sustain it and hence to forward bias the switches. The net

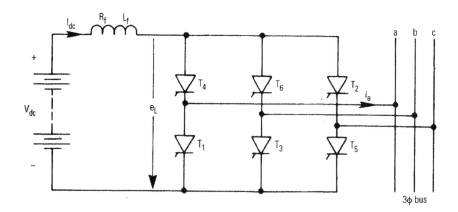

Fig. 2   Three-phase, naturally commutated, bridge rectifier–inverter [20]

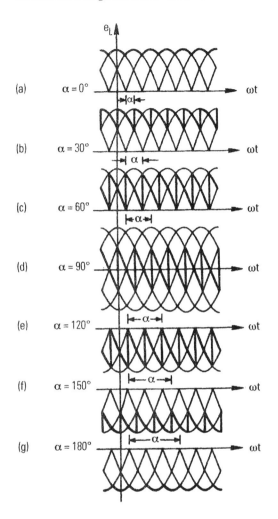

FIG. 3 Instantaneous voltage e_L (ωt) on the dc side of a bridge rectifier–inverter circuit, assuming ideal ac supply (no overlap) [20]

effect is to maintain an SCR in conduction even during segments of the phase voltage waveform when the polarity is such as to reverse bias it. The voltage presented by the ac side is derived from negative segments of individual voltages and has negative average polarity which opposes the flow of current injected by the dc source V_{dc}, If the ac-side effective voltage is smaller than V_{dc}, then I_{dc}

will flow as indicated in Fig. 9.2, and power is transferred from the dc side to the ac side in inverter action.

Voltage $e_L(\omega t)$ in Fig. 9.2 is time varying and does not, in general, coincide with the constant voltage V_{dc} from the dc source. For this reason a filter inductor L_f must be included to absorb the difference or ripple voltage e_{L_f}

$$e_{L_f} = V_{dc} - e_L(\omega t) = L_f \frac{dI_{dc}}{dt} + R_f I_{dc} \tag{9.1}$$

If the dc-side current is very smooth, which occurs when L_f is large, then $dI_{dc}/dt \rightarrow 0$, and the ripple voltage falls largely on the resistance R_f of the filter inductor.

The waveform of the current on the ac side of the bridge depends largely on the magnitude of the filter inductor. If this inductor is large, the line current assumes a rectangular waveform similar to that obtained for rectifier operation with a highly inductive load (Fig. 7.7). The line current pulses have a conduction period of 120° followed by a dwell period of 60°. In the circuit of Fig. 9.2 the average value of the bridge voltage is, from Eq. (7.8),

$$\text{Average value of } e_L(\omega t) = E_{av} = E_{av_0} \cos \alpha$$

$$= \frac{3\sqrt{3}E_m}{\pi} \cos \alpha$$

$$= \frac{3}{\pi}\sqrt{2}\,\frac{\sqrt{3}E_m}{\sqrt{2}} \cos \alpha$$

$$= 1.35E \cos \alpha \tag{9.2}$$

where E_m is the peak phase voltage and E is the rms line voltage.

The average value of the current on the dc side is seen, from Fig. 9.2, to be satisfied by the relation for rectifier operation.

$$I_{dc} = \frac{V_{dc} + E_{av}}{R_f} = \frac{1}{R_f}(V_{dc} + 1.35E \cos \alpha) \tag{9.3}$$

For inverter operation $\alpha > 90°$ and $\cos \alpha$ is negative, so that the filter inductor voltage is $V_{dc} - E_{av}$. At $\alpha = 90°$, $E_{av} = 0$ and the inverter presents a short circuit to the direct current. In order to maintain constant I_{dc} in the presence of adjustable SCR firing angle α, it is necessary to simultaneously vary R_f or V_{dc} or both. If direct voltage V_{dc} is constant, then increase of the retardation angle α would result in a decrease of I_{dc} due to growth of the bucking voltage E_{av}. Current I_{dc} in Fig. 9.2 will become zero, from Eq. (9.3), when

$$\cos \alpha = \frac{\pi V_{dc}}{3\sqrt{3}E_m} = \frac{V_{dc}}{1.35E} \tag{9.4}$$

Since the switches of the rectifier–inverter bridge are presumed to be lossless, there must be a power balance on each side of the circuit. If the rms value of the fundamental component of the inverter in phase a line current is I_{a_1} and the displacement angle is ψ_1, then

$$P = 3\frac{E_m}{\sqrt{2}} I_{a_1} \cos \psi_1 \quad \text{(ac side)}$$

$$= E_{av} I_{dc} \quad \text{(dc side)} \tag{9.5}$$

The power dissipation $I_{dc}^2 R_f$ in the filter comes from the battery, but this is external to the bridge. The real power P becomes zero if $E_{av} = 0$ or $I_{dc} = 0$ or both. To make $P = 0$, the SCR firing conditions are therefore that $\alpha = 90°$, from Eq. (9.2), or $\alpha = \cos^{-1}(V_{dc}/1.35E)$, from Eq. (9.4).

Now the important relationship $\alpha = \psi_1$ of Eq. (7.58) remains true and combining this with Eqs. (9.2) and (9.4) gives

$$I_{av} = \frac{\sqrt{2}\sqrt{3}}{\pi} I_{dc} = \frac{\sqrt{6}}{\pi} I_{dc} = 0.78 I_{dc} \tag{9.6}$$

The fundamental line current therefore has a peak value $\sqrt{2} \times 0.78 I_{dc} = 1.1 I_{dc}$ and is, by inspection, in time phase with (ωt). An example of this is shown in Fig. 9.4a.

Although the real power in watts transferred through the inverter comes from the dc source, the reactive voltamperes has to be provided by the ac supply. The ac-side current can be thought of as a fundamental frequency component lagging its corresponding phase voltage by ψ_1 ($= \alpha$) radians plus a series of higher odd harmonics. In terms of the fundamental frequency component, the inverter action can be interpreted either as drawing lagging current from the ac system or, alternatively, as delivering leading current to the ac system. Rectifier action and inverter action of the bridge circuit are depicted in the equivalent circuits of Fig. 9.5. A notation I_1 is used (rather than I_{a_1}) because the circuits are true for any phase.

The fundamental ac-side components may be represented in phasor form as, for example, in Fig. 9.6. With rectifier operation, both the in-phase component of current $I_1 \cos \alpha$ and the quadrature component $I_1 \sin \alpha$ are drawn from the ac bus. With inverter operation, the in-phase component of current is opposite in sign to the case of rectifier operation and represents active or real power delivered to the ac bus, but the quadrature component still represents reactive voltamperes drawn from the ac bus. As α increases from 90°, with constant current, an increased amount of power is transferred to the ac system and the reactive voltampere requirement is reduced.

Even for the condition of zero power transfer there may still be currents flowing in the inverter. At $\alpha = 90°$, (Fig. 9.4d), the inverter current is in quadra-

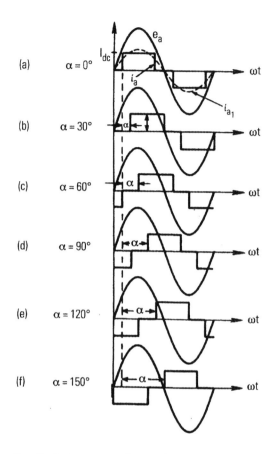

FIG. 4 Instantaneous phase voltage and current on the ac side of a bridge rectifier–inverter assuming ideal ac supply [20]. (Note: It is assumed that V_{dc} is adjusted proportionately to cos α to maintain I_{dc} constant.)

ture lagging its respective phase voltage. The in-phase component of current I_1 sin α is finite, and therefore, from Eq. (9.6), a component of the inverter phase current acts as a ''magnetizing'' current even though there is no magnetic field and no capability of storing energy. In Fig. 7.4, the reactive voltamperes Q is given by an expression complementary to the expression for real power P,

$$Q = \frac{3E_m}{\sqrt{2}} I_{a_1} \sin\alpha = 3\frac{E_m}{\sqrt{2}} I_{a_1} \sin\psi_1 \tag{9.7}$$

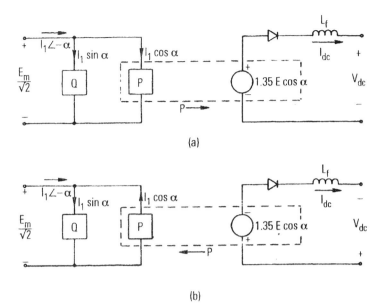

(a)

(b)

FIG. 5   Per-phase equivalent circuits for the three-phase bridge rectifier–inverter: (a) rectifier operation and (b) inverter operation [20]

Combining Eqs. (9.2) and (9.6) with Eq. (9.7) gives

$$Q = E_{av_0} I_{dc} \sin \alpha \tag{9.8}$$

The combination of voltampere components P and Q thus gives

$$\sqrt{P^2 + Q^2} = E_{av_0} I_{dc} \tag{9.9}$$

Since only fundamental components are being considered, the ratio

$$\frac{P}{\sqrt{P^2 + Q^2}} = \frac{E_{av}}{E_{av_0}} = \cos \alpha = \text{displacement factor} \tag{9.10}$$

It can be seen that Eq. (9.10) confirms the earlier result of Eq. (7.58). The rms value of the total ac-side line current is given by Eq. (7.52) if the filter inductance is large. Combining Eq. (9.6) with Eq. (7.52) shows that the ac-side line current distortion factor I_{a_1}/I, has the value $3/\pi$ as in (7.59).

Applications of the use of the fixed-frequency bridge inverter include solar energy systems, wind energy systems and high-voltage direct current (HVDC) interconnections between ac and dc systems.

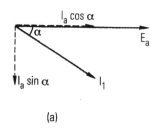

(a)

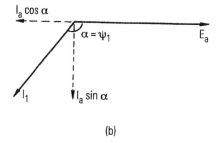

(b)

FIG. 6   Phase diagrams for the fundamental frequency components of the ac-side voltage and current: (a) rectifier operation and (b) inverter operation [20]

9.2   THE EFFECT OF AC SYSTEM INDUCTANCE

The average values of the load side current I_{av} and voltage E_{av} in a three-phase controlled rectifier, in the presence of supply-side reactance ωL_s Ω/phase, are given by Eq. (7.87) and (7.89). These equations remain true for inverter operation, except that the polarity of the voltage is reversed, and are reproduced below.

$$I_{av} = \frac{\sqrt{3}E_m}{2\omega Ls}\left[\cos\alpha - \cos(\alpha+\mu)\right] \tag{9.11}$$

$$E_{av} = -\frac{3\sqrt{3}E_m}{2\pi}\left[\cos\alpha + \cos(\alpha+\mu)\right]$$

$$= -\frac{E_{av_0}}{2}\left[\cos\alpha + \cos(\alpha+\mu)\right] \tag{9.12}$$

where E_m is the peak value of the ac-side phase voltage, α is the switch firing angle, and μ is the overlap angle. The line inductance L_s is usually of the order of a few millihenries, and the line reactance ωL_s, or commutating reactance, is usually a few ohms. For inverter operation, Eq. (9.11) and (9.12) are sometimes rearranged in terms of an extinction angle $\gamma = \pi - \alpha - \mu$, but this does not render any advantage here.

As with rectifier operation, the inverter action can be interpreted in equivalent circuit terms. The form of Eq. (7.90) is still valid but can be most conveniently expressed in terms of extinction angle γ

$$E_{av} = E_{av_0} \cos\alpha - \frac{3}{\pi}\omega L_s I_{av} \tag{9.13}$$

In the presence of overlap

$$\gamma = \pi - \alpha - \mu \tag{9.14}$$

But

$$\cos(\pi - \alpha - \mu) = -\cos(\alpha + \mu) \tag{9.15}$$

Combining Eqs. (9.13)–(9.15) gives

$$E_{av} = -E_{av_0} \cos(\alpha + \mu) - \frac{3}{\pi}\omega L_s I_{av} \tag{9.16}$$

For inverter operation $(\alpha + \mu) > 90°$ and the first term of Eq. (9.16) becomes positive.

The current equation can be written in terms of extinction angle γ by combining Eqs. (9.11), (9.14), and (9.15).

$$I_{av} = \frac{\sqrt{3}E_m}{2\omega L_s}\left(\cos\alpha + \cos\gamma\right) \tag{9.17}$$

9.3   WORKED EXAMPLES

Example 9.1   Power is transferred from a 300-V battery to a three-phase, 230-V, 50-Hz ac bus via a controlled SCR inverter. The inverter switches may be considered lossless and a large filter inductor with resistance 10 Ω is included on the dc side. Calculate the power transferred and the power factor if (1) $\alpha = 90°$, (2) $\alpha = 120°$, and (3) $\alpha = 150°$.

The circuit is represented in Fig. 9.2 with the ac-side current waveform shown in Fig. 9.4. The average voltage on the dc side of the inverter is given by Eq. (7.22),

$$E_{av} = \frac{3\sqrt{3}E_m}{\pi}\cos\alpha$$

In this case the peak phase voltage E_m is

$$E_m = \sqrt{2}\,\frac{230}{\sqrt{3}} = 187.8 \text{ V}$$

$$E_{av} = \frac{3\sqrt{3}}{\pi}\,187.8\cos\alpha = 310.6\cos\alpha$$

$$= \begin{cases} 0 \;\; \text{at } \alpha = 90° \\ -155.3 \text{ V} \;\; \text{at } \alpha = 120° \\ -269 \text{ V} \;\; \text{at } \alpha = 150° \end{cases}$$

The negative sign indicates that E_{av} opposes the current flow that is created by the connection of V_{dc}. The current on the dc side is given by Eq. (9.3).

$$I_{dc} = \frac{V_{dc} + E_{av}}{R_f} = \frac{|V_{dc}| - |E_{av}|}{R_f}$$

At $\alpha = 90°$,

$$I_{dc} = \frac{300}{10} = 30 \text{ A}$$

At $\alpha = 120°$

$$I_{dc} = \frac{300 - 1553}{10} = 14.47 \text{ A}$$

At $\alpha = 150°$,

$$I_{dc} = \frac{300 - 269}{10} = 3.1 \text{ A}$$

The power transferred through the inverter into the ac system is the battery power minus the loss in R_f.

$$P = V_{dc}I_{dc} - I_{dc}^2 R_f$$

At α 90°,

$$P = 300 \times 30 - (30)^2 \times 10 = 0$$

At $\alpha = 120°$,

$$P = 300 \times 14.47 - (14.47)^2 \times 10$$
$$= 4341 - 2093.8 = 2247.2 \text{ W}$$

At $\alpha = 150°$,

$$P = 300 \times 3.1 - (3.1)^2 \times 10$$
$$= 930 - 96.1 = 833.9 \text{ W}$$

The peak height of the ac-side current is also the battery current I_{dc}. The fundamental ac-side current has an rms value given by Eq. (9.6).

$$I_1 = 0.78 I_{dc}$$

At $\alpha = 90°$,

$$I_1 = 23.4 \text{ A}$$

At $\alpha = 120°$,

$$I_1 = 11.3 \text{ A}$$

At $\alpha = 150°$

$$I_1 = 2.42 \text{ A}$$

At $\alpha = 120°$, for example, the power given by the ac-side equation, Eq. (9.5), is

$$P = 3 \times \frac{230}{\sqrt{3}} \times 11.3 \cos 120° = -2251 \text{ W}$$

This agrees very nearly with the calculated power (2247.2 W) on the dc side. The displacement factor DF is defined directly from Eqs. (7.58) and (9.10):

Displacement factor $= |\cos \psi_1| = |\cos \alpha|$

At $\alpha = 90°$,

$$DF = 0$$

At $\alpha = 120°$,

$$DF = |\cos 120°| = 0.5$$

At $\alpha = 150°$,

$$DF = |\cos 150°| = 0.866$$

The distortion factor for a waveform such as the current in Fig. 9.4 was shown in Eq. (7.59) to have the value $3/\pi$. Now the power factor PF is given by

$$PF = \text{displacement factor} \times \text{distortion factor}$$

At $\alpha = 90°$,

$$PF = 0$$

AT $\alpha = 120°$,

$$PF = \frac{3}{\pi} \times 0.5 = 0.477$$

At $\alpha = 150°$,

$$PF = \frac{3}{\pi} \times 0.866 = 0.827$$

Example 9.2 Calculate the switch ratings for the operation of the inverter in Example 9.1

Operation of the inverter of Fig. 9.2 requires that the peak line voltage fall sequentially on each SCR

$$V_{Tmax} = \sqrt{2} \times 230 = 325.3 \text{ V}$$

Each SCR current I_T has one rectangular pulse (positive going only), as in Fig. 9.3, per cycle. This has an rms value

$$I_T = \sqrt{\frac{1}{2\pi} \int_{\pi/6}^{5\pi/6} I_{dc}^2 \, d\omega t} = \sqrt{\frac{I_{dc}^2}{3}}$$

Therefore,

$$I_T = \frac{I_{dc}}{\sqrt{3}}$$

The maximum value of I_{dc} occurs at $\alpha = 90°$, which corresponds to

$$I_T = \frac{30}{\sqrt{3}} = 17.32 \text{ A}$$

The expression above for I_T is confirmed by Eq. (7.52) which defines the rms value of the line current as $\sqrt{2}I_T$.

Practically selected devices might be rated at 400 V, 20 A.

Example 9.3 The current-firing-angle characteristics of Fig. 9.7 were measured from a battery-powered SCR inverter feeding into a three-phase transformer. Deduce the transformer terminal voltage.

A point on one of the characteristics is chosen arbitrarily. Consider the point $V_{dc} = 70$ V, $I_{dc} = 6$ A, when $\alpha = 130°$. The power leaving the battery is $P = 70 \times 6 = 420$ W. Neglecting any power loss in the filter inductor all of this power reappears on the ac side of the inverter. The rms value of the fundamental inverter current is, from Eq. (9.6).

$$I_1 = 0.78I_{dc} = 4.68 \text{ A}$$

In Eq. (9.5),

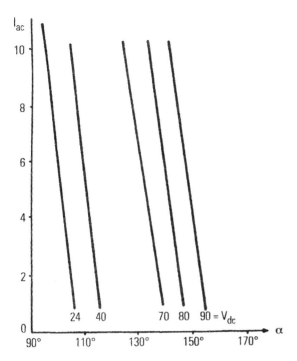

Fɪɢ. 7   Measured dc-side current (amperes) versus firing angle α for a three-phase recti-fier–inverter [20]

$$420 = \frac{3E_m}{\sqrt{2}} \times 4.68 \times \cos 130^\circ$$

which gives

$$E_m = 65.8 \text{ V}$$

The rms line voltage on the ac side was therefore

$$E = \sqrt{3} \times \frac{65.8}{\sqrt{2}} = 80 \text{ V}$$

Example 9.4   The ac line voltage output of a three-phase inverter is 6.6 kV. A dc supply voltage of 7 kV is found to require an ignition angle of 140°. Calculate the associated overlap angle.

The base value E_{av_0} of the dc-side voltage is given by

$$E_{av_0} = \frac{3\sqrt{3}E_m}{\pi}$$

But E_m is the peak value of the ac-side phase voltage

$$E_m = \frac{6600\sqrt{2}}{\sqrt{3}}$$

so that

$$E_{av_0} = \frac{3\sqrt{3}}{\pi} \frac{6600\sqrt{2}}{\sqrt{3}} = 8913 \text{ V}$$

Substituting values into Eq. (9.12) gives

$$7500 = -\frac{8913}{2}\left[\cos 140° + \cos(140° + \mu)\right] \cos(140° + \mu) = -0.9169$$

$$140° + \mu = \cos^{-1}(-0.9169) = 180° - 23.52° = 156.4°$$

Therefore,

$$\mu = 16.4°$$

Example 9.5   The current and voltage on the dc side of a three-phase bridge inverter are 100 A and 30 kV respectively. If the ac line reactance is 15 Ω and the ac bus voltage is 36 kV, calculate the firing angle and the overlap angle.

Base value E_{av_0} of the dc-side voltage is given by

$$E_{av_0} = \frac{3\sqrt{3}E_m}{\pi} = \frac{3}{\pi}\sqrt{3}\frac{36\sqrt{2}}{\sqrt{3}} = 48.62 \text{ kV}$$

Substituting numerical values into Eq. (9.13) gives

$$30000 = 48620 \cos \gamma - \frac{3}{\pi} \times 15 \times 100$$

$$\cos \gamma = \frac{30000 + 1432.4}{48620} = 0.6465$$

$$\gamma = \cos^{-1} 0.6465 = 49.72°$$

Utilizing Eq. (9.14) it is seen that

$$\alpha + \mu = \pi - \gamma = 130.28°$$

A further, independent, relationship is needed to separate α and μ, and this is provided by Eq. (9.12).

$$30000 = \frac{-48620}{2}\left[\cos\alpha + \cos(\alpha+\mu)\right]$$

$$\cos\alpha = \frac{-60000}{48620} - \cos 130.28^\circ = -1.23 + 0.6465 = -0.5835$$

$$\alpha = \cos^{-1}(-0.5835) = 180^\circ - 54.3^\circ = 125.7^\circ$$

Overlap angle μ is therefore

$$\mu = 130.28^\circ - 125.7^\circ = 4.58^\circ$$

PROBLEMS

9.1 A naturally commutated, three-phase inverter contains six ideal SCRs and transfers energy into a 440-V, 50-Hz, three-phase supply from an 800-V dc battery. The battery and the inverter are linked by a smoothing inductor with a resistance of 12.4 Ω. Calculate the power transferred at $\alpha = 90^\circ$, 120°, 150°, and 170°.

9.2 For the inverter application of Problem 9.1 calculate the voltage and rms current ratings required of the switches.

9.3 A large solar energy installation utilizes a naturally commutated, three-phase SCR inverter to transfer energy into a power system via a 660-V transformer. The collected solar energy is initially stored in an 800-V battery that is linked to the inverter through a large filter choke of resistance 14.2 Ω. What is the maximum usable value of the SCR firing angle? Calculate the power transferred at the firing angle of 165°. What is the necessary transformer rating?

9.4 Calculate the necessary SCR voltage and rms current ratings for the inverter application of Problem 9.3.

9.5 Use the inverter characteristics of Fig. 9.7 to deduce the form of the corresponding $V_{dc} - I_{dc}$ characteristics with cos α as the parameter. If the maximum dc-side voltage is 100 V, what is the firing angle required to give a direct current of 10 A if $R = 1\ \Omega$?

9.6 Sketch the main circuit of a naturally commutated, three-phase, controlled bridge inverter. If the ac-side rms line voltage V is fixed, sketch the variation of inverter power transfer with SCR firing angle α and dc-side voltage

V_{dc}. If $\alpha = 120°$, what is the minimum value of ratio V_{dc}/V that will permit inversion?

Sketch the waveform of the current passing between the inverter and the supply and give a phasor diagram interpretation to explain the inverter operation. Why is it often necessary to connect capacitance across the terminals of a naturally commutated inverter of high kVA rating?

9.7   A three-phase, bridge inverter feeds power into an ac supply of 175 kV in which the line reactance is 20 Ω. When the power transferred from the dc side is 50 MW, the extinction angle is 19.2°. Calculate the dc current and voltage. A three-phase bridge inverter delivers power into a 50 kV three-phase bus from a dc supply of 40 kV. Calculate the overlap angle when the switch firing angle is 120°.

9.8   What value of switch firing angle would result in the power transfer specified in Problem 9.7?

9.9   A three-phase bridge inverter delivers power into a 50-kV three-phase bus from a dc supply of 40 kV. Calculate the overlap angle when the switch firing angle is 120°.

Most inverter applications, however, have passive, three-phase loads requiring an adjustable frequency supply. The most common form of load is a three-phase ac motor, which is essentially passive even though the windings contain speed related induced voltages. With passive loads the inverter switches have to be commutated by methods designed into the inverter.

For high power inverters, above 500 kW, the switches are silicon controlled rectifiers (SCRs), and commutation has to be realized by the use of auxiliary commutation circuits, known here as *circuit commutation*. Inverters up to about 500 kW usually use three-terminal, switch-off devices such as power transistors and can be commutated by gate control or extinguished using resonant circuit methods.

Most dc power supplies are good voltage sources characterized by low internal impedance, in which the voltage level is largely maintained in the presence of variable current output. Typical examples are a battery, a dc generator, and a full-wave rectified ac power source. If the direct voltage remains substantially constant, this constitutes an ideal or nearly ideal source, and the associated inverter is known as a *voltage-source inverter* (VSI). In order to maintain a high degree of dc source voltage stability or stiffness a large electrolytic capacitor, of size 2000 μF to 20,000 μF, is connected across the input terminals (Fig. 9.8(a). This item is both bulky and significantly expensive.

There are a few forms of dc power supply in which the source impedance is high and that deliver a direct current of constant level, regardless of the load. These can be used as power supplies for the current-source inverter (CSI), which is completely different in operation and commutation from the VSI. A supply of constant or almost constant current can be approximated by using a voltage source and a large series inductance (Fig. 9.8(b).

Both the VSI and CSI are used in applications where the inverter output frequency needs to be adjustable, such as ac motor speed control. One of the two basic forms of inverter switching action results in output voltage waveforms

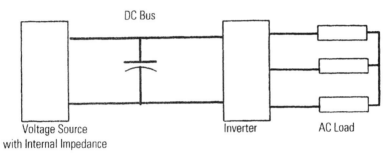

(a) Voltage Stiff Inverter

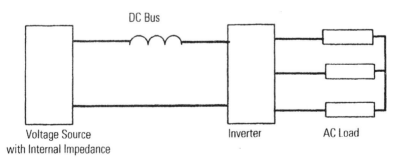

(b) Current Stiff Inverter

FIG. 8   (a) Voltage and (b) current stiff systems [32]

that have the form of square waves or stepped waves, described in Chapter 10. The other basic switching action is known as pulse width modulation (PWM), described in Chapters 8 and 11. For both step-wave and PWM voltage-source inverters the commutation action has to be performed either in the inverter itself, known as *self-commutation*, or by the action of the associated circuitry, known as *circuit commutation*.

10

Three-Phase, Step-Wave Inverter Circuits

10.1 SKELETON INVERTER CIRCUIT

The form of voltage-source inverter (VSI) most commonly used consists of a three-phase, naturally commutated, controlled rectifier providing adjustable direct voltage V_{dc} as input to a three-phase, force-commutated inverter (Fig. 10.1). The rectifier output–inverter input section is known as the *dc link*. In addition to a shunt capacitor to aid direct voltage stiffness the link usually contains series inductance to limit any transient current that may arise.

Figure 10.2a shows the skeleton inverter in which the semiconductor rectifier devices are shown as generalized switches S. The notation of the switching devices in Fig. 10.2 is exactly the same as for the controlled rectifier in Fig. 7.1 and the naturally commutated inverter of Fig. 9.1. In high-power applications the switches are most likely to be SCRs, in which case they must be switched off by forced quenching of the anode voltages. This adds greatly to the complexity and cost of the inverter design and reduces the reliability of its operation.

If the inverter devices are GTOs (Fig. 10.2b), they can be extinguished using negative gate current. Various forms of transistor switches such as BJTs (Fig. 10.2c), and IGBTs (Fig. 10.2d) can be extinguished by control of their base currents, as briefly discussed in Chapter 1. In Fig. 10.2 the commutating circuitry is not shown. It is assumed in the following analysis that each switch can be opened or closed freely.

309

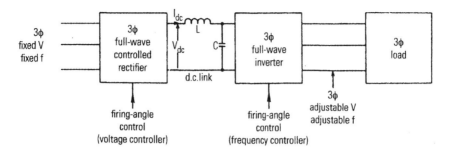

Fɪɢ. 1   Basic form of voltage-source inverter (VSI) [20].

From the power circuit point of view all versions of the skeleton inverter of Fig. 10.2 are identical. In each case the frequency of the generated voltages depends on the frequency of gating of the switches and the waveforms of the generated voltages depend on the inverter switching mode. The waveforms of the associated circuit currents depend on the load impedances.

Many different voltage waveforms can be generated by the use of appropriate switching patterns in the circuit of Fig. 10.2. An invariable requirement in three-phase systems is that the three-phase output voltages be identical in form but phase displaced by 120° electrical from each other. This does not necessarily create a balanced set of load voltages, in the sinusoidal sense of summing to zero at every instant of the cycle, but it reduces the possibility of gross voltage unbalance.

A voltage source inverter is best suited to loads that have a high impedance to harmonic currents, such as a series tuned circuit or an induction motor. The series inductance of such loads often results in operation at low power factors.

10.2 STEP-WAVE INVERTER VOLTAGE WAVEFORMS

For the purpose of voltage waveform fabrication it is convenient to switch the devices of Fig. 10.2 sequentially at intervals of 60° electrical or one-sixth of a period. The use of a dc supply having equal positive and negative voltage values $\pm V_{dc}$ is common. The zero point of the dc supply is known as the *supply zero pole* but is not grounded.

10.2.1 Two Simultaneously Conducting Switches

If two switches conduct at any instant, a suitable switching pattern is defined in Fig. 10.3 for no-load operation. The devices are switched in numerical order, and each remains in conduction for 120° electrical. Phase voltages v_{AN}, v_{BC}, and v_{CN}

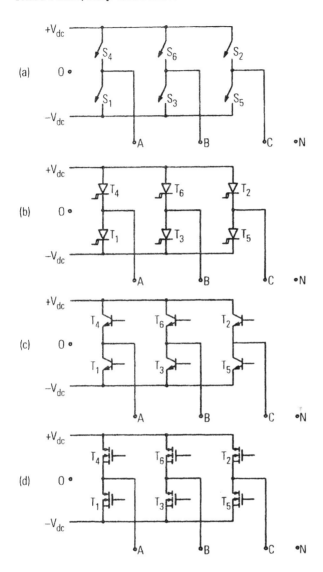

FIG. 2 Skeleton switching circuit of voltage source inverter: (a) general switches, (b) GTO switches, (c) BJT switches, and (d) IGBT switches [20].

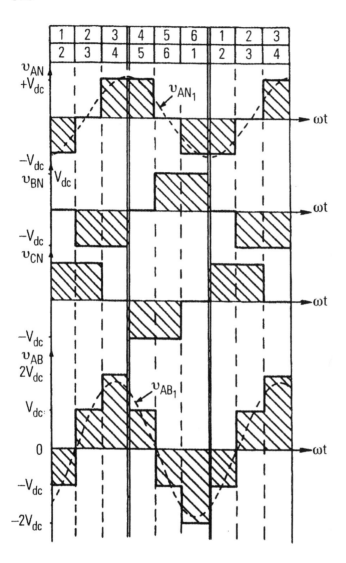

Fig. 3 Load voltage waveforms with two simultaneously conducting switches. No load and resistive load [20].

consist of rectangular pulses of height $\pm V_{dc}$. If equal resistors R are now connected in star to the load terminals A, B, and C of Fig. 10.2, the conduction pattern of Fig. 10.4 ensues for the first half period.

In interval $0 < \omega t < \pi/3$,

$$v_{AN} = -I_L R = -\frac{2V_{dc}}{2R} R = -V_{dc}$$

$$v_{BN} = 0$$

$$v_{CN} = I_L R = \frac{2V_{dc}}{2R} R = +V_{dc}$$

$$v_{AB} = v_{AN} + v_{NB} = v_{AN} - v_{BN} = -V_{dc} \tag{10.1}$$

In the interval $\pi/3 < \omega t < 2\pi/3$,

$$v_{AN} = 0$$

$$v_{BN} = I_L R = +V_{dc}$$

$$v_{CN} = I_L R = +V_{dc}$$

$$v_{AB} = +V_{dc} \tag{10.2}$$

In the interval $2\pi/3 < \omega t < \pi$,

$$v_{AN} = I_L R = +V_{dc}$$

$$v_{BN} = I_L R = -V_{dc}$$

$$v_{CN} = 0$$

$$v_{AB} = 2V_{dc} \tag{10.3}$$

For each interval it is seen that the load current during conduction is

$$I_L = \frac{\pm 2V_{dc}}{2R} = \pm \frac{V_{dc}}{R} \tag{10.4}$$

The results of Eqs. (10.1)–(10.4) are seen to be represented by the waveforms of Fig. 10.3. For this particular mode of switching the load voltage and current waveforms with star-connected resistive load are therefore identical with the pattern of the open-circuit voltages. The potential of load neutral point N is always midway between $+V_{dc}$ and $-V_{dc}$ and therefore coincides with the potential of the supply midpoint 0.

Phase voltage waveform v_{AN} in Fig. 10.3 is given by an expression

$$v_{AN} = (\omega t) = V_{dc} \begin{vmatrix} 240° \\ 120° \end{vmatrix} - V_{dc} \begin{vmatrix} 60°, \ 360° \\ 0°, \ 300° \end{vmatrix} \tag{10.5}$$

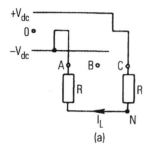

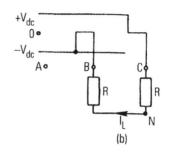

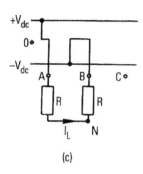

FIG. 4 Current conduction pattern for the case of two simultaneously conducting switches: (a) $0° < \omega t < 60°$, (b) $60° < \omega t < 120°$, and (c) $120° < \omega t < 180°$ [20].

This has the rms value

$$V_{AN} = \sqrt{\frac{1}{2\pi}\int_0^{2\pi} v_{AN}^2(\omega t)\,d\omega t} = \sqrt{\frac{2}{3}}V_{dc} = 0.816V_{dc} \tag{10.6}$$

The fundamental Fourier coefficients of waveform $v_{AN}(\omega t)$ are found to be

$$a_1 = \frac{1}{\pi}\int_0^{2\pi} v_{AN}(\omega t)\cos \omega t\,d\omega t = -\frac{2\sqrt{3}}{\pi}V_{dc} \tag{10.7}$$

$$b_1 = \frac{1}{\pi}\int_0^{2\pi} v_{AN}(\omega t)\sin \omega t\,d\omega t = 0 \tag{10.8}$$

$$c_1 = \sqrt{a_1^2 + b_1^2} = a_1 = -\frac{2\sqrt{3}}{\pi}V_{dc} \tag{10.9}$$

$$\psi_1 = \tan^{-1}\frac{a_1}{b_1} = \tan^{-1}(-\infty) = -90° \tag{10.10}$$

It is seen from Eqs. (10.9) and (10.10) that the fundamental (supply frequency) component of the phase voltages has a peak value $(2\sqrt{3}/\pi)$ V_{dc}, or $1.1V_{dc}$ with its origin delayed by 90°. This $(2\sqrt{3}/\pi)V_{dc}$ fundamental component waveform is sketched in Fig. 10.3.

The distortion factor of the phase voltage is given by

$$\text{Distortion factor} = \frac{V_{AN_1}}{V_{AN}} = \frac{c_{1/\sqrt{2}}}{V_{AN}} = \frac{3}{\pi} \tag{10.11}$$

Line voltage v_{AB} (ωt) in Fig. 10.3 is defined by the relation

$$v_{AB}(\omega t) = V_{dc}\begin{vmatrix}120°, & 240°\\60°, & 180°\end{vmatrix} - V_{dc}\begin{vmatrix}60°, & 300°\\0, & 240°\end{vmatrix} + 2V_{dc}\begin{vmatrix}180°\\120°\end{vmatrix} - 2V_{dc}\begin{vmatrix}360°\\300°\end{vmatrix} \tag{10.12}$$

This is found to have fundamental frequency Fourier coefficients of value

$$a_1 = -\frac{3\sqrt{3}}{\pi}V_{dc}$$

$$b_1 = +\frac{3}{\pi}V_{dc}$$

$$\text{Therefore, } c_1 = \frac{6}{\pi}V_{dc} \qquad \psi_1 = -\tan^{-1}\sqrt{3} = -60° \tag{10.13}$$

The fundamental component of v_{AB} (ωt) is therefore given by

$$v_{AB_1}(\omega t) = \frac{6}{\pi}V_{dc}\sin(\omega t - 60°) \tag{10.14}$$

It is seen in Fig. 10.3 that v_{AB_1} (ωt) leads v_{AN_1} (ωt) by 30°, as in a balanced three-phase system, and comparing Eqs. (10.9) and (10.13), the magnitude $|V_{AB_1}|$ is $\sqrt{3}$ times the magnitude $|V_{AN_1}|$.

With a firing pattern of two simultaneously conducting switches the load voltages of Fig. 10.3 are not retained with inductive load. Instead, the load voltages become irregular with dwell periods that differ with load phase-angle. Because of this, the pattern of two simultaneously conducting switches has only limited application.

10.2.2 Three Simultaneously Conducting Switches

A different load voltage waveform is generated if a mode of switching is used whereby three switches conduct at any instant. Once again the switching devices conduct in numerical sequence but now each with a conduction angle of 180°

electrical. At any instant of the cycle three switches with consecutive numbering are in conduction simultaneously. The pattern of waveforms obtained on no load is shown in Fig. 10.5. With equal star-connected resistors the current conduction patterns of Fig. 10.6 are true for the first three 60° intervals of the cycle, if the load neutral N is isolated.

For each interval,

$$I = \frac{2V_{dc}}{R + R/2} = \frac{4V_{dc}}{3R} \qquad (10.15)$$

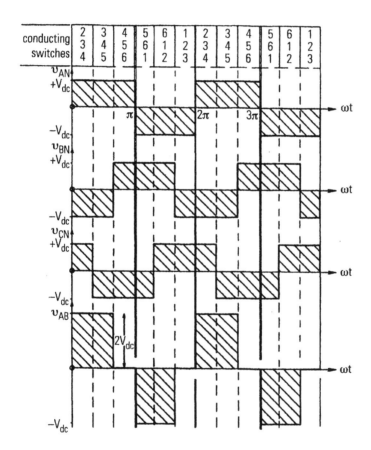

FIG. 5 Output voltage waveforms with three simultaneously conducting switches. No load [20].

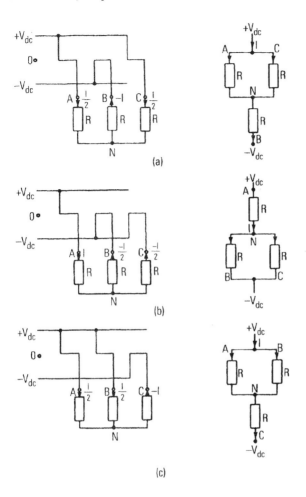

Fig. 6 Current conduction pattern for the case of three simultaneously conducting switches. Star-connected R load: (a) $0° < \omega t < 60°$, (b) $60° < \omega t < 120°$, and (c) $120° < \omega t < 180°$ [20].

In the interval $0 < \omega t < \pi/3$,

$$v_{AN} = v_{CN} = \frac{I}{2}R = \frac{2}{3}V_{dc}$$

$$v_{BN} = -IR = -\frac{4}{3}V_{dc}$$

$$v_{AB} = v_{AN} - v_{BN} = 2V_{dc} \qquad (10.16)$$

In the interval $\pi/3 \leq \omega t \leq \pi$,

$$v_{AN} = v_{BN} = \frac{1}{2}R = \frac{2}{3}V_{dc}$$

$$v_{CN} = -IR = -\frac{4}{3}V_{dc}$$

$$v_{AB} = 2V_{dc} \qquad\qquad (10.17)$$

In the interval $2\pi/3 \leq \omega t \leq \pi$,

$$v_{AN} = v_{BN} = \frac{1}{2}R = \frac{2}{3}V_{dc}$$

$$v_{CN} = IR = -\frac{4}{3}V_{dc}$$

$$v_{AB} = 0 \qquad\qquad (10.18)$$

The load voltage waveforms obtained with star-connected resistive load are plotted in Fig. 10.7. The phase voltages are seen to be different from the corresponding no-load values (shown as dashed lines), but the line voltages remain unchanged. Although the no-load phase voltages do not sum to zero, the load currents, with three-wire star connection, must sum to zero at every instant of the cycle. In Fig. 10.7 the phase voltage v_{AN} is given by

$$v_{AN}(\omega t) = \frac{2}{3}V_{dc}\begin{vmatrix}60°, 180° \\ 0, 120°\end{vmatrix} - \frac{2}{3}V_{dc}\begin{vmatrix}240°, 360° \\ 180°, 300°\end{vmatrix} + \frac{4}{3}V_{dc}\begin{vmatrix}120° \\ 60°\end{vmatrix} - \frac{4}{3}V_{dc}\begin{vmatrix}300° \\ 240°\end{vmatrix}$$
$$(10.19)$$

It can be seen by inspection in Fig. 10.7 that the fundamental frequency component of $v_{AN}(\omega t)$ is in time phase with it, so that

$$\alpha_1 = 0$$

$$\psi_1 = \tan^{-1}\frac{\alpha_1}{b_1} = 0 \qquad\qquad (10.20)$$

Fundamental frequency Fourier coefficient b_1 for the load peak phase voltage is found to be

$$b_1 = c_1 = \frac{4}{\pi}V_{dc} \qquad\qquad (10.21)$$

The corresponding fundamental (supply) frequency Fourier coefficients for line voltage $v_{AB}(\omega t)$ are given by

$$a_1 = \frac{2\sqrt{3}}{\pi}V_{dc}$$

$$b_1 = \frac{6}{\pi}V_{dc}$$

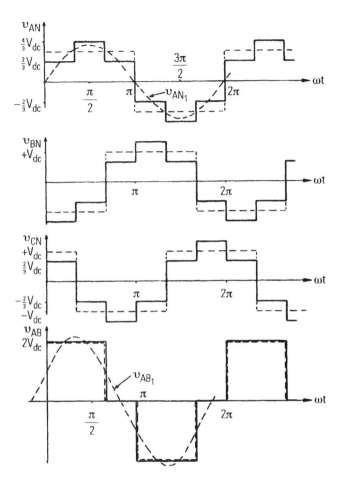

FIG. 7 Output voltage waveforms with three simultaneously conducting switches. Star-connected R load, isolated neutral. No-load waveforms [20].

$$c_1 = \frac{4}{\pi}\sqrt{3V_{dc}} = \sqrt{3} \times \text{the phase value}$$

$$\psi_1 = \tan^{-1}\frac{1}{\sqrt{3}} = 30° \tag{10.22}$$

The positive value $+30°$ for ψ_1 implies that its origin lies to the left of the zero on the scale of Fig. 10.7. Line voltage component $v_{AB}(\omega t)$ is plotted in Fig. 10.7, consistent with Eq. (10.22).

The fundamental components of the load voltages, plotted in Fig. 10.7 show that, as with a three-phase sinusoidal system, the line voltage leads its corresponding phase voltage by 30°. The rms value of phase voltage v_{AN} (ωt) is found to be

$$V_{AN} = \sqrt{\frac{1}{\pi} \int_0^\pi v_{AN}^2 (\omega t) \, d\omega t} = \frac{2\sqrt{2}}{3} V_{dc} = 0.943 V_{dc} \qquad (10.23)$$

Combining Eqs. (10.21) and (10.23) gives the distortion factor of the phase voltage,

$$\text{Distortion factor} = \frac{V_{AN1}}{V_{AN}} = \frac{\frac{c_1}{\sqrt{2}}}{V_{AN}} = \frac{3}{\pi} \qquad (10.24)$$

This is seen to be identical to the value obtained in Eq. (10.11) for the phase voltage waveform of Fig. 10.3 obtained with two simultaneously conducting switches. Although the distortion factors are identical, waveform v_{AN} (ωt) of Fig. 10.7 has a slightly greater fundamental value $(4/\pi)V_{dc}$ than the corresponding value $(2\sqrt{3}/\pi)V_{dc}$ for v_{AN} (ωt) of Fig. 10.3, given by Eq. (10.7). The switching mode that utilizes three simultaneously conducting switches is therefore potentially more useful for motor speed control applications. The properties of relevant step waves and square waves are summarized in Table 10.1.

It can be deduced from the waveforms of Fig. 10.7 that load neutral point N is not at the same potential as the supply neutral point 0. While these points remain isolated, a difference voltage V_{NO} exists that is square wave in form, with amplitude $\pm V_{dc}/3$ and of frequency three times the inverter switching frequency. If the two neutral points are joined, a neutral current will flow that is square wave in form, of amplitude $\pm V_{dc}/R$, and of three times the inverter switching frequency.

10.3 MEASUREMENT OF HARMONIC DISTORTION

The extent of waveform distortion for an alternating waveform can be defined in a number of different ways. The best known of the these, the distortion factor defined by Eq. (10.24), was used in connection with the rectifier circuits of Chapters 2–9.

An alternative measure of the amount of distortion is by means of a property known as the *total harmonic distortion (THD)*, which is defined as

$$THD = \sqrt{\frac{V_{AN}^2 - V_{AN_1}^2}{V_{AN_1}^2}} = \frac{V_{AN_h}}{V_{AN_1}} \qquad (10.25)$$

TABLE 10.1 Properties of Step Waves [20]

Phase voltage wave form	Properties of the phase voltage waveform		Total rms	Distortion factor	THD	Corresponding line voltage waveform
	Peak	RMS				
v_{AN} waveform	$\dfrac{4}{\pi}V_{dc}=1.273V_{dc}$	$\dfrac{4}{\sqrt{2}\pi}V_{dc}=\dfrac{2\sqrt{2}}{\pi}V_{dc}$	V_{dc}	$\dfrac{2\sqrt{2}}{\pi}=0.9$	$\sqrt{\dfrac{\pi^2}{8}-1}=0.483$	v_{AB} waveform
v_{AN} waveform	$\dfrac{2\sqrt{3}}{\pi}V_{dc}=1.1V_{dc}$	$\dfrac{\sqrt{6}}{\pi}V_{dc}=0.78V_{dc}$	$\sqrt{\dfrac{2}{3}}V_{dc}$	$\dfrac{3}{\pi}=0.955$	$\sqrt{\dfrac{\pi^2}{9}-1}=0.311$	v_{AB} waveform
v_{AN} waveform	$\dfrac{6}{\pi}V_{dc}=1.91V_{dc}$	$\dfrac{6}{\sqrt{2}\pi}V_{dc}=1.35V_{dc}$	$\sqrt{2}V_{dc}$	$\dfrac{3}{\pi}=0.955$	$\sqrt{\dfrac{\pi^2}{9}-1}=0.311$	v_{AB} waveform
v_{AN} waveform	$\dfrac{2}{\pi}V_{dc}$	$\dfrac{\sqrt{2}}{\pi}V_{dc}$	$\dfrac{1}{\sqrt{2}}V_{dc}$	$\dfrac{2}{\pi}=0.637$	$\sqrt{\dfrac{\pi^2}{4}-1}=1.212$	v_{AB} waveform

For a pure sinusoid $V_{AN_1} = V_{AN}$, and the *THD* then has the ideal value of zero. The numerator of Eq. (10.25) is seen to represent the effective sum of the nonfundamental or higher harmonic components V_{AN_h}.

A comparison of Eqs. (10.24) and (10.25) shows that for any wave,

$$\text{Distortion factor} = \frac{V_{AN_1}}{V_{AN}} = \sqrt{\frac{1}{1+(THD)^2}} \tag{10.26}$$

10.4 HARMONIC PROPERTIES OF THE SIX-STEP VOLTAGE WAVE

The six-step phase voltage waveforms of Fig. 10.7 are defined by the Fourier series

$$v_{AN}(\omega t) = \frac{4}{\pi} V_{dc} \left(\sin \omega t + \frac{1}{5} \sin 5\omega t + \frac{1}{7} \sin 7\omega t \right.$$
$$\left. + \frac{1}{11} \sin 11\omega t + \frac{1}{13} \sin 13\omega t + \cdots \right) \tag{10.27}$$

It is seen from Eq. (10.27) that the waveform $v_{AN}(\omega t)$ of Fig. 10.7 contains no triplen harmonics and its lowest higher harmonic is of order five with an amplitude equal to 20% of the fundamental. The rms value of the function in Eq. (10.27) is given by

$$V_{AN} = \frac{4V_{dc}}{\sqrt{2}\pi} \sqrt{1 + \frac{1}{5^2} + \frac{1}{7^2} + \frac{1}{11^2} + \frac{1}{13^2} \frac{1}{17^2} + \cdots}$$
$$= \frac{1}{\sqrt{2}} \times \frac{4}{\pi} V_{dc} \sqrt{1.079 + \cdots}$$
$$= 0.935 V_{dc} + \cdots \tag{10.28}$$

which confirms the value obtained by integration in Eq. (10.23).

For the step wave of Fig. 10.7, substituting Eqs. (10.21) and (10.23) into Eq. (10.25) gives

$$THD = \frac{V_{dc} \sqrt{\left(2\sqrt{2}/3\right)^2 - \left(4/\pi\sqrt{2}\right)^2}}{V_{dc}(4/\pi)1/\sqrt{2}}$$
$$= \sqrt{\frac{\pi^2}{9} - 1} = 0.311 \tag{10.29}$$

From Eq. (10.25) harmonic voltage V_{AN_h} is therefore 31.1% of the rms value of the fundamental component and 29.7% of the total rms value. Values of *THD* for other waveforms are given in Table 10.1. In general, if there are N steps/ cycle, each occupying $2\pi/N$ radians, the only harmonics present are of the order $h = nN \pm 1$, where n = 1, 2, 3, $\cdots$ For a six-step waveform Fig. 10.7, for example, $N = 6$ so that $h = 5, 7, 11, 13$, etc., as depicted in Eq. (10.27).

10.5 HARMONIC PROPERTIES OF THE OPTIMUM 12-STEP WAVEFORM

A reduction of the harmonic content can be realized by increase of the number of steps in the phase voltage wave. If a 12-step waveform is used, $N = 12$ and $h = 11, 13, 23, 15, \cdots$ Example 10.4 gives some detail of a certain 12-step waveform calculation. It is found that the optimum 12-step waveform, shown in Fig. 10.8, is represented by the Fourier expression

$$v(\omega t) = \frac{\pi}{3}V(\sin\omega t + \frac{1}{11}\sin 11\omega t + \frac{1}{13}\sin 13\omega t + \frac{1}{23}\sin 23\omega t + \cdots) \qquad (10.30)$$

In each interval of the optimum waveform of Fig. 10.8 the step height corresponds to the average value of the sinusoidal segment. For $0 \le \omega t \le \pi/6$, for example, the average value is

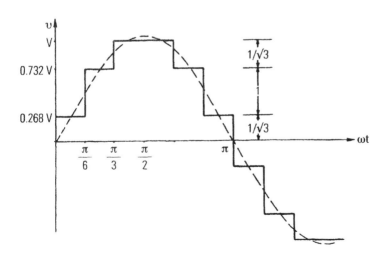

FIG. 8 Twelve-step voltage waveform [20].

$$\text{Step height} = \frac{\pi V}{3} \frac{6}{\pi} \int_0^{\pi/6} \sin \omega t \, d\omega t = 0.268 \text{ V} \qquad (10.31)$$

A 12-step waveform can be fabricated by the use of two six-step inverters with their outputs displaced by 30° or by the series addition of square-wave or PWM voltages.

10.6 SIX-STEP VOLTAGE INVERTER WITH SERIES *R-L* LOAD

When a reactive load is connected to a step-wave inverter, it becomes necessary to include a set of reverse-connected diodes in the circuit to carry return current (Fig. 10.9). The presence of the diodes immediately identifies the circuit as a VSI rather than a current-source inverter (CSI) for which return diodes are unnecessary. In the presence of load inductance with rectifier supply, a shunt capacitor must be connected in the dc link to absorb the reactive voltamperes because there is no path for reverse current in the supply.

10.6.1 Star-Connected Load

In the switching mode where three switches conduct simultaneously, the no-load voltages are given by Fig. 10.5. Let these voltages now be applied to the star-connected *R-L* loads, as in Fig. 10.9. The resulting current undergoes an exponential increase of value. Consider the instant $\omega t = 0$ in the typical steady-state

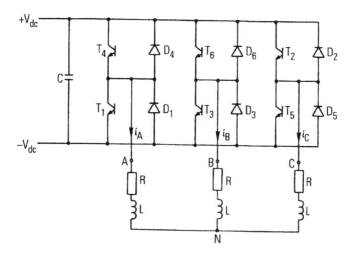

FIG. 9 Voltage-source transistor inverter incorporating return current diodes [20].

cycle shown in Fig. 10.10. Transistor T_1 has been in conduction for 180° and has just switched off. Transistor T_2 has been in conduction for 120° passing positive current I_c. Transistor T_3 is 60° into its conduction cycle resulting in current i_B that is increasing negatively. Transistor T_4 has just switched on, connecting terminal A to $+ V_{dc}$, which will attempt to create positive I_A. The negative current $i_A(0)$ at $\omega t = 0$ is diverted from its previous path through T_1 and passes through diode D_4 to circulate through capacitor C. As soon as $i_A = 0$, diode D_4 switches off, at point t in Fig. 10.10 and T_4 takes up the positive current I_A.

For each interval in Fig. 10.10 the current can be described mathematically by a constant term plus a decaying exponential component. Even if the load is highly inductive the load phase voltages and line voltages largely retain the forms of Fig. 10.7. For example, the diagram of Fig. 10.11 is reproduced from oscillograms of waveforms when a three-phase induction motor is driven from a stepwave, voltage-source inverter. The motor phase voltage is the classical six-step

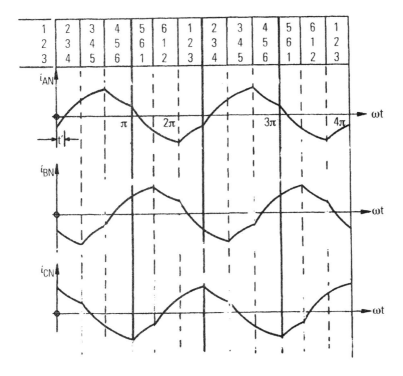

Fɪɢ. 10 Current waveforms for voltage-source six-step inverter with star-connected series R-L load [20].

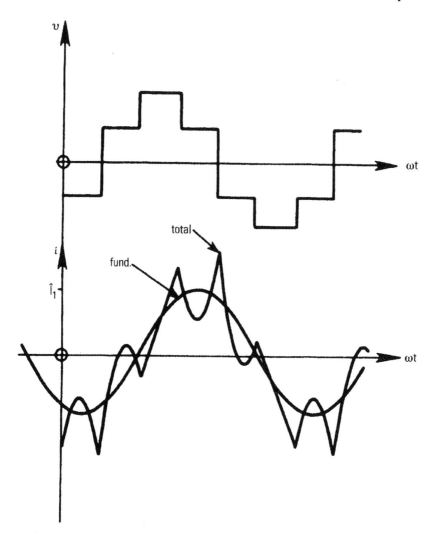

Fig. 11 Waveforms with six-step VSI applied to an induction motor load [20].

waveform. At each switching there is an abrupt change of current slope. A motor input impedance is much more complex than the passive *R-L* load of Fig. 10.9 since the resistance value is speed related and there are magnetically induced voltages in the windings. It can be seen in Fig. 10.11 that the fundamental component of the very "spiky" current lags the voltage by about 60° of phase angle, which is typical of low-speed operation of an induction motor.

10.6.2 Delta-Connected Load

Let the voltages of Fig. 10.7, for the case of three simultaneously conducting switches be applied to a balanced, three-phase, delta-connected load, as in Fig. 10.12. Since the star-connected load of Fig. 10.9 can be replaced by an equivalent delta-connected load, the line current waveforms of Fig. 10.10 remain true. The phase current waveforms can be deduced by the application of classical mathematical analysis or transform methods.

In the interval $0 \leq \omega t \leq 120°$ of Fig. 10.10 a voltage $2V_{dc}$ is impressed across terminals AB so that, with $\cot\phi = R/\omega L$,

$$i_{AB}(\omega t)\bigg|_{0 < \omega t < 120°} = \frac{2V_{dc}}{R}(1 - \varepsilon^{-\cot\phi\omega t}) + i_{AB}(0)\varepsilon^{-\cot\phi\omega t} \tag{10.32}$$

In the interval $120° < \omega t < 180°$ of Fig. 10.10 terminals A and B are coincident and load branch AB is short-circuited so that

$$i_{AB}(\omega t)\bigg|_{120° < \omega t < 180°} = \left[\frac{2V_{dc}}{R}(1 - \varepsilon^{-\cot\phi 2\pi/3}) + i_{AB}(0)\varepsilon^{-\cot\phi 2\pi/3}\right]\varepsilon^{-\cot\phi(\omega t - 2\pi/3)} \tag{10.33}$$

Since the current wave possesses half-wave inverse symmetry, $i_{AB}(0) = i_{AB}(\pi) = i_{AB}(2\pi)$. Putting $\omega t = \pi$ in Eq. (10.33) and utilizing the inverse-symmetry identity give

$$i_{AB}(0) = -\frac{2V_{dc}}{R}\frac{\varepsilon^{-\cot\phi\pi/3} - \varepsilon^{-\cot\phi\pi}}{1 + \varepsilon^{-\cot\phi\pi}} \tag{10.34}$$

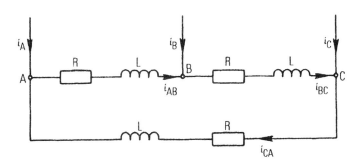

FIG. 12 Delta-connected series R-L load [20].

Combining Eq. (10.34) with Eqs. (10.32) and (10.33), respectively, gives

$$
i_{AB}(\omega t)\bigg|_{0 < \omega t < 120°} = \frac{2V_{dc}}{R}\left(1 - \frac{1 + \varepsilon^{-\cot\phi\pi/3}}{1 + \varepsilon^{-\cot\phi\pi}}\varepsilon^{-\cot\phi\omega t}\right) \tag{10.35}
$$

$$
i_{AB}(\omega t)\bigg|_{120° < \omega t < 180°} = \frac{2V_{dc}}{R}\left[1 - \frac{1 - \varepsilon^{-\cot\phi 2\pi/3}}{1 + \varepsilon^{-\cot\phi\pi}}\varepsilon^{-\cot\phi(\omega t - 2\pi/3)}\right] \tag{10.36}
$$

Current $i_{CA}(\omega t)$ in Fig. 10.12 is given by expressions corresponding to those of Eqs. (10.35) and (10.36) but with the time delayed by $4\pi/3$ radians. The rms value of the branch current is defined by the expression

$$
I_{AB} = \sqrt{\frac{1}{2\pi}\int_0^{2\pi} i_{AB}^2(\omega t)\,d\omega t} \tag{10.37}
$$

In elucidating Eq. (10.37) it is convenient to use the substitutions

$$
K_1 = \frac{1 + \varepsilon^{-\cot\phi\pi/3}}{1 + \varepsilon^{-\cot\phi\pi}} \qquad K_2 = \frac{1 - \varepsilon^{-\cot\phi 2\pi/3}}{1 + \varepsilon^{-\cot\phi\pi}} \tag{10.38}
$$

An examination of K_1 and K_2 above shows that

$$
K_2 = 1 - K_1\varepsilon^{-\cot\phi 2\pi/3} \tag{10.39}
$$

Substituting Eqs. (10.35) and (10.36) into Eq. (10.37) gives

$$
\begin{aligned}
I_{AB}^2 &= \frac{2}{2\pi}\times\frac{4V_{dc}^2}{R^2}\left[\int_0^{120°}(1 - K_1\varepsilon^{-\cot\phi\omega t})^2\,d\omega t\right.\\
&\quad\left. + \int_{120°}^{180°}\left(K_2\varepsilon^{-\cot\phi(\omega t - 2\pi/3)}\right)^2\,d\omega t\right]\\
&= \frac{4V_{dc}^2}{\pi R^2}\left[\left(\omega t + \frac{2K_1}{\cot\phi}\varepsilon^{-\cot\phi\omega t} - \frac{K_1^2\varepsilon^{-\cot\phi 2\omega t}}{2\cot\phi}\right)\bigg|_0^{120°}\right.\\
&\quad\left. + \left(-\frac{K_2^2\varepsilon^{-\cot\phi 2(\omega t - 2\pi/3)}}{2\cot\phi}\right)\bigg|_{120°}^{180°}\right]\\
&= \frac{4V_{dc}^2}{\pi R^2}\left[\frac{2\pi}{3} + \frac{2K_1}{\cot\phi}(\varepsilon^{-\cot\phi 2\pi/3} - 1) - \frac{K_1^2}{2\cot\phi}(\varepsilon^{-\cot\phi 4\pi/3} - 1)\right.\\
&\quad\left. - \frac{K_2^2}{2\cot\phi}(\varepsilon^{-\cot\phi 2\pi/3} - 1)\right]
\end{aligned} \tag{10.40}
$$

Eliminating the explicit exponential terms between Eqs. (10.38) and (10.40) gives

$$I_{AB}^2 = \frac{4V_{dc}^2}{\pi R^2}\left[\frac{2\pi}{3} + \frac{1}{\cot\phi}\left(\frac{3}{2} - K_2 - 2K_1 + \frac{K_1^2}{2} - \frac{K_2^2(1-K_2)}{2K_1}\right)\right]$$ (10.41)

Line current $i_A(\omega t)$ in Fig. 10.12 changes in each $60°$ interval of conduction. In general, $i_A(\omega t) = i_{AB}(\omega t)$, so that

$$i_A(\omega t)\bigg|_{0 < \omega t < 60°} = \frac{2V_{dc}}{R}\left(1 - \frac{(1+\varepsilon^{-\cot\phi\pi/3})(2-\varepsilon^{-\cot\phi\pi/3})}{1+\varepsilon^{-\cot\phi\pi}}\varepsilon^{-\cot\phi\omega t}\right)$$ (10.42)

$$i_A(\omega t)\bigg|_{60° < \omega t < 120°} = \frac{2V_{dc}}{R}\left(2 - \frac{(1+\varepsilon^{-\cot\phi\pi/3})^2}{1+\varepsilon^{-\cot\phi\pi}}\varepsilon^{-\cot\phi(\omega t-\pi/3)}\right)$$ (10.43)

$$i_{AB}(\omega t)\bigg|_{120° < \omega t < 180°} = \frac{2V_{dc}}{R}\left(1 - \frac{(1+2\varepsilon^{-\cot\phi\pi/3})(1+\varepsilon^{-\cot\phi\pi/3})}{1+\varepsilon^{-\cot\phi\pi}}\right.$$
$$\left.\varepsilon^{-\cot\phi(\omega t-2\pi/3)}\right)$$ (10.44)

A typical pattern of waves consistent with Eqs. (10.42) to (10.44) is shown in Fig. 10.13. At any instant the current $i_A(\omega t)$ must be flowing through one of the devices T_1, T_4, D_1, or D_4 in the inverter of Fig. 10.9. In the interval $0 \le \omega t \le 60°$, the negative part of $i_A(\omega t)$, up to $\omega t = \pi$, is conducted via transistor T_4. For $\omega t > 180°$, the positive current $i_A(\omega t)$ reduces to zero through diode D_1 and then goes negative via T_1. The properties of both the transistor and the diode currents can be calculated by use of the appropriate parts of Eqs. (10.35)–(10.44). The oscillating unidirectional current in the dc link (Fig. 10.13) consists of a repetition of the current $i_A(\omega t)$ in the interval $60° \le \omega t \le 120°$. For the interval, $0 \le \omega t \le 60°$, $i_{dc}(\omega t)$ is defined by

$$i_{dc}(\omega t) = \frac{2V_{dc}}{R}\left(2 - K_3\varepsilon^{-\cot\phi\omega t}\right)$$ (10.45)

where

$$K_3 = \frac{(1+\varepsilon^{-\cot\phi\pi/3})^2}{1+\varepsilon^{-\cot\phi\pi}}$$ (10.46)

This link current will become negative for part of the cycle if the load is sufficiently inductive. The boundary condition for the start of negative link current is $i_{dc}(\omega t) = 0$ at $\omega t = 0$, which occurs when $K_3 = 2$. This happens for loads

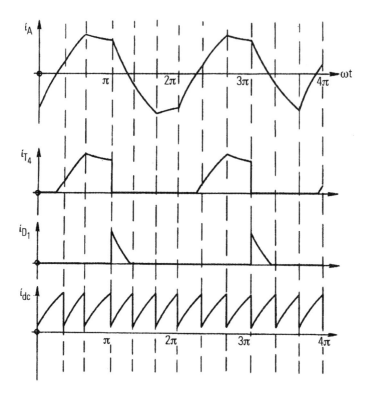

FIG. 13   Current waveforms for six-step VSI with delta-connected, series R-L load [20].

with a power factor smaller than 0.525 lagging. The average value of I_{dc} (ωt) in the interval $0 \le \omega t \le 60°$ and therefore in all the intervals is given by

$$I_{dc} = \frac{3}{\pi} \int_0^{60°} \frac{2V_{dc}}{R} \left(2 - K_3 \varepsilon^{-\cot\phi\omega t}\right) d\omega t$$

$$= \frac{3}{\pi} \frac{2V_{dc}}{R} \left(2\omega t + \frac{K_3}{\cot\phi} \varepsilon^{-\cot\phi\omega t}\right)\Bigg|_0^{60°}$$

$$= \frac{3}{\pi} \frac{2V_{dc}}{R} \left[\frac{2\pi}{3} + \frac{K_3}{\cot\phi}\left(\varepsilon^{-\cot\phi\pi/3} - 1\right)\right] \qquad (10.47)$$

10.7   WORKED EXAMPLES

Example 10.1   An ideal dc supply of constant voltage V supplies power to a three-phase force-commutated inverter consisting of six ideal transistor switches. Power is thence transferred to a delta-connected resistive load of R Ω per branch. The mode of inverter switching is such that two transistors are in conduction at any instant of the cycle. Deduce and sketch waveforms of the phase and line currents.

The load is connected so that the system currents have the notation shown in Fig. 10.12. The triggering sequence is given at the top of Fig. 10.5. At any instant of the cycle two of the three terminals A, B, and C will be connected to the supply, which has a positive rail $+V$ while the other rail is zero potential. The load effectively consists of two resistors R in series shunted by another resistor R. In the interval $0 \leq \omega t \leq \pi/3$, for example, transistors T_1 and T_2 are conducting so that

$$i_C = -i_A = \frac{V}{2R/3} = \frac{3}{2}\frac{V}{R}$$

$$i_B = 0$$

$$i_{CA} = \frac{2}{3}i_C = \frac{V}{R}$$

$$i_{BC} = i_{AB} = -\frac{1}{3}i_C = -\frac{1}{2}\frac{V}{R}$$

In the interval $\pi < \omega t < 2\pi/3$, transistors T_2 and T_3 are conducting, resulting in the isolation of terminal A so that

$$i_C = -i_B = \frac{3}{2}\frac{V}{R}$$

$$i_A = 0$$

$$i_{BC} = -\frac{V}{R}$$

$$i_{CA} = i_{AB} = +\frac{1}{2}\frac{V}{R}$$

In the interval $2\pi/3 \leq \omega t \leq \pi$, transistors T_3 and T_4 are in conduction so that terminal B has the negative rail potential of zero while terminal A is connected to the $+V$ rail, so that

$$i_C = 0$$

$$i_A = -i_B = \frac{3}{2}\frac{V}{R}$$

$$i_{AB} = \frac{V}{R}$$

$$i_{CA} = i_{BC} = -\frac{1}{2}\frac{V}{R}$$

The pattern of waveforms so produced (Fig. 10.14) is that of a six-step phase (i.e., branch) current but a square-wave line current. In fact, the pattern of waveforms is identical in form, but with different amplitude scaling, to that obtained with a star-connected load of R Ω/phase in Fig. 10.7 when three transistors conduct simultaneously.

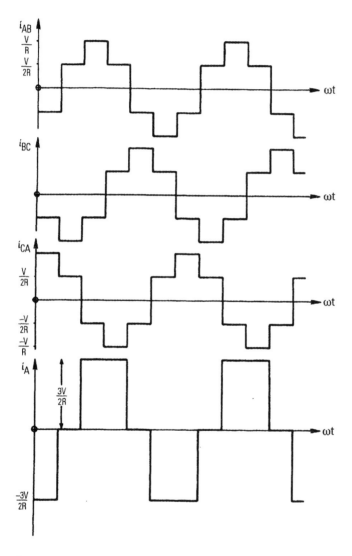

FIG. 14   Voltage waveforms of VSI with delta-connected R load (Example 10.1) [20].

Example 10.2 The voltage waveform of a certain type of 12-step inverter is given in Fig. 10.15. For this waveform calculate the fundamental value, the total rms value, and the distortion factor.

The waveform of Fig. 10.15 is defined by the relation

$$
e(\omega t) = \frac{E_m}{3}\bigg|_{0,\,4\pi/5}^{\pi,\,5\pi} + \frac{2E_m}{3}\bigg|_{\pi/5,\,3\pi/5}^{2\pi/5,\,4\pi/5} + E_m\bigg|_{2\pi/5}^{3\pi/5}
$$

For the interval $0 \le \omega t \le \pi$ the rms value E is given by

$$
E = \sqrt{\frac{1}{\pi}\int_0^\pi e^2(\omega t)\,d\omega t}
$$

$$
E^2 = \frac{1}{\pi}\left(\frac{E_m^2}{9}\,\omega t\right)\bigg|_{0,\,4\pi/5}^{\pi,\,5\pi} + \frac{1}{\pi}\left(\frac{4E_m^2}{9}\,\omega t\right)\bigg|_{\pi/5,\,3\pi/5}^{2\pi/5,\,4\pi/5} + \frac{1}{\pi}\left(E_m^2\,\omega t\right)\bigg|_{2\pi/5}^{3\pi/5}
$$

$$
= \frac{E_m^2}{\pi}\left[\frac{1}{9}\left(\frac{\pi}{5} - 0 + \pi - \frac{4\pi}{5}\right) + \frac{4}{9}\left(\frac{2\pi}{5} - \frac{\pi}{5} + \frac{4\pi}{5} - \frac{3\pi}{5}\right) + \left(\frac{3\pi}{5} - \frac{2\pi}{5}\right)\right]
$$

$$
= \frac{E_m^2}{\pi}\left(\frac{1}{9}\frac{2\pi}{5} + \frac{4}{9}\frac{2\pi}{5} + \frac{\pi}{5}\right)
$$

$$
= E_m^2\left(\frac{2}{45} + \frac{8}{45} + \frac{9}{45}\right) = E_m^2\,\frac{19}{45} = 0.65E_m
$$

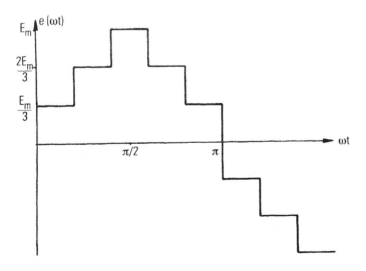

Fig. 15 Voltage waveform of 12-step VSI in Example 10.2 [20].

It is obvious that the fundamental component of waveform $e(\omega t)$ in Fig. 10.15 is symmetrical with respect to the waveform itself. Therefore,

$$b_1 = \frac{1}{\pi} \int_0^{2\pi} e(\omega t) \sin \omega t \, d\omega t$$

$$= \frac{2}{\pi} \int_0^{\pi} e(\omega t) \sin \omega t \, d\omega t$$

In this case

$$b_1 = \frac{2}{\pi} \left(-\frac{E_m}{3} \cos \omega t \Big|_{0,\ 4\pi/5}^{\pi,\ 5\pi} - \frac{2E_m}{3} \Big|_{\pi/5,\ 3\pi/5}^{2\pi/5,\ 4\pi/5} - E_m \cos \omega t \Big|_{2\pi/5}^{3\pi/5} \right)$$

$$= \frac{2E_m}{\pi} \left[-\frac{1}{3} \left(\cos \frac{\pi}{5} - \cos 0 + \cos \pi - \cos \frac{4\pi}{5} \right) \right.$$

$$\left. -\frac{2}{3} \left(\cos \frac{2\pi}{5} - \cos \frac{\pi}{5} + \cos \frac{4\pi}{5} - \cos \frac{3\pi}{5} \right) - \cos \frac{3\pi}{5} + \cos \frac{2\pi}{5} \right]$$

$$= \frac{2E_m}{\pi} \left(+\frac{2}{3} + \frac{1}{3} \cos \frac{\pi}{5} - \frac{1}{3} \cos \frac{4\pi}{5} + \frac{1}{3} \cos \frac{2\pi}{5} - \frac{1}{3} \cos \frac{3\pi}{5} \right)$$

$$= \frac{2E_m}{\pi} (+2 + 0.809 + 0.809 + 0.309 + 0.309)$$

$$= \frac{2E_m}{\pi} (4.24) = \frac{2.82 E_m}{\pi} = 0.9 E_m$$

$$\text{Distortion factor} = \frac{E_1}{E} = \frac{0.9}{\sqrt{2} \times 0.65} = 0.98$$

Example 10.3 A six-step voltage source inverter is supplied with power from an ideal battery of constant voltage $V = 150$ V. The inverter has a delta-connected series R-L load, where $R = 15\Omega$, $X_L = 25\Omega$ at 50 Hz. Calculate the rms current in the load, the power transferred, and the average value of the supply current at 50 Hz.

In this example an inverter of the form of Fig. 10.9 supplies power to a load with the connection of Fig. 10.12. The pattern of phase or branch currents $i_{AB}(\omega t)$, $i_{BC}(\omega t)$, $i_{CA}(\omega t)$ is similar in form to the load currents with star-connected load shown in Fig. 10.10. The line currents have the typical form $i_A(\omega t)$ given in Fig. 10.13. The branch current i_{AB} (ωt) is defined by Eqs. (10.35) and (10.36), where the voltage is now V (rather than $2V_{dc}$)

$$\phi = \tan^{-1}\frac{\omega L}{R} = \tan^{-1}1.67 = 59.1°$$

$$\cot\phi = \cot 59.1° = 0.6$$

$$\varepsilon^{-\cot\phi\pi/3} = \varepsilon^{-0.63} = 0.533$$

$$\varepsilon^{-\cot\phi 2\pi/3} = \varepsilon^{-1.26} = 0.285$$

$$\varepsilon^{-\cot\phi\pi} = \varepsilon^{-1.88} = 0.152$$

$$\varepsilon^{-\cot\phi 4\pi/3} = \varepsilon^{-2.51} = 0.081$$

Now in Eq. (10.38)

$$K_1 = \frac{1+\varepsilon^{-\cot\phi\pi/3}}{1+\varepsilon^{-\cot\phi\pi}} = \frac{1.533}{1.152} = 1.33$$

$$K_2 = \frac{1-\varepsilon^{-\cot\phi 2\pi/3}}{1+\varepsilon^{-\cot\phi\pi}} = \frac{0.715}{1.152} = 0.621$$

Substituting into Eq. (10.41) gives

$$I_{AB}^2 = \frac{150^2}{\pi\times15^2}\left[2.094+1.67\left(1.5-0.621-2.66+\frac{1.77}{2}-\frac{0.386\times0.379}{2.66}\right)\right]$$

$$I_{AB} = 10\sqrt{\frac{1}{\pi}(2.094-1.67\times0.951)} = 4.014\ \text{A}$$

The total power dissipation is

$$P = 3I^2R = 3\times(4.014)^2\times15 = 725\ \text{W}$$

The average value of the link current may be obtained by integrating Eq. (10.45) between the limits 0 and $\pi/3$:

$$I_{dc} = \frac{3V}{\pi R}\left[\frac{2\pi}{3}+\frac{K_3}{\cot\phi}\left(\varepsilon^{-\cot\phi\pi/3}-1\right)\right]$$

In this case, from Eq. (10.38),

$$K_3 = \frac{\left(1+\varepsilon^{-\cot\phi\pi/3}\right)^2}{1+\varepsilon^{-\cot\phi\pi}} = \frac{(1.533)^2}{1.152} = 2.04$$

Therefore,

$$I_{dc} = \frac{3}{\pi}\frac{150}{15}\left[2.094 + \frac{2.04}{0.6}(0.533 - 1)\right]$$

$$= \frac{30}{\pi}(2.094 - 1.588) = 4.83 \text{ A}$$

The power entering the inverter through the link is

$$P_{in} = VI_{dc} = 150 \times 4.83 = 725 \text{ W}$$

which agrees with the value of the load power.

PROBLEMS

10.1   Sketch the circuit diagram of a three-phase, force-commutated inverter incorporating six SCRs and six diodes. The commutation system should not be shown. Two SCRs only conduct at any instant, and power is transferred from the dc source voltage $\pm V$ into a balanced three-phase resistive load. Explain the sequence of SCR firing over a complete cycle and sketch a resulting per-phase load voltage waveform consistent with your firing pattern.

10.2   Sketch the skeleton circuit of the basic six-switch, force-commutated inverter with direct supply voltage $\pm V$. The switching mode to be used is that where three switches conduct simultaneously at every instant of the cycle. Deduce and sketch consistent waveforms of the output phase voltages v_{AN}, v_{BN}, v_{CN} (assuming phase sequence ABC) and the line voltage v_{AB} on open circuit over a complete time cycle, indicating which switches are conducting through each 60° interval. What is the phase difference between the fundamental component v_{AB_1} of the line voltage v_{AB} and the fundamental component v_{AN_1} of the phase voltage v_{AN}? In what ways would a phasor diagram of the fundamental, open-circuit phase voltages give a misleading impression of the actual operation?

10.3   The basic circuit of a six-switch, force-commutated inverter with supply voltage $\pm V$ is shown in Fig. 10.2. The triggering mode to be used is where three switches conduct simultaneously. Deduce and sketch waveforms of the instantaneous phase voltages v_{AN}, v_{BN}, v_{CN} and the instantaneous line voltage v_{AB} for open-circuit operation with phase sequence ABC. Indicate which of the six switches are conducting during each 60° interval of the cyclic period. If equal resistors R are connected to terminals A, B, C as a star-connected load, deduce and sketch the waveform of phase current i_{AN}.

10.4   In the inverter circuit of Fig. 10.2 the triggering mode to be used is where three switches conduct simultaneously. The load consists of three identical resistors R connected in wye (star).

1.   If the load neutral point N is electrically isolated from the supply neutral point O, deduce the magnitude, frequency, and waveform of the neutral–neutral voltage V_{NO}.
2.   If the two neutral points N and O are joined, deduce the magnitude, frequency, and waveform of the neutral current.

10.5   The stepped waveform of Fig. 10.16 is typical of the phase voltage waveform of a certain type of inverter. Use Fourier analysis to calculate the magnitude and phase angle of the fundamental component of this waveform. Sketch in correct proportion, the waveform and its fundamental component. What is the half-wave average value of the stepped wave compared with the half-wave average value of its fundamental component?

10.6   A set of no-load, phase voltage waveforms v_{AN}, v_{BN}, v_{CN} produced by a certain type of inverter is given in Fig. 10.5. Sketch, on squared paper, the corresponding no-load line voltages v_{AN}, v_{BN}, v_{CA}. Calculate the magnitude and phase-angle of the fundamental component v_{AN_1} of the line voltage v_{AN} and sketch v_{AN_1} in correct proportion to v_{AN}. What is the half-wave average value of v_{AN} compared with the corresponding half-wave average value of v_{AN_1}? The set of voltages in Fig. 10.5 is applied to a set of equal star-connected resistors of resistance r. Deduce and

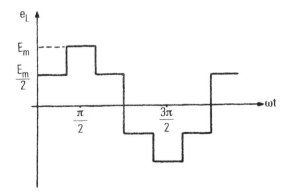

FIG. 16   Motor phase voltage waveform in Problem 10.5 [20].

sketch the waveform of the current in phase A with respect to the open-circuit voltage v_{AN}.

10.7   An ideal dc supply of constant voltage V supplies power to a three-phase, force-commutated inverter consisting of six ideal transistors. Power is then transferred to a delta-connected resistive load of R Ω per branch (Fig. 10.17). The mode of inverter switching is such that three transistors are conducting simultaneously at every instant of the cycle. Show that the line current waveforms are of six-step form with a peak height of $2V/R$. Further show that the phase (branch) currents are square waves of height V/R.

10.8   For the periodic voltage waveform of Fig. 10.18 calculate the fundamental component, the total rms value, the distortion factor, and the displacement factor.

10.9   For the 12-step waveform of Fig. 10.8 show that the step height for the interval $\pi/6 < \omega t < \pi/3$ is given by 0.732 V. Also show that the fundamental component of this waveform has a peak height of $\pi/3$ V and a displacement angle $\psi_1 = 0$.

10.10   For the 12-step voltage waveform of Fig. 10.8 calculate the rms value and hence the distortion factor.

10.11   A six-step voltage source inverter is supplied from an ideal battery with terminal voltage $V = 200$ V. The inverter supplies a delta-connected load with a series $R-L$ impedance in each leg consisting of $R = 20$ Ω,

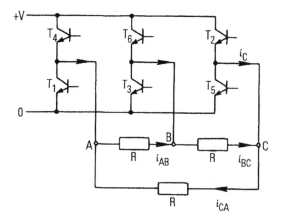

FIG. 17   Inverter circuit connection in Problem 10.7 [20].

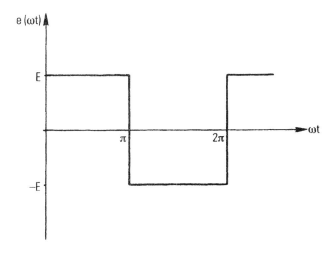

FIG. 18   Voltage waveform of Problem 10.10 [20].

$X_L = 30 \ \Omega$ at the generated frequency. Calculate the rms load current and the average value of the supply current. Check that, within calculation error, the input power is equal to the load power.

10.12   Repeat Problem 10.11 if the load inductance is removed.

10.13   For the inverter operation of Problem 10.11 calculate the maximum and minimum values of the time-varying link current.

11

Three-Phase, Pulse-Width Modulation, Controlled Inverter Circuits

The skeleton three-phase inverter circuits of Fig. 10.2 apply not only for step-wave operation but also for the switching technique known as *pulse width modulation* (PWM). Section 8.4 gives a brief account of the principles of PWM—this is subsumed within the present chapter.

11.1   SINUSOIDAL PULSE WIDTH MODULATION

A periodic (carrier) waveform of any wave shape can be modulated by another periodic (modulating) waveform of any other wave shape, of lower frequency. For most wave-shape combinations of a carrier waveform modulated by a modulating waveform, however, the resultant modulated waveform would not be suitable for either power applications or for information transmission.

For induction motor speed control the motor voltage waveforms should be as nearly sinusoidal as possible. If nonsinusoidal voltages are used, as with most inverter drives, it is preferable to use waveforms that do not contain low-order harmonics such as the fifth and the seventh because these can cause torque ripple disturbances, especially at low speeds. The lower order harmonics of a modulated voltage wave can be greatly reduced if a sinusoidal modulating signal modulates a triangular carrier wave. The pulse widths then cease to be uniform as in Fig. 8.5 but become sinusoidal functions of the angular pulse position, as in Fig. 11.1. With sinusoidal PWM the large look-up tables of precalculated values needed

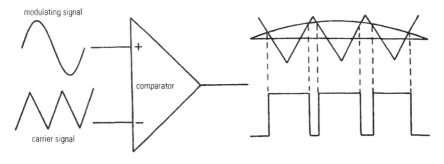

modulating signal

comparator

carrier signal

output modulated PWM signal

FIG. 1   Principle of sinusoidal modulation of triangular carrier wave [20].

for harmonic elimination in step-wave inverters (Sec. 8.4.3) are avoided. Inverter control becomes complex with sinusoidal PWM, and the switching losses are much greater than for conventional six-step operation of step-wave inverters. For SCR and GTO devices the switching frequencies are usually in the range 500–2500 Hz, and for power transistors the frequency can be as high as 10 kHz.

11.1.1 Sinusoidal, Modulation with Natural Sampling

The principle of sinusoidal PWM was discussed in Sec. 8.4 and is followed up in this present section, with reference to Fig. 11.2. A sinusoidal modulating signal $v_m (\omega t) = V_m \sin \omega t$ is applied to a single-sided triangular carrier signal $v_c (\omega_c t)$, of maximum height V_c, to produce the output modulated wave $v_o (\omega t)$ of the same frequency as the modulating wave. For the example shown, the modulation ratio M, defined as V_m/V_c, is 0.75 and the frequency ratio p, defined as f_c/f_m or f_c/f, has the value 12 (Fig. 11.2). The peak value $\pm V$ of the output wave is determined by the dc source voltage.

11.1.2 Three-Phase Sinusoidal PWM with Natural Sampling

The basic skeleton inverter circuit, with a wye-connected load, is shown in Fig. 11.3. This has exactly the same general form as the inverter circuits of Chapter 10, except that a midpoint of the dc supply is shown explicitly. For three-phase operation the triangular carrier wave is usually doubled sided, being symmetrical and without dc offset. Each half wave of the carrier is then an identical isosceles

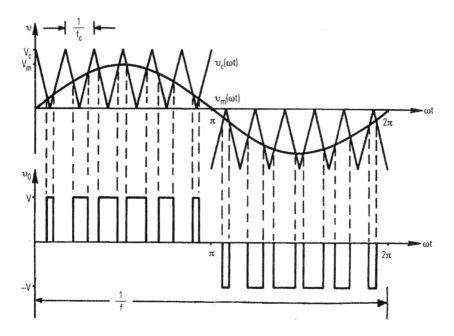

FIG. 2   PWM voltage waveform obtained by sinusoidal modulation using natural sampling: $p = 12$, $M = 0.75$ [20].

triangle. Waveforms for a three-phase system are shown in Fig. 11.4 in which frequency ratio $p = 9$ and modulation ratio M is almost unity. For balanced three-phase operation, p should be an odd multiple of 3. The carrier frequency is then a triplen of the modulating frequency so that the output-modulated waveform does not contain the carrier frequency or its harmonics. In Fig. 11.4 the three-phase, modulated output voltages are seen to be identical but with a mutual phase displacement of 120° or one-third of the modulating voltage cycle, and the peak level is determined by the dc source level.

The order of harmonic components k of the modulated waveform are given by

$$k = np \pm m \tag{11.1}$$

where n is the carrier harmonic order and m is the carrier side band. The major harmonic orders are shown in Table 11.1 for several values of p. At $p = 15$, for example, the lowest significant harmonic is $k = p - 2 = 13$, and this is of much higher order than the harmonics $k = 5, 7$, obtained with a six-step waveform. It is found that the $2p \pm 1$ harmonics are dominant in magnitude for values of

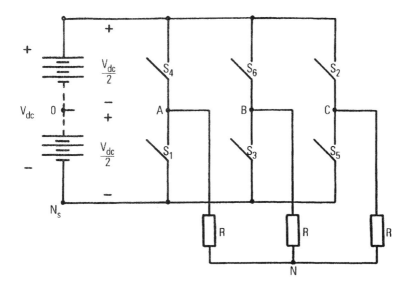

FIG. 3   Basic skeleton inverted circuit [20].

modulation ratio up to about $M = 0.9$. When $p > 9$, the harmonic magnitudes at a given value of M are independent of p.

Fourier analysis of a sinusoidally pulse width modulated waveform is very complex. The nth harmonic phase voltage component for a waveform such as those of Fig. 11.4 is given by an expression of the form

$$
v_n = \frac{MV_{dc}}{2} \cos \omega_m t
$$
$$
+ \frac{2V_{dc}}{\pi} \sum_{m=1}^{\infty} J_0\left(mM\frac{\pi}{2}\right) \sin\left(\frac{m\pi}{2}\right) \cos(m\omega_c t)
$$
$$
+ \frac{2V_{dc}}{\pi} \sum_{m=1}^{\infty} \sum_{n=\pm 1}^{\pm\infty} \frac{J_n\left(mM\frac{\pi}{2}\right)}{m} \sin\left[(m+n)\frac{\pi}{2}\right] \cos(m\omega_c t + n\omega_m t) \quad (11.2)
$$

In Eq. (11.2) the terms J_0, and J_n represent first-order Bessel functions, ω_m is the fundamental frequency of the modulating and output waves, and ω_c is the carrier frequency.

The first term of Eq. (11.2) gives the amplitude of the fundamental frequency output component, which is proportional to modulation index M, in the range $0 \leq M \leq 1$, for all values of $p \geq 9$.

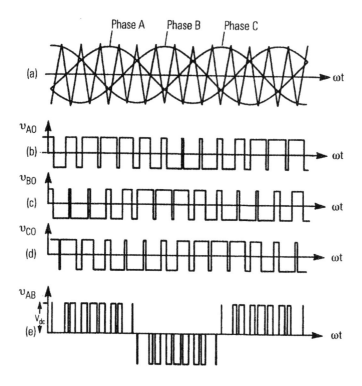

FIG. 4 Voltage waveforms for a three-phase sinusoidal PWM inverter, $m \approx 1$ $p = 9$: (a) timing waveforms, (b)–(d) pole voltages, and (e) output line voltage [21].

TABLE 11.1   Major Harmonics in a PWM Waveform with Naturally Sampled Sinusoidal Modulation [20]

Frequency ratio p	$n = 1$ $m = 2$	$n = 2$ $m = 1$	$n = 3$ $m = 2$
	$p \pm 2$	$2p \pm 1$	$3p \pm 2$
3	5, 1	7, 5	11, 7
9	11, 7	19, 17	29, 25
15	17, 13	31, 29	47, 43
21	23, 19	43, 41	65, 61

$$V_{1(\text{peak})} = \frac{MV_{dc}}{2} \tag{11.3}$$

The corresponding rms value of the fundamental component of the modulated line-to-line voltage V_{L_1} (rms) is given by

$$V_{L_{1(\text{r.m.s})}} = \frac{\sqrt{3}V_{1(\text{peak})}}{\sqrt{2}}$$

$$= \frac{\sqrt{3}}{2\sqrt{2}} MV_{dc} = 0.612 MV_{dc} \tag{11.4}$$

There is no straightforward analytical expression for the related values of the higher harmonic voltage components. Calculated values of these, in the range $0 \leq M \leq 1$ with $p > 9$, are given in Fig. 11.5 [21]. If the modulating wave peak amplitude is varied linearly with the modulating frequency f_m, then ratio M/f_m is constant. A waveform having a constant ratio of fundamental voltage to frequency is particularly useful for ac motor speed control.

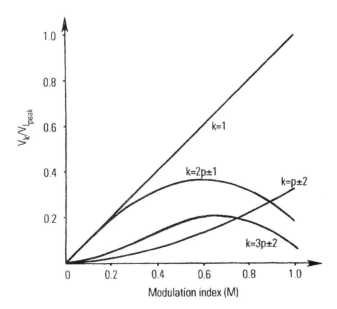

FIG. 5   Harmonic component voltages (relative to peak fundamental value) for sinusoidal PWM with natural sampling $p > 9$ [21].

The second term of Eq. (11.2) describe harmonics at the carrier frequency and its multiples, while the third term refer to sidebands around each multiple of the carrier frequency.

11.1.3 Overmodulation in Sinusoidal PWM Inverters

Increase of the fundamental component of the modulated output voltage V_1, beyond the $M = 1$ value, is possible by making $M > 1$, but V_1 is then no longer proportional to M (Fig. 11.6). In this condition of overmodulation the process of natural sampling no longer occurs. Some intersections between the carrier wave and the modulating wave are lost, as illustrated in Fig. 11.7. The result is that some of the pulses of the original PWM wave are dropped in the manner shown in Fig. 11.8.

In the extreme, when M reaches the value $M = 3.24$, the original forms of PWM waveform in Fig. 11.4 are lost. The phase voltages then revert to the quasi-square wave shape of Table 10.1, Fig. 10.5, and Fig. 10.7 in which harmonics of orders 5 and 7 reappear. Variation of the fundamental output voltage versus modulation ratio M is shown in Fig. 11.6. For pulse voltage V_{dc} (i.e., twice the value given in Fig. 11.3) the rms fundamental line value of the quasi-square wave is

$$V_{1_{(rms)}} = \frac{4}{\pi} \frac{\sqrt{3}}{2\sqrt{2}} V_{dc} = \frac{\sqrt{6}}{\pi} V_{dc} = 0.78 V_{dc} \tag{11.5}$$

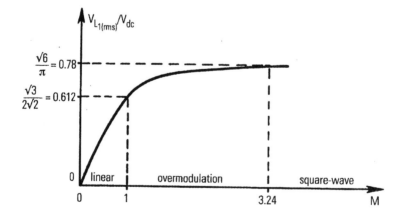

FIG. 6   RMS fundamental line voltage (relative to V_{dc}) versus modulation ratio M for sinusoidal modulation [20].

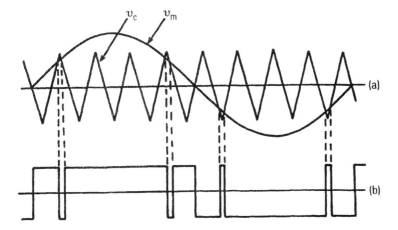

FIG. 7 Overmodulation of triangular carrier wave by a sinusoidal modulating wave, M = 1.55 [21].

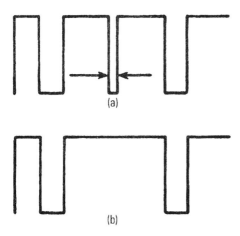

FIG. 8 Example of pulse dropping due to overmodulation: (a) containing a minimum pulse and (b) minimum pulse dripped [20].

Overmodulation increases the waveform harmonic content and can also result in undesirable large jumps of V_1, especially in inverter switches with large dwell times.

Other options for increase of the fundamental output voltage beyond the $M = 1$ value, without increase of other harmonics, are to use a nonsinusoidal reference (modulating) wave such as a trapezoid or a sine wave plus some third harmonic component.

11.1.4 Three-Phase Sinusoidal PWM with Regular Sampling

As an alternative to natural sampling, the sinusoidal reference wave can be sampled at regular intervals of time. If the sampling occurs at instants corresponding to the positive peaks or the positive and negative peaks of the triangular carrier wave, the process is known as *uniform* or *regular sampling*. In Fig. 11.9a sample value of the reference sine wave is held constant until the next sampling instant when a step transition occurs, the stepped version of the reference wave becomes, in effect, the modulating wave. The resulting output modulated wave is defined by the intersections between the carrier wave and the stepped modulating wave.

When sampling occurs at carrier frequency, coincident with the positive peaks of the carrier wave (Fig. 11.9a), the intersections of adjacent sides of the carrier with the step wave are equidistant about the non sampled (negative) peaks. For all values of M the modulated wave pulse widths are then symmetrical about the lower (nonsampled) carrier peaks, and the process is called *symmetrical regular sampling*. The pulse centers occur at uniformly spaced sampling times.

When sampling coincides with both the positive and negative peaks of the carrier wave (Fig. 11.9b), the process is known as *asymmetrical regular sampling*. Adjacent sides of the triangular carrier wave then intersect the stepped modulation wave at different step levels and the resultant modulated wave has pulses that are asymmetrical about the sampling point.

For both symmetrical and asymmetrical regular sampling the output modulated waveforms can be described by analytic expressions. The number of sine wave values needed to define a sampling step wave is equal to the frequency ratio p (symmetrical sampling) or twice the frequency ratio, $2p$ (asymmetrical sampling). In both cases the number of sample values is much smaller than in natural sampling, which requires scanning at sampling instants every degree or half degree of the modulating sine wave.

The use of regular sampled PWM in preference to naturally sampled PWM requires much less ROM-based computer memory. Also, the analytic nature of regular sampled PWM waveforms makes this approach feasible for implementation using microprocessor-based techniques because the pulse widths are easy to calculate.

sample-and-hold —→
at f_c

v_c—→

v_r

v_m

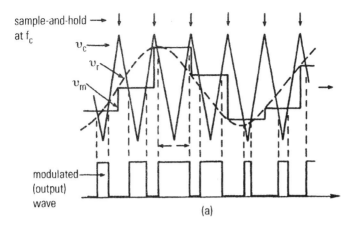

modulated—
(output)
wave

(a)

sample-and-hold—→
at $2f_c$

v_c

v_r

v_m

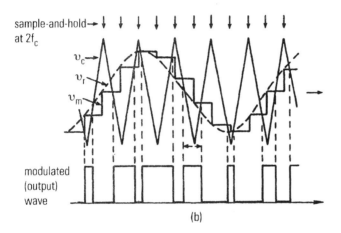

modulated
(output)
wave

(b)

FIG. 9   Sinusoidal modulation of a triangular carrier wave using regular sampling, $M =$ 0.75, $p = 4.5$: (a) symmetrical sampling and (b) asymmetrical sampling [27].

Some details of various forms of pulse width modulation, using a symmetrical triangular carrier wave, are given in Table 11.2.

11.2   PWM VOLTAGE WAVEFORMS APPLIED TO A BALANCED, THREE-PHASE, RESISTIVE-INDUCTIVE LOAD

A double-sided triangular carrier wave modulated by a sinusoid results in the pulse waveforms v_A and v_B of Fig. 11.10. If modulating signal v_{mB} is delayed

TABLE 11.2   Techniques of Pulse Width Modulation (PWM) Using a Symmetrical Triangular Carrier Wave [20]

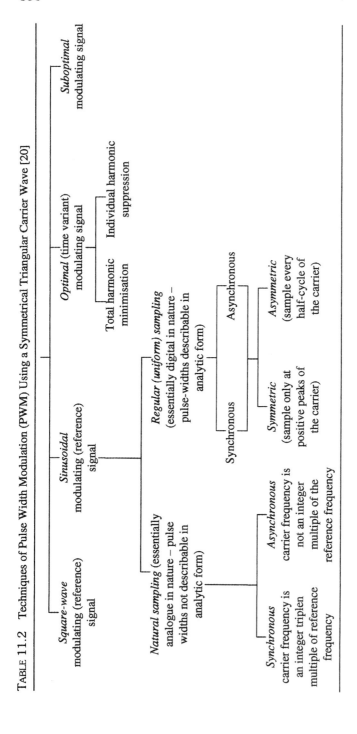

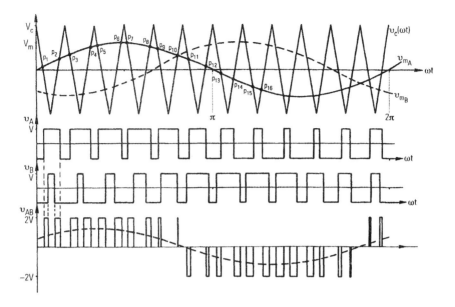

Fig. 10 PWM voltage waveforms with sinusoidal natural sampling instants, $M = 0.75$, $p = 12$ [20].

$120°$ with respect to v_{mA} the resulting modulated wave v_B is identical in form to v_A but is also delayed by $120°$. The corresponding line voltage v_{AB} ($= v_A - v_B$) has a fundamental component that leads the fundamental component of v_A by $30°$, as in a sinusoidal balanced set of voltages. Note that the positive pulse pattern of v_{AB} (ωt) is not quite the same as the negative pulse pattern, although the two areas are the same and give zero time average value. This issue is the subject of Example 11.1.

The application of a PWM voltage waveform to an inductive load results in a current that responds (very nearly) only to the fundamental component. The harmonics of a PWM waveform, including the fundamental, are a complicated function of the carrier frequency ω_c, the modulating (output) frequency ω_m, the carrier amplitude V_c, and the modulating wave amplitude V_m (combined in the modulation index M), as indicated in Eq. (11.2).

Harmonic components of the carrier frequency are in phase in all three load phases and therefore have a zero sequence nature. With a star-connected load there are no carrier frequency components in the line voltages.

An approximate method of calculating the harmonic content of a PWM waveform is to use graphical estimation of the switching angles, as demonstrated in Example 11.1. A precise value of the intersection angles between the triangular carrier wave and the sinusoidal modulating wave can be obtained by equating the appropriate mathematical expressions. In Fig. 11.10, for example, the modulating wave is synchronized to the peak value of the carrier wave. The first intersection p_1 between the carrier v_c ($\omega_m t$) and modulating wave v_{mA} ($\omega_m t$) occurs when

$$\frac{V_m}{V_c} \sin \omega_m t = 1 - \frac{24}{\pi} \omega_m t \tag{11.6}$$

Intersection p_2 occurs when

$$\frac{V_m}{V_c} \sin \omega_m t = -3 + \frac{24}{\pi} \omega_m t \tag{11.7}$$

This oscillating series has the general solution, for the Nth intersection,

$$p_N = (2N - 1)(-1)^{N+1} + (-1)^N \frac{2p}{\pi} \omega_m t \tag{11.8}$$

where $N = 1, 2, 3, \cdots, 24$.

Expressions similar to Eqs. (11.6) and (11.7) can be obtained for all of the intersections, as shown in Example 11.1. Equations of the form Eqs. (11.6)–(11.8) are transcendental and require being solved by iteration.

11.3 PWM VOLTAGE WAVEFORMS APPLIED TO A THREE-PHASE INDUCTION MOTOR

The basic differences of structure between the voltage source, step-wave inverter, such as that of Fig. 10.1, and the voltage source, PWM inverter are given in Fig. 11.11. A step-wave inverter uses a controlled rectifier to give a direct-voltage source of adjustable level at the input to the dc link. The voltage level of the inverter output is controlled by the adjustable V_{dc} link voltage, whereas the frequency is controlled independently by the gating of the inverter switches.

A PWM inverter uses a diode bridge rectifier to give a fixed level of V_{dc} at the dc link. Both the voltage and frequency of the inverter are simultaneously controlled by gating of the inverter switches (Fig. 11.11b). The complete assembly of rectifier stage, dc link, and inverter stage is shown in Fig. 11.12. Since the output voltage of the diode bridge rectifier is not a pure direct voltage, a filter inductor is included to absorb the ripple component.

The use of a fixed dc rail voltage means that several independent inverters can operate simultaneously from the same dc supply. At low power levels the use of transistor (rather than thyristor) switches permits fast switching action and fast current and torque transient response, compared with step-wave inverters.

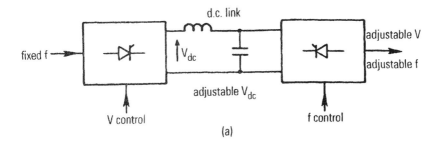

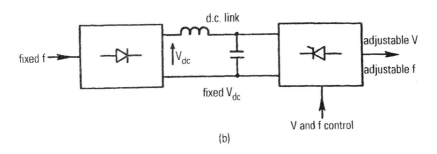

FIG. 11 Basic forms of voltage source inverter (VSI): (a) step wave (or quasi-square wave) and (b) PWM [20].

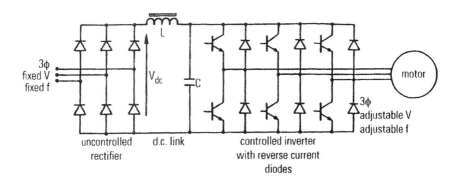

FIG. 12 Main circuit features of PWM VSI with motor load [20].

Because the harmonic currents are small and can be made of relatively high order, compared with single-pulse or multiple-pulse modulation, and because the fundamental component is easily controlled, PWM methods are becoming increasingly popular for ac motor control. Although the harmonic currents may be small, however, the harmonic heating losses may be considerable through increase of the motor resistances due to the skin effect. PWM switching techniques are better suited to power transistor inverters than to thyristor inverters because the commutation losses due to the many switchings are then less significant. Above about 100 Hz the commutation losses with PWM switching become unacceptably large, and stepped-wave techniques are used in ac motor drives.

When PWM voltage waveforms are applied to an induction motor, the motor torque responds largely to the fundamental frequency component. Motor current harmonics are usually small and of high harmonic order, depending on the frequency ratio p.

The harmonics of the PWM applied voltage are often more significant than those of the consequent motor current. This has the result that the eddy-current and hysteresis iron losses, which vary directly with flux and with frequency, are often greater than the copper losses in the windings. The total losses due to harmonics in a PWM driven motor may exceed those of the comparable step-wave driven motor. It is a common practice that a PWM driven motor is derated by an amount of 5–10%.

Torque pulsations in a PWM drive are small in magnitude and are related to high harmonic frequencies so that they can usually be ignored. The input current waveform to a dc link-inverter drive is determined mostly by the rectifier action rather than by the motor operation. This has a wave shape similar to that of a full-wave, three-phase bridge with passive series resistance–inductance load, so that the drive operates, at all speeds, at a displacement factor near to the ideal value of unity.

11.4   WORKED EXAMPLES

Example 11.1   A double-sided triangular carrier wave of height V_c is natural sampling modulated by a sinusoidal modulating signal $v_m(\omega t) = V_m \sin \omega t$, where $V_m = 0.6V_c$. The carrier frequency ω_c is 12 times the modulating frequency ω_m. Sketch a waveform of the resultant modulated voltage and calculate its principal harmonic components and its rms value.

The waveforms are shown in Fig. 11.10. The phase voltage $v_A(\omega t)$ is symmetrical about $\pi/2$ radians and contains only odd harmonics. Since $v_A(\omega t)$ is antisymmetrical about $\omega t = 0$, the Fourier harmonics $a_n = 0$, so that the fundamental output component is in phase with the modulating voltage $v_m(\omega t)$.

It is necessary to determine the intersection points p_1 to p_6.

Point p_1: From Eq. (12.31), $p = 12$, $V_m/V_c = 0.6$ so that

$$0.6 \sin \omega t = 1 - \frac{24}{\pi} \omega t$$

which gives

$$\omega t = 7°$$

Point p_2:

$$\frac{V_m}{V_c} \sin \omega t = \frac{4N}{\pi} \omega t - 3$$

or

$$\omega t = 24°$$

Point p_3:

$$\frac{V_m}{V_c} \sin \omega t = 5 - \frac{4N}{\pi} \omega t$$

$$\omega t = 34.5°$$

Point p_4:

$$\frac{V_m}{V_c} \sin \omega t = \frac{4N}{\pi} \omega t - 7$$

$$\omega t = 56°$$

Point p_5:

$$\frac{V_m}{V_c} \sin \omega t = 9 - \frac{N}{\pi} \omega t$$

$$\omega t = 63°$$

Point p_6:

$$\frac{V_m}{V_c} \sin \omega t = \frac{4N}{\pi} \omega t - 11$$

$$\omega t = 87°$$

For the first quarter cycle in Fig. 11.10 waveform v_A is given by

$$v_A(\omega t) = V \begin{vmatrix} 24°, 56°, 87° \\ \\ 7°, 34.5°, 63° \end{vmatrix} - V \begin{vmatrix} 7°, 34.5°, 63°, 90° \\ \\ 0°, 24°, 56°, 87° \end{vmatrix}$$

Fourier coefficients b_n are given by

$$b_n = \frac{4}{\pi} \int_0^{\pi/2} v_A(\omega t) \sin n\omega t \, d\omega t$$

in this case,

$$b_n = \frac{4V}{n\pi}\left(-\cos n\omega t \begin{vmatrix} 24°, & 56°, & 87° \\ 7°, & 34.5°, & 63° \end{vmatrix} + \cos n\omega t \begin{vmatrix} 7°, & 34.5°, & 63°, & 90° \\ 0°, & 24°, & 56°, & 87° \end{vmatrix} \right)$$

$$= \frac{8V}{n\pi}(\cos n7° + \cos n34.5° + \cos n63° - \cos n24° - \cos n56° - \cos n87° - 0.5)$$

It is found that the peak values of Fourier coefficient b_n are

$$b_1 = 0.62 \text{ V}$$

$$b_3 = \frac{8V}{3\pi}(0.038) = 0.0323 \text{ V}$$

$$b_5 = \frac{8V}{5\pi}(0.102) = 0.052 \text{ V}$$

$$b_7 = \frac{8V}{7\pi}(0.323) = 0.118 \text{ V}$$

$$b_9 = \frac{8V}{9\pi}(0.876) = 0.248 \text{ V}$$

$$b_{11} = \frac{8V}{11\pi}(2.45) = 0.567 \text{ V}$$

$$b_{13} = \frac{8V}{13\pi}(2.99) = -0.585 \text{ V}$$

$$b_{15} = \frac{8V}{15\pi}(3.44) = 0.548 \text{ V}$$

$$b_{17} = \frac{8V}{17\pi}(1.52) = -0.23 \text{ V}$$

$$b_{19} = \frac{8V}{19\pi}(1.22) = -0.164 \text{ V}$$

The rms value of the waveform is

$$V_A = \frac{V}{\sqrt{2}} \sqrt{b_1^2 + b_3^2 + b_5^2 + \cdots + b_{19}^2}$$

$$= \frac{V}{\sqrt{2}} \sqrt{0.598} = 0.547 \text{ V}$$

The distortion factor is

$$\text{Distortion factor} = \frac{\dfrac{b_1}{\sqrt{2}}}{V_A} = \frac{0.626}{\sqrt{2}\times0.547} = 0.809$$

Note that the highest value harmonics satisfy an $n = p \pm 1$ relation and that the low-order harmonics have small values. This enhances the suitability of the waveform for ac motor speed control.

Example 11.2   The PWM voltage waveform $v_A(\omega t)$ of Fig. 11.10 is generated by an inverter that uses a modulating frequency of 50 Hz. If the dc supply is 200 V, calculate the rms current that would flow if $v_A(\omega t)$ was applied to a single-phase series R-L load in which $R = 10\ \Omega$ and $L = 0.01$ H.

At the various harmonic frequencies the load impedance is

$$|Z_h| = \sqrt{R^2 + (n\omega L)^2}$$

which gives

$$Z_1 = \sqrt{10^2 + (2\pi\times50\times0.01)^2} = 10.48\ \Omega$$
$$Z_3 = 13.74\ \Omega \quad Z_{13} = 42\ \Omega$$
$$Z_5 = 18.62\ \Omega \quad Z_{15} = 48.17\ \Omega$$
$$Z_7 = 24.16\ \Omega \quad Z_{17} = 54.33\ \Omega$$
$$Z_9 = 30\ \Omega \qquad\ Z_{19} = 60.52\ \Omega$$
$$Z_{11} = 35.97\ \Omega$$

If the harmonic voltages of Example 11.1 are divided by the respective harmonic impedances above, one obtains the following peak current harmonics:

$$I_1 = \frac{V_1}{Z_1} = \frac{0.626\times200}{10.48} = 11.95\ \text{A}$$
$$I_3 = 0.47\ \text{A} \qquad I_{13} = 2.97\ \text{A}$$
$$I_5 = 0.56\ \text{A} \qquad I_{15} = 2.42\ \text{A}$$
$$I_7 = 0.977\ \text{A} \qquad I_{17} = 0.85\ \text{A}$$
$$I_9 = 1.65\ \text{A} \qquad I_{19} = 0.54\ \text{A}$$
$$I_{11} = 3.15\ \text{A}$$

The harmonic sum is

$$\sum I_n^2 = I_1^2 + I_3^2 + I_5^2 + \cdots + I_{19}^2 = 172\ \text{A}^2$$

This has an r.m.s. value

$$I = \sqrt{\frac{\sum I_n^2}{2}} = 9.27 \text{ A}$$

which compares with a fundamental rms current of $11.95/\sqrt{2} = 8.45$ A.

The current distortion factor is therefore $8.45/9.27 = 0.912$, which is greater (i.e., better) than the corresponding voltage distortion factor of 0.809.

Example 11.3   The PWM voltage waveform v_A (ωt) of Fig. 11.10 is applied to a single-phase series R-L circuit with $R = 10 \, \Omega$ and $L = 0.01$ H. Voltage v_A (ωt) has a frequency of 50 Hz and an amplitude $V = 200$ V. Deduce the waveform of the resulting current.

The waveform v_A (ωt) is reproduced in Fig. 11.13. If a direct voltage V is applied across a series R-L circuit carrying a current I_o, the subsequent rise of current satisfies the relation

$$i(\omega t) = \frac{V}{R}\left(1 - \varepsilon^{-t/\tau}\right) + I_o\varepsilon^{-t/\tau} = \frac{V}{R} - \left(\frac{V}{R} - I_o\right)\varepsilon^{-t/\tau}$$

where I_o is the value of the current at the switching instant. In this case $\tau = L/R = \omega L/\omega R = \pi/10\omega$ so that

$$i(\omega t) = 20 - (20 - I_o)\varepsilon^{-10/\pi\omega t}$$

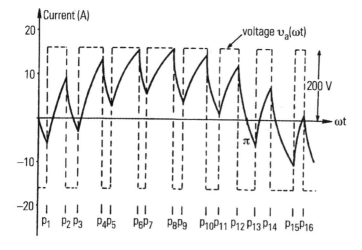

Fig. 13   PWM waveforms with series R-L load, from Example 11.3 [20].

At switch-on $I_o = 0$ and the current starts from the origin. Consider the current values at the voltage switching points in Fig. 11.13. These time values are given in Example 11.1.

At point p_1,

$$\omega t = 7° = 0.122 \text{ radian}$$
$$i(\omega t) = -20(1 - 0.678) = -6.44 \text{ A}$$

At point p_2,

$$\omega t = 24° - 7° = 0.297 \text{ radian}$$
$$i(\omega t) = 20 - (20 + 6.44) \times 0.39 = 9.8 \text{ A}$$

At point p_3,

$$\omega t = 34.5° - 24° = 0.183 \text{ radian}$$
$$i(\omega t) = -20 - (-20 - 9.8) \times 0.56 = -3.3 \text{ A}$$

At point p_4,

$$\omega t = 56° - 34.5° = 0.375 \text{ radian}$$
$$i(\omega t) = 20 - (20 + 3.3) \times 0.303 = 12.94 \text{ A}$$

At point p_5,

$$\omega t = 63° - 56° = 0.122 \text{ radian}$$
$$i(\omega t) = -20 - (-20 - 12.94) \times 0.678 = 2.33 \text{ A}$$

At point p_6,

$$\omega t = 87° - 63° = 0.419 \text{ radian}$$
$$i(\omega t) = 20 - (20 - 2.33) \times 0.264 = 15.34 \text{ A}$$

Similarly,

At point p_7, $\omega t = 93° - 87° = 0.105$ radian, $i(\omega t) = 5.3$ A.
At point p_8, $\omega t = 116° - 93° = 0.401$ radian, $i(\omega t) = 15.9$ A.
At point p_9, $\omega t = 124.5° - 116° = 0.148$ radian, $i(\omega t) = 2.4$ A.
At point p_{10}, $\omega t = 145.5° - 124.5° = 0.367$ radian, $i(\omega t) = 14.53$ A.
At point p_{11}, $\omega t = 155° - 145.5° = 0.166$ radian, $i(\omega t) = 0.3$ A.
At point p_{12}, $\omega t = 172.5° - 155° = 0.305$ radian, $i(\omega t) = 12.5$ A.
At point p_{13}, $\omega t = 187.5° - 172.5° = 0.262$ radian, $i(\omega t) = -5.9$ A.
At point p_{14}, $\omega t = 201° - 187.5° = 0.236$ radian, $i(\omega t) = 7.78$ A.
At point p_{15}, $\omega t = 220.5° - 201° = 0.34$ radian, $i(\omega t) = -10.6$ A.

At point p_{16}, $\omega t = 229.5° - 220.5° = 0.157$ radian, $i(\omega t) = 1.43$ A.

The time variation of the current, shown in Fig. 11.13, is typical of the current waveforms obtained with PWM voltages applied to inductive and ac motor loads.

Example 11.4 The PWM waveforms of Fig. 11.4 have a height of 240 V and are applied as the phase voltage waveforms of a three-phase, four-pole, 50-Hz, star-connected induction motor. The motor equivalent circuit parameters, referred to *primary*, are $R_1 = 0.32 \, \Omega$, $R = 0.18 \, \Omega$, $X_1 = X_2 = 1.65 \, \Omega$, and X_m = large. Calculate the motor rms current at 1440 rpm. What are the values of the main harmonic currents?

A four-pole, 50-Hz motor has a synchronous speed

$$N_s = \frac{f120}{p} = \frac{50 \times 120}{4} = 1500 \text{ rpm}$$

At a speed of 1440 rpm the per-unit slip is given by Eq. (9.3)

$$S = \frac{1500 \times 1440}{1500} = 0.04$$

It is seen from Fig. 11.4a that the PWM line voltage waveform of Fig. 11.4 is the value of the battery (or mean rectified) supply voltage V_{dc} in the circuit of Fig. 11.3. The rms value of the per-phase applied voltage is therefore, from Eq. (11.3),

$$V_{1_{rms}} = 0.9 \times \frac{240}{2\sqrt{2}} = 76.4 \text{ V/phase}$$

The motor per-phase equivalent circuit is a series R-L circuit. At 1440 rpm this has the input impedance

$$Z_{in_{1440}} = \left(0.32 + \frac{0.18}{0.04} \right) + j1.65$$

$$= 4.82 + j1.65 = 5.095\angle 18.9° \ \Omega/\text{phase}$$

Therefore,

Therefore, $I_1 = \dfrac{76.4\angle 0}{5.095\angle 18.9} = 15\angle -18.9° \text{ A/phase}$

With $M = 0.9$ the dominant harmonics are likely to be those of order $2p \pm 1$ and $p \pm 2$. In this case, therefore, with $p = 9$, the harmonics to be considered are $n = 7, 11, 17,$ and 19. The harmonic slip values are

$$S_7 = \frac{7 - (1 - 0.04)}{7} = \frac{6.04}{7} = 0.863$$

$$S_{11} = \frac{11 + (1 - 0.04)}{11} = \frac{11.96}{11} = 1.087$$

$$S_{17} = \frac{17 + (1 - 0.04)}{17} = \frac{17.96}{17} = 1.056$$

$$S_{19} = \frac{19 - (1 - 0.04)}{19} = \frac{18.04}{19} = 0.95$$

For the nth harmonic currents the relevant input impedances to the respective equivalent circuit are

$$Z_{in_7} = \left(0.32 + \frac{0.18}{0.863}\right) + j7 \times 1.65 = 0.53 + j11.55 = 11.56\ \Omega$$

$$Z_{in_{11}} = \left(0.32 + \frac{0.18}{1.087}\right) + j11 \times 1.65 = 0.486 + j18.15 = 18.16\ \Omega$$

$$Z_{in_{17}} = \left(0.32 + \frac{0.18}{1.056}\right) + j17 \times 1.65 = 0.291 + j28.1 = 28.2\ \Omega$$

$$Z_{in_{19}} = \left(0.32 + \frac{0.18}{0.949}\right) + j19 \times 1.65 = 0.51 + j31.35 = 31.35\ \Omega$$

The 7, 11, 17, and 19 order harmonic voltage levels have to be deduced from Fig. 11.5. For $M = 0.9$ the $2p \pm 1$ and $p \pm 2$ levels have the value 0.26 of the peak fundamental value. The $M = 1$ value of the r.m.s. component of the fundamental phase voltage, from Eq. (11.3), is

$$V_{1_{rms}} = \frac{240}{2\sqrt{2}} = 84.86\ \text{V/phase}$$

Therefore, the rms harmonic voltage values

$$V_7 = V_{11} = V_{17} = V_{19} = 0.26 \times 84.86 = 22\ \text{V/Phase}$$

The appropriate rms harmonic phase currents are

$$I_7 = \frac{22.1}{11.56} = 1.91 \text{A} \qquad I_{17} = \frac{22.1}{28.2} = 0.78 \text{A}$$

$$I_{11} = \frac{22.1}{18.16} = 1.22 \text{A} \qquad I_{19} = \frac{22.1}{3.135} = 0.7 \text{A}$$

The total rms current is obtained by the customary square-law summation

$$I_{rms} = \sqrt{I_1^2 + I_7^2 + I_{11}^2 + I_{17}^2 + I_{19}^2 + \cdots}$$
$$I_{rms}^2 = 15^2 + (1.91)^2 + (1.22)^2 + (0.78)^2 + (0.7)^2$$
$$= 225 + 3.65 + 1.49 + 0.6 + 0.49$$
$$= 231.23 \text{ A}^2$$
$$I_{rms} = \sqrt{231.23} = 15.25 \text{ A}$$

which is about 2% greater than the fundamental value.

PROBLEMS

11.1   A single-sided triangluar carrier wave of peak height V_c contains six pulses per half cycle and is modulated by a sine wave, $v_m\ (\omega t) = V_m \sin \omega t$, synchronized to the origin of a triangular pulse. Estimate, graphically, the values of ωt at which intersections occur between v_c $(\omega_c t)$ and $v_m\ (\omega t)$ when $V_m = V_c$. Use these to calculate values of the harmonics of the modulated wave up to $n = 21$ and thereby calculate the rms value.

11.2   The modulated voltage waveform described in Problem 11.1 is applied to a series R-L load in which $R = 25\ \Omega$ and $X_L = 50\ \Omega$ at 50 Hz. If the constant height of the PWM voltage wave is 400 V, calculate the resulting current harmonics up to $n = 21$. Calculate the resultant rms current. Compare the value of the current distortion factor with the voltage distortion factor.

11.3   Calculate the power dissipation in the R-L series circuit of Problem 11.2. Hence calculate the operating power factor.

11.4   The PWM voltage waveform $v_o\ (\omega t)$ if Fig. 11.2 is applied to the series R-L load, $R = 25\ \Omega$ and $X_L = 50\ \Omega$ at 50 Hz. If $V = 250$ V and $f = 50$ Hz, reduce the waveform of the resulting current.

11.5   The PWM waveform $v_o\ (\omega t)$ of Fig. 11.2 has a height $V = 240$ V and is applied as the phase voltage waveform of a three-phase, four-pole, 50 Hz, star-connected induction motor. The motor equivalent circuit parameters, referred to primary turns, are $R_1 = 0.32\ \Omega$, $R_2 = 0.18\ \Omega$, $X_1 = X_2 = 1.65\ \Omega$, $X_m =$ large. Calculate the motor rms current at 1440 rpm. What are the values of the main harmonic currents?

11.6   For the induction motor of Problem 11.5 calculate the input power and hence the power factor for operation at 1440 rpm.

11.7   If only the fundamental current component results in useful torque pro-
        duction, calculate the efficiency of operation for the motor of Problem
        11.5 at 1440 rpm. (a) neglecting core losses and friction and windage,
        (b) assuming that the core losses plus friction and windage are equal to
        the copper losses.

12

Phase-Controlled Cycloconverters

12.1 INTRODUCTION

A cycloconverter is a static frequency changer that converts an ac supply of fixed input frequency to an ac output of adjustable but lower frequency. The converter consists of an array of back-to-back or inverse-parallel connected switches, usually silicon controlled rectifiers (SCRs). By the controlled opening and closing of the switches it is possible to fabricate output voltage waveforms having a fundamental component of the desired output frequency.

Cycloconverters fall into two broad categories:

1. Phase-controlled cycloconverters, in which the firing angle is controlled by adjustable gate pulses as in controlled rectifier circuits such as those of Chapters 5 and 7. These are discussed in this present chapter.
2. Envelope cycloconverters, in which the switches remain fully on like diodes and conduct for consecutive half cycles. This form of cycloconverter is discussed in the following chapter.

All cycloconverters are line commutated, like controlled rectifiers. It is because of the necessity of producing natural commutation of the current between successive switchings that the realizable output frequency is lower than the input frequency. Theoretically the fundamental component of the highly distorted output voltage waveform is likely to be notionally between 1/3 and 2/3 of the input frequency.

364

For phase-controlled cycloconverters with 60-Hz supply the practical maximum output frequency is likely to be about 36 Hz for three-phase (six pulse) bridges and 18 Hz for single-way, three-phase (three pulse) rectifiers.

In addition to severe output voltage distortion the input current to a cyclo-converter is usually significantly distorted. Also, the fundamental component of the input current lags the supply voltage, resulting in poor input power factor, irrespective of whether the load current is lagging, leading, or of unity power factor. Very often both input and output filters may be needed.

It is possible to implement a cycloconverter in several different forms. The single-phase to single-phase version (Fig. 12.1a) will be used to illustrate the principle. Practical uses of the cycloconverter include the speed control of adjustable frequency ac induction motors and in aircraft variable-speed, constant-frequency (VSCF) supply systems.

A phase-controlled cycloconverter has exactly the same circuit topology as a dual converter. For a dual converter the firing angles of the converter switches are constant in time, as in bridge rectifier circuits, to result in rectifier operation with dc output. On the other hand, cycloconverter operation utilizes circuits in which the switch firing-angles are functions of time and the output is ac.

12.2 SINGLE-PHASE, PHASE-CONTROLLED DUAL CONVERTER

12.2.1 Circulating current free mode

The single-phase, full-wave bridge rectifier of Fig. 3.21 was discussed in Examples 3.8 and 3.9. Two such bridges can be connected back to back, as shown in Fig. 12.2, to form a single-phase dual converter. Each bridge is connected to the same ac source but no thyristor firing overlap can be permitted between them as this would short-circuit the supply. By the gating of diagonal pairs of switches in the positive (P) group, a positive load voltage is developed, as in Fig. 3.19, with the polarity indicated in Fig. 12.2. To establish negative load voltage (not possible with a single bridge), it is necessary to deploy the negative (N) group of switches in diagonal pairs.

In an ideal dual converter the firing angles of the two converters are controlled so that their dc output voltages are exactly equal. If the positive bridge is gated at an angle α_P to produce a rectified output voltage V_{dc}, then the negative bridge is simultaneously gated at α_N to also produce the same output voltage V_{dc}, of consistent polarity.

Thus the two bridges are delivering identical output dc voltages at any instant and the two firing angles must satisfy the relation

$$\alpha_p + \alpha_N = \pi \tag{12.1}$$

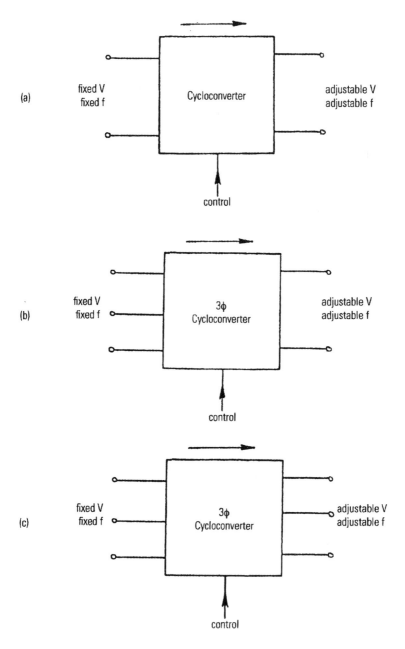

FIG. 1 Basic forms of cycloconverters: (a) one phase to one phase, (b) three phase to one phase, and (c) three phase to three phase.

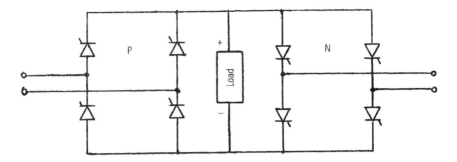

FIG. 2   Single-phase, full-wave dual converter in a noncirculating current connection.

Since the two bridges are acting alternately only one bridge carries load current at any given time. There is then no current circulating between the two bridges, and the mode of operation is usually called *circulating current free* operation or *noncirculating current* operation. The average value of the rectified load voltage was developed in Sec. 3.2.2. For the voltage waveform of Fig. 3.19, which applies here,

$$E_{av} = \frac{2E_m}{\pi} \cos \alpha_p \tag{12.2}$$

where E_m is the peak value of the single-phase supply voltage.

The actions of the two bridges are illustrated in Fig. 12.3, which shows the load voltage and current for a highly inductive load, assuming sinusoidal load current (i.e., neglecting the load current ripple). While the P bridge is causing both positive load voltage and current, it is in rectifying mode. But with inductive load the load current continues positive after the load voltage has reversed. For this interval the P bridge is inverting. While the load voltage and current are both negative, the N bridge is rectifying. When the load current continues negative after the voltage has again become positive, the N bridge is inverting.

Figure 12.4 describes the complementary nature of the P and N group firing angles for a thyristor switch dual converter. The interval $0 \le \alpha_p \le 90°$, $180° \le \alpha_N \le 90°$ represents rectification by the P bridge and inversion by the N bridge. Conversely, the interval $90° \le \alpha_p < 180°$ and $90° \le \alpha_N \le 0$ represents rectification by the N group and inversion by the P group. In practise, allowance has to be made for supply system reactance, which causes commutation overlap, and for finite thyristor turn-off time. The combined effect is to restrict the converter delay angle to less than $180°$.

In order to maintain the circulating current free condition, the gating pulses to the thyristors must be controlled so that only the bridge carrying current is kept in conduction. The temporarily idle converter must be blocked, either by removing its gating pulses altogether or by retarding the pulses away from the

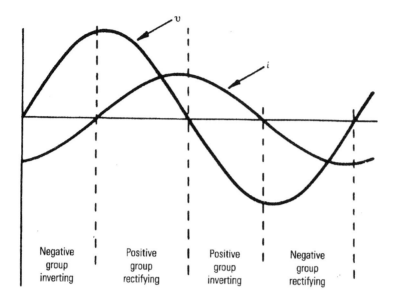

FIG. 3 Load voltage and current for a single-phase dual converter inductive load, $0 <$ $\alpha < 90°$.

condition of Eq. (12.1). Even then some severe control problems arise when the load current is discontinuous (e.g., with resistive loads).

Example voltage waveforms are given in Fig. 12.5 for the case of $\alpha_p =$ 60°. The firing pattern chosen is such as to give an ac output voltage v_o of double wavelength or one-half frequency of the input. If the supply frequency is 60 Hz, then the output voltage has a fundamental component of 30 Hz, plus higher harmonics. Because this switching produces an ac output, this is an example of the dual converter used as a cycloconverter frequency changer.

12.2.1.1 RMS Load Voltage

The rms value V of the output voltage waveform v_o (ωt) in Fig. 12.5 is equal to the rms value of any of the half sinusoid sections. This is shown, in Example 12.1, to be

$$V = V_m \sqrt{\frac{1}{2\pi}\left[\left(\pi - \alpha_p\right) + \sin 2\alpha_p\right]} \tag{12.3}$$

12.2.1.2 Fundamental Load Voltage

The peak fundamental component of the load voltage V_1 is expressed in terms of the a and b terms of its Fourier definition, consistent with the Appendix. If there are m half cycles of input in each half period of the output, then

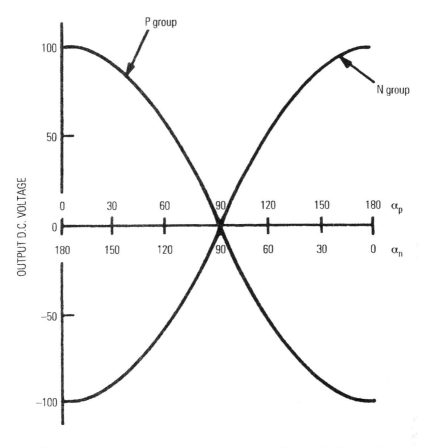

Fig. 4 Relationship of output dc voltage to thyristor firing angles for a dual converter.

$$V_{1_b} = \frac{2V_m}{m\pi}\left[\int_{\alpha_p}^{\pi} \sin\omega_o t\ \sin\left(\frac{\omega_o t}{m}\right)d\omega_o t + \int_{\alpha_p}^{\pi} \sin\omega_o t\ \sin\left(\frac{\omega_o t + \pi}{m}\right)d\omega_o t\right.$$

$$\left. + \int_{\alpha_p}^{\pi} \sin\omega_o t\ \sin\left(\frac{\omega_o t + (m-1)\pi}{m}\right)d\omega_o t\right]$$

$$= \frac{V_m}{m\pi}\sum_{n=1}^{m}\left\{\frac{m}{m-1}\left[\sin\frac{n\pi}{m} - \sin\left(\alpha_p - \frac{\alpha_p + (n-1)\pi}{m}\right)\right]\right.$$

$$\left. + \frac{m}{m+1}\left[\sin\frac{n\pi}{m} + \sin\left(\alpha_p - \frac{\alpha_p + (n-1)\pi}{m}\right)\right]\right\}$$

$$\tag{12.4}$$

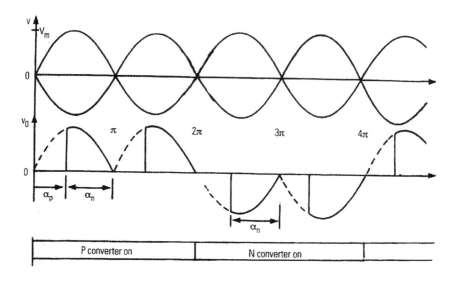

FIG. 5   Voltage waveforms of a single-phase dual converter with resistive load.

Similarly,

$$V_{1_a} = \frac{V_m}{m\pi} \sum_{n=1}^{m} \left\{ \frac{m}{m+1} \left[\cos\frac{n\pi}{m} + \cos\left(\alpha_p + \frac{\alpha_p + (n-1)\pi}{m} \right) \right] \right.$$
$$\left. + \frac{m}{m-1} \left[\cos\frac{n\pi}{m} + \cos\left(\alpha_p - \frac{\alpha_p + (n-1)\pi}{m} \right) \right] \right\}$$

$$(12.5)$$

The total peak fundamental load voltage is then given by

$$V_1 = \sqrt{V_{1_a}^2 + V_{1_b}^2}$$

$$(12.6)$$

Higher harmonic components can be calculated by dividing the parameter m inside the square brackets of Eqs. (12.4) and (12.5) by the appropriate order of harmonic.

The single-phase, center-tapped or push–pull circuit of Fig. 12.6 performs the same circuit function as that of Fig. 12.2. It uses only one-half as many thyristor switches, but they must have twice the voltage rating. Figure 12.6a demonstrates an essential feature of cycloconverters which is that they use inverse-parallel or back-to-back combinations of switches. The circuit of Fig. 12.6b is electrically identical to that of Fig. 12.6a but is topologically rearranged to

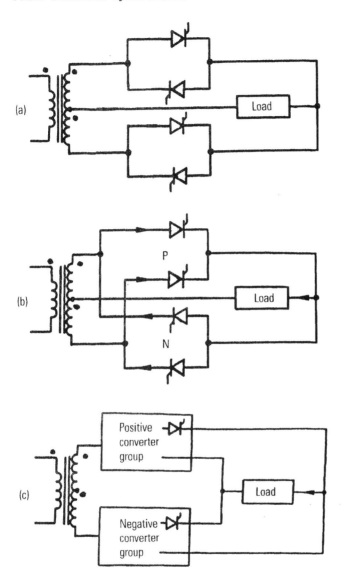

FIG. 6   Single-phase dual converter with center-tapped supply (a) demonstrating the inverse-parallel feature, (b) demonstrating the positive and negative group topology, and (c) generalizing the circuit for a circulating current free cycloconverter.

show the positive (*P*) and negative (*N*) switch groups, also identified in Fig. 12.2. A more general representation of the two converter groups constituting a single-phase dual converter is shown in Fig. 12.6(c).

12.2.2 Single-Phase Dual Converter in the Circulating Current Mode

In the dual bridge converter of Fig. 12.2 it is possible to exactly match the two dc voltages across the load by observing the firing restriction of Eq. (12.1). But the rectified voltages are not purely dc. In addition to the predominant dc components, they also contain ''ripple'' voltages that consist of ac harmonic components, as described in Sec. 4.3. Although the dc components from the two bridges can be precisely matched, the ac ripple voltages are not equal instantaneously and cannot be matched by a direct circuit connection.

For satisfactory circuit operation it is necessary to absorb the ripple voltage and thereby to account for differences between the instantaneous voltages of the two bridges. This can be realized by the connection of a current limiting reactor between the dc terminals of the two bridges, as in Fig. 12.7. Current can now circulate between the two bridge circuits without passing through the load, and the ripple voltages fall across the reactor coils, not across the load. In the presence of such interphase or intergroup reactors, the switches from both bridges can operate, and current can flow simultaneously. If the load current remains continuous, the average (dc) value of the load voltage can be smoothly varied from positive maximum to negative maximum.

Any single-phase arrangement is essentially of a two-pulse nature. Every control cycle involves two firing pulses in the circuit configuration, whether it

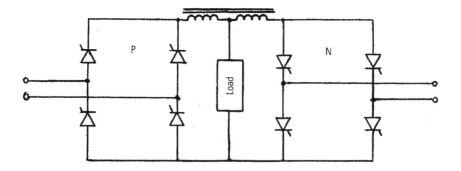

FIG. 7 Single-phase, full-wave dual converter with interphase reactor to facilitate circulating current operation.

is the dual-converter bridge or the center-tapped, midpoint connection. With two-pulse control the ripple voltage level of a single-phase cycloconverter is relatively large, requiring the necessity of large smoothing reactors.

In addition, the distorted input current to a single-phase dual converter contains subharmonics of the supply frequency. With large rated loads this may not be acceptable to the electricity supply authority.

For these reasons single-phase supplied cycloconverters are not greatly used. More versatile control is found to be available via the use of multiphase supplies that, in effect, increase the pulse number. The ideal situation, of sinusoidal output current without filtering, would require an infinitely large pulse number.

12.2.3 Examples on Single-Phase Converters

Example 12.1   Show that the rms value V of a waveform of the form v_o (ωt) in Fig. 12.5 is given by Eq. (12.3).

The rms value of a function is independent of polarity and is equal to the rms value of the smallest repetitive section. In Fig. 12.5

$$v_o(\omega t) = V_m \sin \omega t \Big|_{\alpha_p}^{\pi}$$

but, in general,

$$V^2 = \frac{1}{\pi} \int_{\alpha_p}^{\pi} v_o^2(\omega t)\, d\omega t$$

Substituting v_o (ωt) into the rms equation gives

$$V^2 = \frac{1}{\pi} \int_{\alpha_p}^{\pi} V_m^2 \sin^2(\omega t)\, d\omega t$$

$$= \frac{V_m^2}{\pi} \int_{\alpha_p}^{\pi} \frac{1 - \cos 2\omega t}{2}\, d\omega t$$

$$= \frac{V_m^2}{2\pi} \left[\left(\omega t - \frac{\sin 2\omega t}{2} \right) \right]_{\alpha_p}^{\pi} d\omega t$$

$$= \frac{V_m^2}{2\pi} \left(\pi - \alpha_p - \frac{\sin 2\pi}{2} + \frac{\sin 2\alpha_p}{2} \right)$$

$$= \frac{V_m^2}{2\pi} \left[(\pi - \alpha_p) + \frac{1}{2} \sin 2\alpha_p \right]$$

Taking the square roots of each side

$$V = V_m \sqrt{\frac{1}{2\pi}\left[(\pi - \alpha_p) + \frac{1}{2}\sin 2\alpha_p\right]}$$

Example 12.2   When operating as a cycloconverter a single-phase, circulating current free dual converter with input 110 V, 60 Hz is used to produce an output voltage waveform of the general form (Fig. 12.5). The output frequency is one-third of the input frequency, and $\alpha_p = 45°$. Calculate the amplitude of the fundamental current.

From Eq. (12.4), with $m = 3$,

$$V_{1_b} = \frac{110\sqrt{2}}{3\pi} \sum_{n=1}^{3} \left\{ \frac{3}{2}\left[\sin\frac{n\pi}{3} - \sin\left(\frac{\pi}{4} - \frac{\frac{\pi}{4} + (n-1)\pi}{3}\right)\right]\right.$$

$$\left. + \frac{3}{4}\left[\sin\frac{n\pi}{3} + \sin\left(\frac{\pi}{4} - \frac{\frac{\pi}{4} + (n-1)\pi}{3}\right)\right]\right\}$$

$$V_{1_a} = \frac{110\sqrt{2}}{3\pi} \sum_{n=1}^{3} \left\{ \frac{3}{4}\left[\cos\frac{n\pi}{3} + \cos\left(\frac{\pi}{4} + \frac{\frac{\pi}{4} + (n-1)\pi}{3}\right)\right]\right.$$

$$\left. + \frac{3}{2}\left[\cos\frac{n\pi}{3} + \cos\left(\frac{\pi}{4} - \frac{\frac{\pi}{4} + (n-1)\pi}{3}\right)\right]\right\}$$

The sections of these equations are evaluated in Table 12.1

$$V_{1_b} = \frac{110\sqrt{2}}{3\pi}\left(\frac{3\sqrt{3}}{2} + \frac{3}{2}\right) = 67.6 \text{ V}$$

$$V_{1_a} = \frac{110\sqrt{2}}{3\pi}\left(\frac{3\sqrt{3}}{2} - 3\right) = -6.64 \text{ V}$$

$$V_1 = \sqrt{V_{1_a}^2 + V_{1_b}^2} = \sqrt{67.6^2 + 6.64^2} = 67.92 \text{ V}$$

The rms value of the fundamental load voltage is therefore

TABLE 12.1

n	$\sin\dfrac{n\pi}{3}$	$\cos\dfrac{n\pi}{3}$	$\dfrac{\pi/4+(n-1)\pi}{3}$	$\dfrac{\pi}{4}+\dfrac{\pi/4+(n-1)\pi}{3}$	$\dfrac{\pi}{4}-\dfrac{\pi/4+(n-1)\pi}{3}$
1	$\sqrt{\dfrac{3}{2}}$	$\dfrac{1}{2}$	$\dfrac{\pi}{12}$	$\dfrac{\pi}{3}=60°$	$\dfrac{\pi}{6}=30°$
2	$\sqrt{\dfrac{3}{2}}$	$-\dfrac{1}{2}$	$\dfrac{5\pi}{12}=75°$	$\dfrac{2\pi}{3}=120°$	$-\dfrac{\pi}{6}=-30°$
3	0	-1	$\dfrac{9\pi}{12}=135°$	$\pi=180°$	$-\dfrac{\pi}{2}=-90°$

$$V_{1(\text{rms})} = \frac{67.92}{\sqrt{2}} = 48 \text{ V}$$

12.3 THREE-PHASE TO SINGLE-PHASE CYCLOCONVERTERS

12.3.1 Three-Phase, Half-Wave (Three-Pulse) Cycloconverter Circuits

12.3.1.1 Circulating Current Free Circuit

The three-phase, four-wire, half-wave controlled rectifier circuit was extensively described in Chapter 5.

In general the average output voltage of a p pulse rectifier is given by

$$E_{av} = E_m \frac{\sin(\pi/p)}{\pi/p} \cos\alpha \tag{12.7}$$

For the three-phase, four-wire, half-wave circuit $p = 3$ and

$$E_{av} = \frac{3\sqrt{3}}{2\pi} E_m \cos\alpha \tag{12.8}$$

where E_m is the peak value of the supply voltage per phase. For a single-phase supply $p = 2$ and Eq. (12.7) reduces to Eq. (12.2).

More comprehensive control of the load voltage waveform can be obtained by a corresponding dual converter of the type shown in Fig. 12.8. The positive (P) and negative (N) bridges must operate alternately without overlap in the

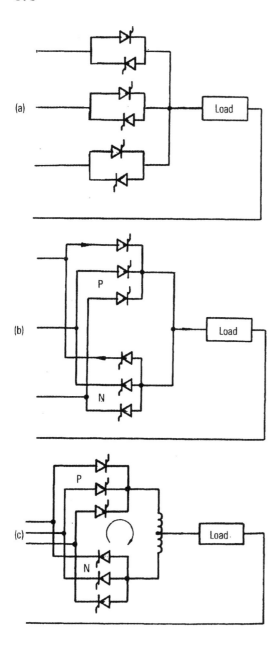

FIG. 8 Three-phase-to-single-phase, half-wave (three-pulse) cycloconverter (a) demonstrating the inverse-parallel feature, (b) demonstrating the positive (*P*) and negative (*N*) groups, and (c) showing the midpoint circulating current connection.

electrically identical circuits of Fig. 12.8a and b to result in circulating current free operation. The output is then no longer restricted to the function of rectification with unidirectional output current. Load current can flow in both directions so that ac operation is realizable and regeneration action is possible.

The most important feature of dual converter ac operation is that the firing angles of the converter switches can be time varied, cycle by cycle. If the time-varying firing angle is described as $\alpha(t)$, then the output voltage (no longer dc) has a peak amplitude V_o given by

$$V_o = \frac{3\sqrt{3}}{2\pi} E_m \cos\alpha(t) \tag{12.9}$$

When the firing angle $\alpha(t)$ is time varied in a periodic manner, the effect is to produce time-phase modulation, usually just called *phase modulation*. This permits the output (modulated) voltage of the converter to be controlled in both amplitude and frequency independently. The supply frequency acts as a carrier frequency, and the output frequency ω_o is also the modulating frequency.

A form of modulating function commonly used is

$$\alpha(t) = \pm \cos^{-1}(M \cos \omega_0 t) = \pm \cos^{-1}(M \cos 2\pi f_0 t) \tag{12.10a}$$

or, since, $\cos^{-1}x = \pi/2 - \sin^{-1}x$, it may be written in the form

$$\alpha(t) = \pm \sin^{-1}(M \sin \omega_0 t) = \pm \sin^{-1}(M \sin 2\pi f_0 t) \tag{12.10b}$$

The inverse cosine function limits the range of $\alpha(t)$ and can be taken to give $90° \le \alpha(t) \le 180°$, necessary for rectifier operation. The same technique can be employed by using the inverse cosine of a sine function. In Eqs. (12.10) the parameter M is the modulation ratio (ratio peak modulating voltage/peak carrier voltage) described in Chapter 11, and f_0 is the desired output frequency.

For example, to convert a three-phase supply into a single-phase 20-Hz output, it is necessary to use the modulating function (i.e., firing angle)

$$\alpha(t) = \pm \cos^{-1}(M \cos 40\pi t) \tag{12.11a}$$

or

$$\alpha(t) = \pm \sin^{-1}(M \sin 40\pi t) \tag{12.11b}$$

Equations (12.11) describe a pair of triangular modulating functions, attributable, respectively, to the P and N bridges of Fig. 12.8b, with periodicity 0.50 s. The output voltage waveform Fig. 12.9, with $M = 1$, is highly distorted, but its fundamental (20Hz) component is clearly discernible.

It is inherent in cycloconverter operation that the output voltage, made up of sections of sine waves, is highly distorted. The amount of distortion increases as the ratio output frequency/input frequency (i.e., modulating frequency/carrier

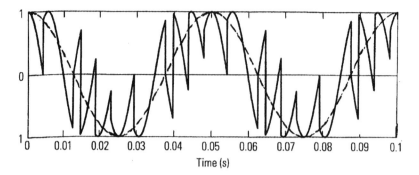

FIG. 9   Normalized output voltage waveform (θ/E_m) of a three-phase to single-phase (three-pulse) circulating current free cycloconverter with inductive load. $M = 1, f_{in} = 60$ Hz, $f_o = 20$ Hz (fundamental) [32].

frequency) increases. In terms of the PWM frequency ratio, defined in Eq. (8.15), it can be said that minimum output voltage distortion requires cycloconversion using the maximum possible value of frequency ratio f_{in}/f_o or ω_{in}/ω_o. A higher value of frequency ratio ω_{in}/ω_o can be realized by increase of the converter circuit pulse number. As in single-phase-to-single-phase conversion, control problems arise when the load current is discontinuous (e.g., with resistive load).

12.3.1.2 Circulating current circuit

To alleviate the control problems arising with discontinuous load current an inter-group reactor may be connected, as in Fig. 12.8c. The firing angles of the two converters are regulated so that both bridges conduct simultaneously and a con-trolled amount of current I_c is allowed to circulate between them unidirectionally. The ideal cosine relationship of Fig. 12.4 is preserved so that the output average voltages of the two bridges are identical, satisfying Eq. (12.1). Although the reactor absorbs the ripple voltages, as described in Sec. 12.2.2, its size and cost represent significant disadvantages. Figure 12.10 shows waveforms for the no-load operation of the circuit of Fig. 12.8c at a particular setting of firing angles. The reactor voltage (Fig. 12.10d) has a time-average value of zero, and the circu-lating current Fig. 12.10e is unidirectional. With a highly inductive load imped-ance both of the bridge circuits deliver (ideally) sinusoidal currents with zero offsets. The dc components (i.e., zero offsets) then cancel in the load to give a substantially sinusoidal current of the modulating frequency, as shown by the dotted line of Fig. 12.10c.

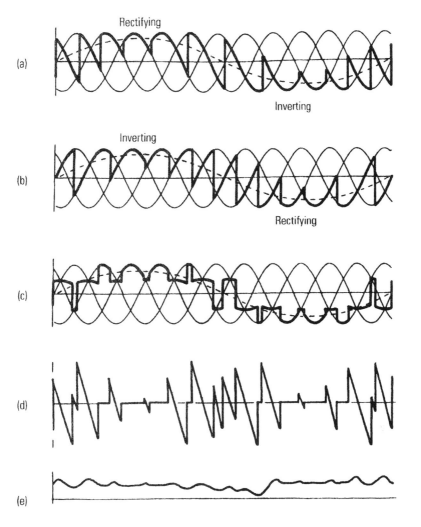

FIG. 10 Voltage waveforms for three-phase to single-phase circulating current cyclo-converter. No load. $M = 0.87$, $p = 1:4$, $\alpha_p = 30°$, $\alpha_N = 150°$: (a) P group voltages, (b) N group voltages, (c) total load voltage, (d) reactor voltage, and (e) circulating current [3].

12.3.2   Three-Phase, Full-Wave (Six-pulse) Cycloconverter Circuits

Two three-phase, full-wave controlled bridge rectifier circuits of the form of Fig. 7.1 can be used to form a dual converter, shown in Fig. 12.11. The two separate P and N converters are shown in Fig. 12.11a and redrawn to show the back-to-back nature of the connection in Fig. 12.11b, representing circulating current free operation. Figure 12.11b is seen to be the full-wave version of the half-wave circuit of Fig. 12.8a. Similarly, the full-wave circulating current circuit of Fig. 12.11c is the full-wave version of the half-wave circuit of Fig. 12.8c.

For the three-phase, full-wave converter $p = 6$, and the peak load voltage may be inferred from Eq. (12.4) to be

$$V_o = \frac{3\sqrt{3}}{\pi} E_m \cos \alpha \tag{12.12}$$

This is seen to be twice the value of the half-wave, three-pulse converter given in Eq. (12.5).

12.3.3   Examples on Three-Phase to Single-Phase Converters

Example 12.3   A three-phase-to-single-phase, three-pulse cycloconverter delivers power to a load rated at 200 V, 60 A, with power factor 0.87 lagging. Estimate the necessary input voltage and power factor.

From Eq. (12.9) the peak value of the required input voltage is given by

$$200\sqrt{2} = \frac{3\sqrt{3}}{2\pi} E_m \cos \alpha(t)$$

To accommodate the worst case, when $\cos \alpha = 1$,

$$E_m = \frac{200 \times \sqrt{2} \times 2\pi}{3\sqrt{3}} = 342 \text{ V}$$

so that the rms supply voltage E is given by

$$E = \frac{342}{\sqrt{2}} = 242 \text{ V/phase}$$

The rms value of the specified single-phase load current is 60 A. If this is shared equally between the three input phases, each input line (assuming sinusoidal operation) has an rms current

$$I_{in} = \frac{I_L}{\sqrt{3}} = \frac{60}{\sqrt{3}} = 34.64 \text{ A/phase}$$

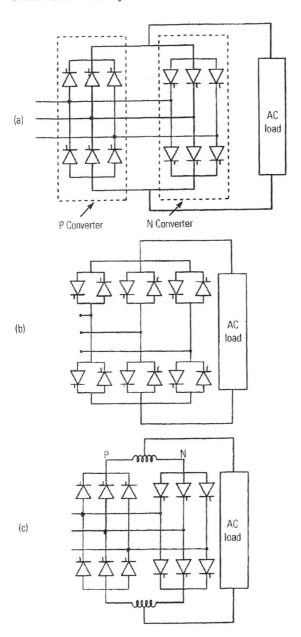

(a)

P Converter   N Converter

(b)

(c)

P   N

FIG. 11   Three-phase-to-single-phase, full-wave (six-pulse) cycloconverter: (a) double-bridge, noncirculating current configuration, (b) inverse-parallel topological form of (a), and (c) double-bridge, circulating current configuration.

The load power is seen to be

$$P_L = 200 \times 60 \times 0.87 = 10.44 \text{ kW}$$

Assuming that the input power is shared equally by the three phases and is equal to the output power,

$$P_{in} = \frac{P_L}{3} = \frac{10.44}{3} = 3.48 \text{ kW/phase}$$

The input power factor is then

$$PF = \frac{P_{in}}{EI_{in}} = \frac{3480}{242 \times 24.64} = 0.415 \text{ lagging}$$

Note that this value of input power factor is the maximum possible value, with $\alpha = 0$. If α is increased, the power factor will decrease.

Example 12.4   A three-phase-to-single-phase, full-wave cycloconverter of the type shown in Fig. 12.11a supplies a series R-L load, where $R = 1.2\ \Omega$ and $\omega_o L = 1.6\ \Omega$ at the specified output of 110 V (rms) at frequency 4 Hz. If the input is 220-V line, 60 Hz, calculate (1) the necessary firing angle, (2) load current, (3) load power, (4) input current, (5) thyristor current, and (6) input power factor.

The specified output voltage wave is

$$v_o(\omega_o t) = \sqrt{2 V_{o(\text{rms})}} \sin \omega_o t$$
$$= \sqrt{2} \times 110 \sin 8\pi t$$

But the peak load voltage $\sqrt{2 V_{o(\text{rms})}}$ is defined in terms of Eq. (12.12), where E_m is the peak phase voltage

$$V_o = \frac{3\sqrt{3}}{\pi} E_m \cos \alpha$$

1. Therefore,

$$\sqrt{2} \times 110 = \frac{3\sqrt{3}}{\pi} \times \frac{220}{\sqrt{3}} \times \cos \alpha$$

giving $\cos \alpha = 0.74$ or $\alpha = 42.2°$.

2. The load impedance to currents of 4-Hz frequency is

$$Z = \sqrt{1.2^2 + 1.6^2} = 2\ \Omega$$

The rms load current is therefore $I_o = 110/2 = 55$ A at a load power factor

$$PF_L = \frac{R}{Z} = \frac{1.2}{2} = 0.6 \text{ lagging}$$

3. The load power is therefore

$$P_L = V_{o(\text{rms})}I_o \cos\phi_o = 110 \times 55 \times 0.6 = 3630 \text{ W}$$

4. With symmetrical operation the load current is supplied equally by the three input currents. Each rms input current is therefore

$$I_{in} = \frac{I_o}{\sqrt{3}} = \frac{55}{\sqrt{3}} = 31.76 \text{ A}$$

5. If each presumed sinusoidal line rms current is divided equally between two thyristors, in Fig. 12.11b, then

$$I_{thyr} = \frac{I_{in}}{\sqrt{2}} = \frac{31.76}{\sqrt{2}} = 22.5 \text{ A (rms)}$$

6. It is assumed that the converter is lossless so that the output power is equal to the input power, giving

$$P_{in} = P_L = 3630 \text{ W}$$

The input power factor is

$$PF_{in} = \frac{P_{in}}{3EI_{in}} = \frac{3630}{3 \times \left(220/\sqrt{3}\right) \times 31.76} = 0.3 \text{ lagging}$$

The input power factor in this case, is only one-half the value of the load power factor.

12.4 THREE-PHASE TO THREE-PHASE CYCLOCONVERTER

12.4.1 Load Voltage Waveforms of the Half-Wave (Three-Pulse) Cycloconverter

The three-phase, four-wire cycloconverter circuits of Fig. 12.12 are of three-pulse nature and use 18 semiconductor switches. They each represent three circuits of the single-phase forms of Fig. 12.8. With balanced three-phase loads the neutral connection can be omitted. Most practical cycloconverter circuits consist of combinations of the basic three-pulse group, connected so as to realize six-pulse or 12-pulse performance. The standard method of determining the switch firing pattern is called the *cosine crossing method* and is similar to the principle described in Chapters 8 and 11 for pulse-width-modulated rectifiers and inverters. Detailed mathematical derivations of the output voltage waveform and the input current waveform for multipulse cycloconverters are very lengthy and complex. The interested reader is referred to the excellent and authoritative standard text by Pelly [3].

If the load current is continuous and the modulating function of Eq. (12.10b) is used, then

$$\sin \alpha(t) = M \sin \omega_o t \tag{12.13}$$

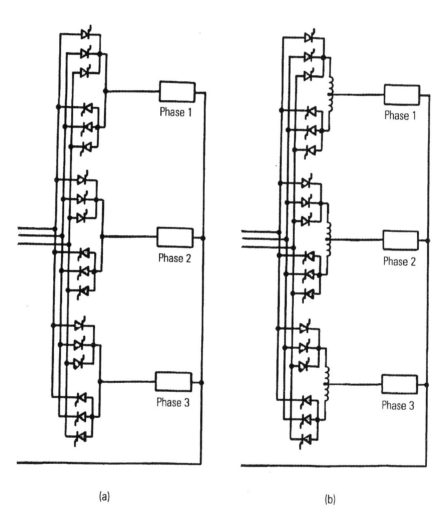

(a)   (b)

FIG. 12 Three-phase-to-three-phase, half-wave (3 pulse) cycloconverter circuits: (a) circulating current free and (b) circulating current connection.

and this implies the cosine crossing method of switch firing control. For the three-pulse circulating current circuit of Fig. 12.12b it is found that the output voltage may be written [3]

$$v_o = \frac{3\sqrt{3}E_m}{2\pi} \left[M \sin \omega_o t + \frac{1}{2} \sum_{p=1}^{\infty} \left([A][B] + [C][D] \right) \right]$$

(12.14)

where

$$[A] = \sum_{n=0}^{2n=3(2p-1)+1} \left(\frac{a_{[3(2p-1)-1]2n}}{3(2p-1)-1} + \frac{a_{[3(2p-1)+1]2n}}{3(2p-1)+1} \right)$$

$$[B] = \sin\left[3(2p-1)\omega_{in}t + 2n\omega_o t \right] + \sin\left[3(2p-1)\omega_{in}t - 2n\omega_o t \right]$$

$$[C] = \sum_{n=0}^{2n+1=6p+1} \left(\frac{a_{(6p-1)(2n+1)}}{6p-1} + \frac{a_{(6p+1)(2n+1)}}{6p+1} \right)$$

$$[D] = \sin\left[6p\omega_{in}t + (2n+1)\omega_o t \right] - \sin\left[6p\omega_{in}t - (2n+1)\omega_o t \right]$$

The first term in Eq. (12.14) is the term representing the desired fundamental output frequency ω_o. In Fig. 12.9, for example, this would be the 20-Hz component, shown dotted. When the modulation index M is unity in Eq. (12.14) the higher harmonic terms in the summation section include values

$$\sin(3\omega_{in}t \pm 2\omega_o t), \sin(3\omega_{in}t \pm 4\omega_o t), \sin(6\omega_{in}t \pm 5\omega_o t), \text{etc.}$$

For the three-pulse circulating current free circuit, (Fig. 12.12a), all of the terms of Eq. (12.14) are present plus a further series of higher harmonic terms. The harmonic component frequencies f_h expressed as multiples of the input frequency f_{in} may be written in terms of output frequency f_o and integers p and n for two separate harmonic families

$$\frac{f_h}{f_{in}} = 3(2p-1) \pm 2n \frac{f_o}{f_{in}}$$

(12.15a)

and

$$\frac{f_h}{f_{in}} = 6p \pm (2n+1) \frac{f_o}{f_{in}}$$

(12.15b)

where $1 \leq p \leq \infty$ and $0 \leq n \leq \infty$.

With $p = 1, n = 0$ in Eq. (12.15b), for example, it is seen that the harmonic frequencies are $f_h = 5f_o$ and $f_h = 7f_o$. The lowest order harmonic frequency is

$f_h = 3f_o$ when $p = 1$ and $n = 1$ in Eq. (12.15b) and $p = 1$ and $n = 0$ in Eq. (12.15a).

12.4.2 Load Voltage Waveforms of the Full-Wave (Six-Pulse) Cycloconverter

A six-pulse cycloconverter can be implemented by doubling the number of switches from 18 to 36. Each load phase current is shared equally between two switch groups, resulting in symmetrical three-phase operation. The midpoint or push–pull center-tapped circuit of Fig. 12.13 facilitates circulating current operation. This requires a double-star supply of two separate secondary windings on the input transformer, feeding the six input terminals in two separate three-phase groups.

The output voltages v_0 with six-pulse circulating current operation may be written [3]

$$v_o = k\frac{3\sqrt{3}E_m}{2\pi}\left(M\sin\omega_o t + \frac{1}{2}\sum_{p=1}^{\infty}\sum_{n=0}^{2n+1=6p+1}[A][B] \right) \tag{12.16}$$

where

$$[A] = \left[\frac{a_{(6p-1)(2n+1)}}{6p-1} + \frac{a_{(6p+1)(2n+1)}}{6p+1} \right]$$

$$[B] = \sin\left[6p\omega_{in}t + (2n+1)\omega_o t \right] - \sin\left[6p\omega_{in}t - (2n+1)\omega_o t \right]$$

$k = 1$ for the six-pulse midpoint circuit

$k = 2$ for the six-pulse bridge circuit

If a three-phase-to-three-phase bridge connection (Fig. 12.14) is used, only one input transformer secondary winding is required. For this noncirculating current circuit with continuous load current, the output voltage for phase 1 in Fig. 12.14 is given in Fig. 12.15 for three different types of load impedance. The output voltage expression then combines all of the terms of Eq. (12.16) plus a large family of additional higher harmonic terms. The large number of unwanted harmonic terms typically includes sum and difference combinations of the input frequency ω_{in} and output frequency ω_o, such as $3\omega_{in} \pm 2\omega_o$, $6\omega_{in} \pm 5\omega_o$ etc. As in three-pulse operation, the load voltage with the circulating current free mode contains harmonics of orders

$$\frac{f_h}{f_{in}} = 6p(2n+1)\frac{f_o}{f_{in}} \tag{12.17}$$

where $1 \le p \le \infty$ and $0 \le n \le \infty$.

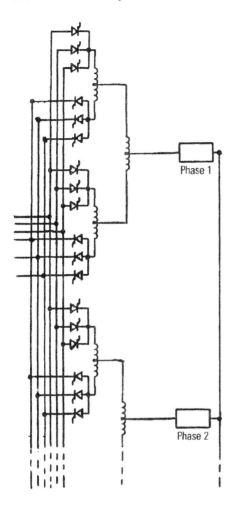

Fɪɢ. 13 Three-phase, full-wave (six pulse) center-tapped, circulating current cycloconverter.

The $3(2p - 1) \pm 2n(f_o / f_{in})$ group harmonics associated with three-pulse operation in Eq. (12.15a) are not present in six-pulse operation. For 12-pulse operation in the circulating current free mode (not considered in this text) the frequencies of the harmonic components of the output voltage are found to be

$$\frac{f_h}{f_{in}} = 12p \pm (2n+1)\frac{f_o}{f_{in}} \tag{12.18}$$

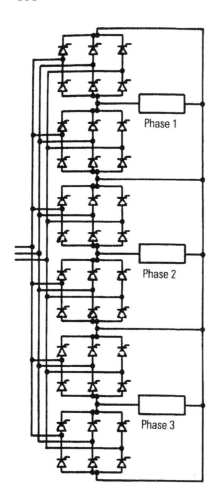

FIG. 14   Three-phase (six pulse) circulating current free bridge cycloconverter [11].

12.4.3   Input Current, Power, and Displacement Factor

Expressions for the input current waveforms to both three-pulse and six-pulse cycloconverters are, if anything, even more lengthy and complex than corresponding expressions for the load voltage waveforms. Theoretical waveforms are shown in Fig. 12.16 for the input currents with three different loads in the circulating

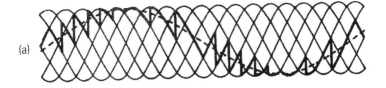

(a)

(b)

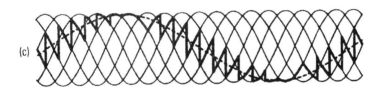

(c)

FIG. 15 Load phase voltage for three-phase-to-three-phase (six-pulse) bridge cycloconverter. $M = 1, f_0 / f_{in} = 1/3$. (a) R load, $\cos\phi = 1$. (b) R-L load, $\cos\phi = 0.5$ lag. (c) R-
C load, $\cos\phi = 0.5$ lead.

current free mode of operation. With resistive load the output displacement angle
ϕ_o is zero, and the output displacement factor $\cos\phi_o$ is unity. But the fundamental
component of the input current (shown dotted) lags its phase voltage by about
$32°$ giving an input displacement factor $\cos\phi_{in} = 0.85$ lagging. For both the
inductive and capacitive loads shown (Fig. 12.16b and c), the load impedance
displacement factor $\phi_o = 60°$. In both cases the fundamental component of the
input current lags its respective phase voltage by more than $60°$, so that $\cos\phi_{in}$
$< \cos\phi_o$. In other words, the power factor of the total burden of cycloconverter
plus load is lower (i.e., worse) than that of the load alone.

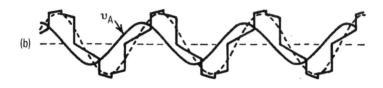

FIG. 16   Input phase voltage V and current i for three-phase-to-three-phase (six-pulse) bridge cycloconverter. $M = 1, f_o/f_{in} = 1/3$. (a) R load, $\cos\phi = 1$. (b) R-L load, $\cos\phi = 0.5$ lag. (c) R-C load, $\cos\phi = 0.5$ lead [3].

12.4.4   Components of the Fundamental Input Current

12.4.4.1 In-Phase Component (I_p)

In all cases the fundamental input current lags the phase supply voltage. The rms in-phase component of the fundamental current I_p, can be can be expressed in terms of the rms output current I_o by [3]

$$I_p = qsM \frac{\sqrt{3}}{2\pi} I_o \cos\phi_o \qquad\qquad (12.19)$$

where

q = number of output phases
s = number of three-pulse switch groups in series per phase (integer)
M = modulation ratio (peak modulating voltage/E_m)

ϕ_o = output (load) displacement angle

I_o = total rms output (load) current (noting that this current is usually substantially sinusoidal)

For the three-phase-to-three-phase systems, $q = 3$ so that

$$I_p = sM\frac{3\sqrt{3}}{2\pi}I_o \cos\phi_o \tag{12.20}$$

For the six-pulse circuits of Fig. 12.13 and Fig. 12.14, parameter $s = 1$ or 2 in Eq. (12.20).

12.4.4.2 Quadrature Lagging Component (I_Q)

Due both to any reactive load impedance and to the switching nature of the cycloconverter, the input current always lags its respective phase voltage. This corresponds with the description of power factor for the controlled, three-phase bridge rectifier in Sec. 7.1.3. For three-phase-to-three-phase systems the rms value of the quadrature component of the fundamental input current may be written [3]

$$I_Q = 2s \times \frac{3\sqrt{3}}{\pi^2} \times I_o \sum_{n=0}^{\infty} \frac{-a_{1_{2n}} \cos 2n\phi_o}{(2n-1)(2n+1)} \tag{12.21}$$

Parameter $a_{1_{2n}}$ in Eq. (12.21) represents the values of an infinite series of negative harmonic terms defined by

$$\cos\left[(6p+1)\sin^{-1} M \sin\omega_o t\right] = a_{(6p+1)_0} + a_{(6p+1)_2} \cos 2\omega_o t + \cdots$$
$$+ a_{(6p+1)_{2n}} \cos 2n\omega_o t + \cdots \tag{12.22}$$

where

$$a_{(6p\pm1)_0} = \frac{1}{2\pi}\int_0^{2\pi} \cos\left[(6p\pm1)\sin^{-1} M \sin\omega_o t\right] d\omega_o t$$

$$a_{(6p\pm1)_{2n}} = \frac{1}{\pi}\int_0^{2\pi} \cos\left[(6p\pm1)\sin^{-1} M \sin\omega_o t\right](\cos 2n\omega_o t)\, d\omega_o t$$

Some values of the coefficients $a_{1_{2n}}$ are given in Example 12.6. For instance, with $M = 1$ and $n = 0$, it is found that $a_{1_0} = 0.637$. In general, the ratio of the in-phase and quadrature components I_p/I_Q is independent of the pulse number, the number of output phases, and the frequency ratio f_o/f_{in}. These proportion values depend only on the load displacement factor $\cos\phi_o$ and the output voltage ratio M.

The rms value of the total fundamental input current component is therefore

$$I_{1in} = \sqrt{I_p^2 + I_Q^2} \qquad (12.23)$$

The input displacement factor is, correspondingly,

$$\cos\phi_{in} = \frac{I_p}{I_{1in}} = \frac{I_p}{\sqrt{I_p^2 + I_Q^2}} \qquad (12.24)$$

The total rms input current I_{in} contains the two components I_p and I_Q of the fundamental harmonic plus the summation of the higher harmonic terms I_h:

$$I_{in}^2 = I_p^2 + I_Q^2 + \sum_{h=2}^{\infty} I_h^2 \qquad (12.25)$$

With a three-phase output the rms input current I_{in} does not have any fixed analytical relationship with the rms output current I_o.

12.4.5 Distortion Factor of the Input Current Waveform

The distortion factor defined Eq. (2.26) and used throughout this book is equally true for cycloconverters.

$$\text{Current distortion factor} = \frac{I_{1in}}{I_{in}} \qquad (12.26)$$

With a balanced three-phase load the input current distortion is determined largely by the circuit switching action rather than by load reactance effects. Input currents with low distortion are available if $p \geq 6$.

12.4.6 Input Power (P_{in})

It is a basic property of ac electric circuits that real power in watts can only be transferred by the combination of voltage and current components of the same frequency. Since the supply voltage is of single frequency, it is possible to transfer power from the supply to the converter only via the fundamental (supply frequency) component of current. For a three-phase supply

$$P_{in} = 3EI_{1in} \cos \angle_{I_{1in}}^{E} = 3EI_{1in} \cos \phi_{in} \qquad (12.27)$$

where E is the rms phase voltage. But the in-phase component of the fundamental current I_p is given by

$$I_p = I_{1in} \cos \phi_{in}$$ (12.28)

Combining Eqs. (12.27) and (12.28) gives

$$P_{in} = 3EI_p$$ (12.29)

If Eq. (12.29) is now combined with Eq. (12.20), it is seen that

$$P_{in} = 3sM \frac{3\sqrt{3}}{2\pi} EI_o \cos\phi_o$$ (12.30)

If the switching action of the cycloconverter thyristors causes negligible loss, then the output power P_o equals the input power:

$$P_o = P_{in}$$ (12.31)

An alternative interpretation is to assume sinusoidal output current so that

$$P_o = 3E_oI_o \cos\phi_o$$ (12.32)

The rms value of the desired output frequency voltage E_o may be therefore obtained by equating Eqs. (12.30) and (12.32)

$$E_o = sM \frac{3\sqrt{3}}{2\pi} E$$ (12.33)

12.4.7 Power Factor

The power factor is given by

$$PF = \frac{P_{in}}{S_{in}} = \frac{P_{in}}{EI_{in}}$$ (12.34)

Combining Eqs. (12.25)(12.30), and (12.34) gives a very cumbersome expression that is not helpful in calculating the power factor and is not reproduced here.

12.4.8 Worked Examples

Example 12.7   A three-phase-to-three-phase circulating current free, six-pulse cycloconverter is connected as a bridge, as in Fig. 12.14. The input supply is rated at 440 V, 60 Hz. The (presumed sinusoidal) output current is 100 A when the modulation ratio M is unity. Calculate the load power and the input displacement factor when the phase angle of the balanced load impedance is $\phi_o = 0°$, representing resistive load.

In the circuit of Fig. 12.14 there are effectively two three-pulse switch groups in series per phase so that $s = 2$ in Eqs. (12.19) and (12.20). With $M = 1$, the rms in-phase component I_p of the fundamental supply current is

$$I_p = 2 \times \frac{3\sqrt{3}}{2\pi} \times 100 \times \cos \phi_o = 165.4 \cos \phi_o$$

For $\phi_o = 0$, $\cos\phi_o = 1.0$ and $I_p = 165.4$ A. The rms quadrature component of the fundamental current I_Q is given in Eq. (12.21) with the summation term evaluated in Table 12.2, for $M = 1$

$$I_Q = 2 \times 2 \times \frac{3\sqrt{3}}{\pi^2} \times 100 \times 0.652 = 137.3 \text{ A}$$

The fundamental component of the input current I_{1in} is then given by

$$I_{1in} = \sqrt{I_p^2 + I_Q^2}$$

At $\phi_o = 0$,

$$I_{1in} = \sqrt{165.4^2 + 137.3^2} = \sqrt{27237 + 18851} = \sqrt{46208} = 214.96 \text{ A}$$

The per-phase input voltage is $E = 440/\sqrt{3} = 254$ V. The input power to the cycloconverter is then, from Eq. (12.27),

$$P_{in} = 3 \times 254 \times 214.96 \times \cos \phi_{in} = 148.91 \times 10^3 \times \cos \phi_{in}$$

But the input power can also be obtained from Eq. (12.29),

$$P_{in} = 3EI_p = 3 \times 254 \times 165.4 = 126 \text{ kW}$$

The input displacement factor is therefore

$$\text{Displacement factor} = \cos \phi_{in} = \frac{126}{163.81} = 0.77 \text{ lagging}$$

The load power P_o is equal to the input power P_{in}

$$P_o = P_{in} = 126 \text{ kW}$$

Example 12.8   Calculate the order of the lowest higher harmonic terms in the output voltage wave of a six-pulse, circulating current free cycloconverter if $f_{in} = 60$ Hz and the frequency ratio $\omega_o / \omega_{in} = 1/4$.

For a circuit of the form of Fig. 12.14 the output voltage harmonic order is defined by Eq. (12.17).

$$f_h = 6pf_{in} \pm (2n + 1)f_o$$

where $1 \le p \le \infty$ and $0 \le n \le \infty$, $f_{in} = 60$Hz, and $f_o = 15$Hz. Some values are given in Table 12.3

Example 12.9   Evaluate some of the coefficients a_{12n} in Eq. (12.21) for $M = 1, 0.8, 0.5, 0.1$ and $n = 0, 1, 2, 3$.

TABLE 12.2

M	a_{1_0} ($n = 0$)	a_{1_2} ($n = 1$)	a_{1_4} ($n = 2$)	a_{1_6} ($n = 3$)
1	0.637	0.424	−0.085	0.036
0.8	0.813	0.198	−0.012	0.002
0.5	0.934	0.067	−0.001	0
0.3	0.977	0.023	0	0
0.1	0.997	0.003	0	0

Using the terms in Eq. (12.22), it is found that the coefficients shown in Table 12.2 can be calculated. Substituting values from the Table 12.2 into Eq. (12.21) gives coefficients for the summation term (using a square-law relationship) in Table 12.4. For $M \leq 0.6$ and $n > 1$ the higher harmonic terms are negligibly small. With reactive load impedance the summation terms are subject to sign changes due to the $\cos 2n\phi_n$ term in Eq. (12.21).

PROBLEMS

12.1   For the waveform v_o (ωt) in Fig. 12.5, with peak voltage 163 V, calculate the rms voltage V_o for the values α (a) 30°, (b) 60°, (c) 90°, (d) 120°, (e) 150°.

12.2   Show that the fundamental component of waveform v_o (ωt) in Fig. 12.5 is zero. What is the frequency of the lowest order harmonic?

TABLE 12.3

p	n	f_h (order)	f_h (Hz)
1	0	$6f_{in} \pm f_o$	375, 345
	1	$6f_{in} \pm 3f_o$	405, 315
	2	$6f_{in} \pm 5f_o$	435, 285
	3	$6f_{in} \pm 7f_o$	465, 255
2	0	$f_{in} \pm f_o$	735, 705
	1	$12f_{in} \pm 3f_o$	765, 675
	2	$12f_{in} \pm 5f_o$	795, 645
	3	$12f_{in} \pm 7f_o$	825, 615

TABLE 12.4

M	n = 0	n = 1	n = 2	n = 3	Σ
1	0.637	−0.141	+0.0057	−0.001	0.652
0.5	0.934	−0.026	+0.00067	0	0.934
0.1	0.997	−0.001	0	0	0.997

12.3 A three-phase, four-wire, half-wave rectifier dual converter operated from a 230-V, 60-Hz three-phase supply. Calculate the average value of the load voltage for firing angles (a) 0°, (b) 30°, (c) 60°, (d) 90°.

12.4 A three-phase-to-single-phase cycloconverter delivers an output of 250 V, 70 A to a load of 0.9 power factor lagging. Assume that the switches of the converter are ideal and that the output current is substantially sinusoidal. Calculate the input power factor.

12.5 A full-wave, three-phase to single-phase circulating current free cyclo-converter has an output of 120 V (rms) when the output frequency is 6.667 Hz. The input supply is 220 V, 60 Hz. The load impedance consists effectively of a series R-L circuit with $R = 1\ \Omega$ and $\omega L = 2\ \Omega$ (at the specified output frequency). Calculate (a) firing angle, (b) load current, (c) load power, (d) input current, (e) thyristor switch current, and (f) input displacement factor.

12.6 The switching angle of a cycloconverter is given by the function

$$\alpha(t) = \cos^{-1}(\cos 20\pi t)$$

to convert a three-phase 60-Hz input to a single-phase 10-Hz output. Show that this switching function is double triangular with peak values ± 180° at 0.05 s.

12.7 A three-phase to three-phase, three-pulse, circulating current free cyclo-converter operates with an input frequency ratio $\omega_{in} / \omega_o = 9$. What are typical frequencies of the low-order higher harmonics?

12.8 A three-phase-to-three-phase, full-wave, six-pulse, circulating current free cycloconverter is supplied from 60-Hz mains. If the switching ratio in $\omega_{in} / \omega_o = 9$, calculate typical frequencies of the low-order higher harmonics.

12.9 A three-phase-to-three-phase, six-pulse, circulating current free cyclo-converter is connected as a full-wave bridge. It delivers a current of 120 A to a resistive load when operating with unity modulation ratio. The three-phase supply is rated at 440 V, 60 Hz. Calculate the load power, input power, input displacement factor, and load voltage.

13

Envelope Cycloconverters

As an alternative to the adjustable gate control of phase-controlled cycloconverters described in Chapter 12, the switches can operate continuously, like diodes. This requires that the gate control signal must fully be on during the whole of one half cycle and fully off during the other half cycle. In effect the rectifier valves, usually thyristors, act as on–off switches for the whole of a half cycle. As in phase control, the commutation of the conducting switches occurs naturally. The output voltage waveform follows the profile of the ac supply voltage, as in Fig. 13.1, leading to the name *envelope cycloconverter*. In Fig. 13.1 there are three half cycles of the supply voltage, which defines the frequency ratio ω_{in}/ω_o = 3. As with rectifier circuit operation an increase of pulse number reduces the voltage ripple and the higher harmonic content. In envelope cycloconverter operation it is invariably necessary to use some form of input transformer connection. Conduction in successive phases of the transformer secondary windings has to occur, with respect to the supply voltages, so as to normally provide an output voltage envelope that is approximately a stepped waveform of the desired output frequency.

13.1 SINGLE-PHASE CYCLOCONVERTER OPERATION

In the circuit of Fig. 13.2a the voltage applied across the load can be the full supply voltage of either polarity or some fraction A, depending on the transformer

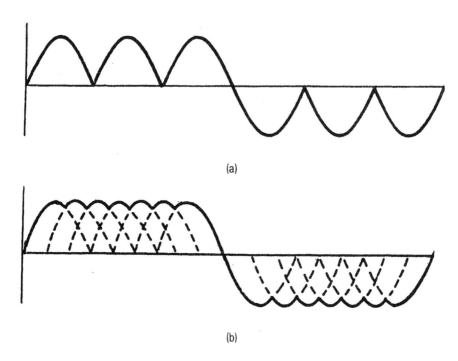

(a)

(b)

FIG. 1 Output voltage waveforms of an envelope cycloconverter, $\omega_o/\omega_{in} = 1/3$: (a) single-phase (two pulse) and (b) three-phase (six pulse).

secondary tap setting. The switches are shown as triacs but can equally well be pairs of inverse-parallel connected silicon controlled rectifiers. With an ideal supply and ideal switches the characteristic waveform of Fig. 13.2(b) can be obtained, having three peaks in an overall half period.

13.2 THREE-PHASE-TO-SINGLE-PHASE CYCLOCONVERTER

Bidirectional switches from each line of a three-phase supply can be combined to form a single-phase output as in Fig. 13.3. This connection is seen to be a bidirectional extension of the various half-wave rectifier circuits of Chapter 5. With a cycloconverter, however, the load voltage is shaped to produce frequency conversion, and the load current is bidirectional to give an ac output.

An important parameter defining the performance is the number of peaks M of the supply voltage in a half period of the output voltage. In Figs. 13.1 and

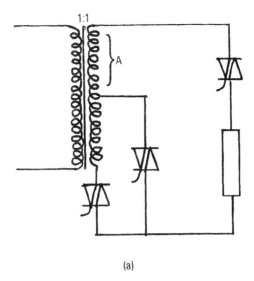

(a)

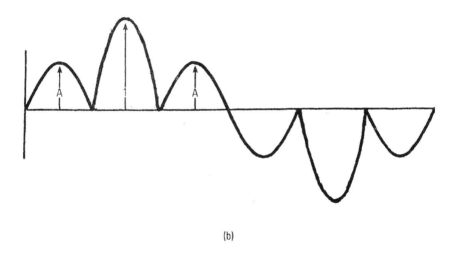

(b)

FIG. 2   Single-phase cycloconverter: (a) circuit diagram for two-pulse operation and (b) load voltage waveforms, $\omega_o/\omega_{in} = 1/3$ [9].

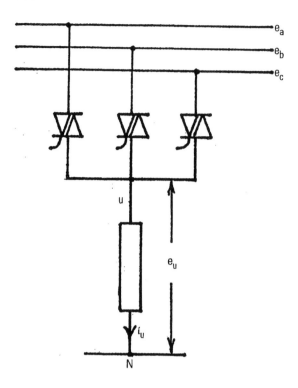

FIG. 3   Three-phase-to-single-phase cycloconverter circuit [9].

13.2 it is seen that $M = 3$. The number of peaks in a one-fourth period m depends on the periodicity and is defined by

$$m = \frac{(M+1/2)\pm1/2}{2} \quad (-\text{for } M \text{ even}; \; + \text{ for } M \text{ odd})\tag{13.1}$$

Typical voltage waveforms with resistive load are given in Fig. 13.4. Commutation occurs naturally, but not precisely at the waveform crossovers due to overlap caused by supply reactance. The frequency ratio does not have to be integer and, in this case, is plotted for $\omega_{in}/\omega_o = 10/3$.

   With an inductive load, such as induction motor, there is within the same half cycle both a direct rectifier mode in which energy is transferred from the input power supply through the transformer to the load and an inverse rectifier mode in which energy stored in the inductive load is recovered and returned back to the supply. The triggering angles of the switches to ensure the desired

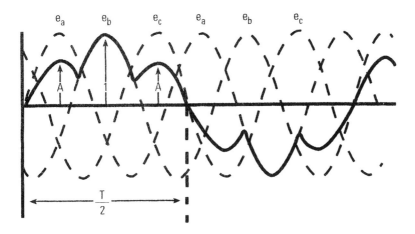

FIG. 4   Voltage waveforms for a three-phase-to-single-phase cycloconverter. R load, $\omega_{in}/\omega_o = 10/3$, $M = 3$ [9].

commutation margin are vitally important to avoid commutation failure. The relation between the output voltage e_u and corresponding lagging current (presumed sinusoidal) i_u for the circuit of Fig. 13.3 is shown in Fig. 13.5. It can be seen that the fundamental component of the load voltage (not shown) is symmetric with the waveform itself so that the displacement angle $\theta_1 - \theta_3$ is about 60°. Prior to instant θ_1 and after instant θ_3, the converter operates in direct rectification mode. During the period $\theta_1 - \theta_3$, operation occurs in the inverse rectification mode. To commutate from the input voltage e_a to voltage e_b, for example, the commutation process must initiate at an angle γ prior to the point of intersection θ_2 in Fig. 13.5. The lead angle γ is related to the overlap angle μ, which is changed proportionately with the instantaneous current value I_γ the instant when $\gamma > \mu$. Failure to observe the inequality $\gamma > \mu$ would mean that the outgoing thyristor could not be turned off within the period γ and would continue in conduction beyond θ_2. This would result in operational failure and system shut down.

13.3   THREE-PHASE-TO-THREE-PHASE FOUR-WIRE ENVELOPE CYCLOCONVERTER

13.3.1   Load Voltage Waveforms

The load side connection of the three-phase, 4-wire, cycloconverter circuit of Fig. 13.6 is the same as the three-pulse circulating current free circuit of Fig.

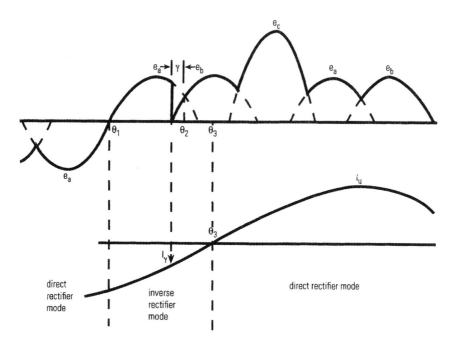

Fig. 5 Voltage and current waveforms for three-phase-to-single-phase cycloconverter. R-L load, $M = 5$, $\Phi_o = 60°$ [9].

12.12(a). In addition, however, the wye-connected transformer secondary windings are tapped to give either a fraction A or the whole input voltage to each phase of the load. For four-wire operation switch sw remains closed. Variation of the load phase voltage e_U, is shown in Fig. 13.7 for the range $M = 2$ to $M = 6$. The phase voltage waveform is identical in shape and equal in magnitude in the three phases, as shown in Fig. 3.8, for the case $M = 6$. It should be noted, however, that the three voltages do not constitute a symmetrical, balanced set as would be the case for sinusoidal operation. Although equal in magnitude, they are not, in general, mutually displaced in phase by precisely 120°. This is further discussed in Sec. 13.3.5. It is also clear from Fig. 13.8 that the three corresponding transformer secondary voltages e_a, e_b, and e_c are unbalanced in magnitude and different in waveform. Variation of the transformer secondary voltage waveforms e_a for different values of M is shown in Fig. 13.9. All of the waveforms are free of a dc component, but each different value of M has its own wave shape for e_a, with the different wave shapes for e_b and e_c.

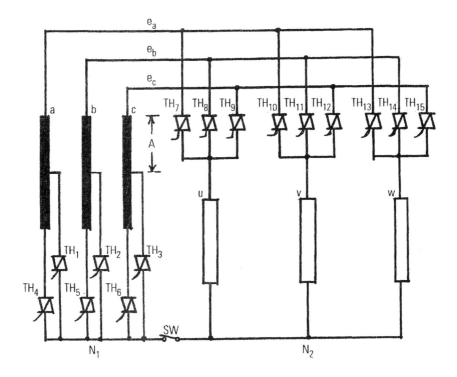

FIG. 6   Three-phase-to-three-phase cycloconverter [9].

13.3.2   Fourier Analysis of the Load Voltage Waveform

The load voltage waveforms e_U, e_V, and e_W in the circuit of Fig. 13.6 are always mathematically odd functions so that the cosine terms a_n of the Fourier series are zero. Each output phase voltage e_o can be analyzed as a Fourier summation

$$e_o = \sum_{n=1}^{\alpha} b_n \sin n\omega t \tag{13.2}$$

In terms of the period of the output voltage waveform T,

$$b_n = \frac{2}{T} \int_0^T e_o(t) \sin 2\pi n \frac{t}{T} \, dt \tag{13.3}$$

When there are m peaks in one-fourth period of the output voltage waveform, then

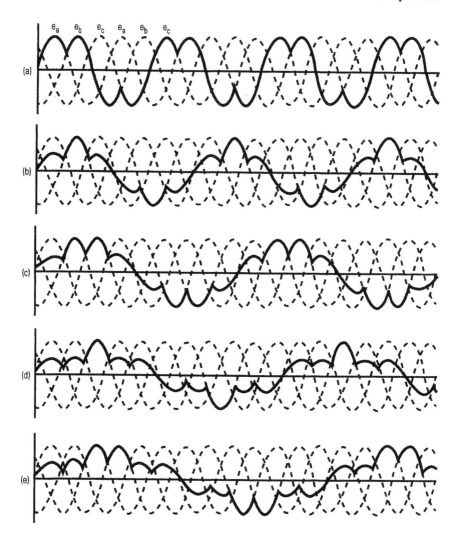

FIG. 7 Load voltage waveform e_u for three-phase envelope cycloconverter of Fig. 13.6, $A = 0.5$: (a) $M = 2$, (b) $M = 3$, (c) $M = 4$, (d) $M = 5$, and (e) $M = 6$ [9].

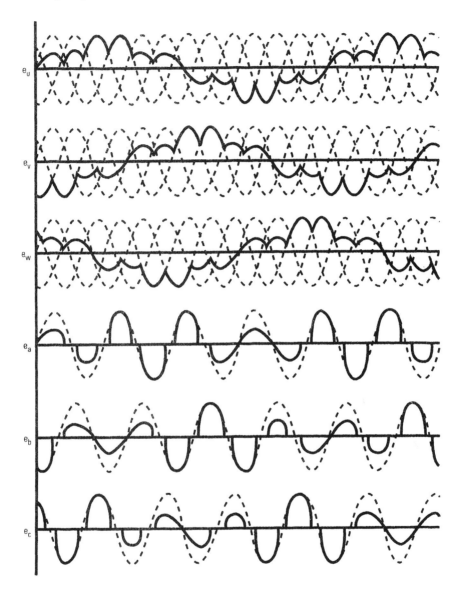

FIG. 8 Voltage waveforms for the three-phase envelope cycloconverter of Fig. 13.6, *A* = 0.5, *M* = 6 [9].

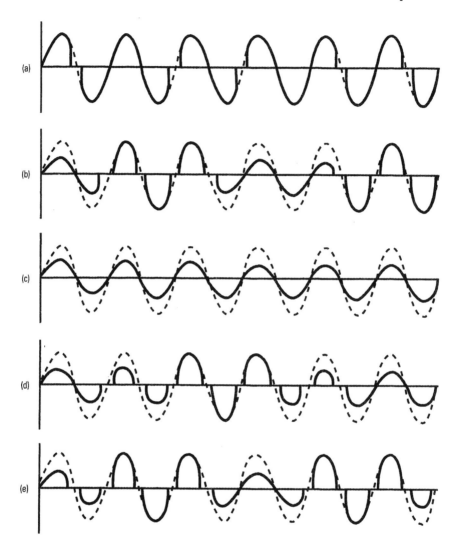

FIG. 9   Transformer secondary voltage e_a for the three-phase envelope cycloconverter of Fig. 13.6. A = 0.5: (a) $M = 2$, (b) $M = 3$, (c) $M = 4$, (d) $M = 5$, and (e) $M = 6$ [9].

$$b_n = \frac{8}{T} \sum_{i=0}^{m} \int_{T_i}^{T_i+1} e_i(t) \sin 2\pi n \frac{t}{T} \, dt \tag{13.4}$$

Parameter m in Eq. (13.4) is related to the waveform identifying parameter M of Figs 13.7–13.9 by Eq. (13.1). The envelope voltage $e_i(t)$ existing between the instants T_i and T_{i+1} is

$$e_i(t) = A_i \sin(2\pi p \frac{t}{T} - \theta_i) \tag{13.5}$$

where phase displacement θ_i is defined as

$$\theta_i = \frac{2i\pi}{3} \tag{13.6}$$

Substituting Eqs. (13.5) and (13.6) into Eq. (13.4) gives

$$b_n = \frac{8}{T} \sum_{i=0}^{M} \int_{T_i}^{T_i+1} A_i \sin\left(2\pi p \frac{t}{T} - \frac{2i\pi}{3}\right) \sin \frac{2\pi n t}{T} \, dt \tag{13.7}$$

Factor A_i in Eq. (13.7) is given by

$$A_i = \begin{cases} 1 & \text{central waves} \\ A & \text{both sets of side waves} \end{cases} \tag{13.8}$$

The frequency ratio p defined in Chapters 8 and 11 is used here:

$$p = \frac{\text{input frequency}}{\text{output frequency}} = \frac{f_{in}}{f_o} = \frac{\omega_{in}}{\omega_o} \tag{13.9}$$

The wavelength of the output voltage waveform envelope T is related to the output frequency by

$$\omega_o = 2\pi f_o = \frac{2\pi}{t} \tag{13.10}$$

Substituting Eqs. (13.9) and (13.10) into Eq. (13.7) permits the coefficients b_n to be expressed in terms of the input and output angular frequencies ω_{in} and ω_o, respectively.

$$b_n = \frac{8}{T} \sum_{i=0}^{M} \int_{T_i}^{T_i+1} A_i \sin\left(\omega_{in} t - \frac{2i\pi}{3}\right) \sin n\omega_o t \, dt \tag{13.11}$$

Because the Fourier coefficients b_n are sinusoidal components, the displacement angles θ_i are notionally multiples of $2\pi/3$. But in this case it is found that the three-phase voltage waves that they represent are not symmetrical in phase displacement, as discussed in Sec. 13.3.5. There will therefore be asymmetry between the fundamental components b_1. The integral in Eq. (13.11) can be evaluated to give a form involving sum and difference terms

$$
\begin{aligned}
b_n &= \frac{8}{T} \sum_{i=0}^{M} \frac{A_i}{2} \int_{T_i}^{T_{i+1}} \left\{ \cos\left[(p-n)\omega_o t - \frac{2i\pi}{3} \right] - \cos\left[(p+n)\omega_o t - \frac{2i\pi}{3} \right] \right\} \\
&= \frac{A_i T}{4(p-n)\pi} \left\{ \sin\left[(p-n)\omega_o T_{i+1} - \frac{2i\pi}{3} \right] - \sin\left[(p-n)\omega_o T_i - \frac{2i\pi}{3} \right] \right\} \\
&\quad - \frac{A_i T}{4(p+n)\pi} \left\{ \sin\left[(p+n)\omega_o T_{i+1} - \frac{2i\pi}{3} \right] - \sin\left[(p+n)\omega_o T_i - \frac{2i\pi}{3} \right] \right\}
\end{aligned}
$$

(13.12)

13.3.3   RMS Value and Distortion Factor of the Load Voltage Waveform

The rms value c of the resultant output voltage waveform can be obtained by summation of the rms values of the Fourier components b_n

$$
c = \sqrt{\sum_{n=1}^{\infty} b_n^2} = \sqrt{\frac{8}{T} \sum_{i=0}^{m} \int_{T_i}^{T_{i+1}} e_i^2(t)\, dt}
$$

(13.13)

The voltage distortion factor, used throughout this book, is the ratio of the fundamental component of voltage to the total rms value. It is a measure of the departure of the voltage waveform from the ideal sinusoidal wave shape, having a maximum value of unity.

$$
\text{Voltage distortion factor} = \frac{b_i}{c}
$$

(13.14)

Comparison is made in Fig. 13.10 between the rectangular waveform v_{AN} of Fig. 10.5, the six-step waveform v_{AN} of Fig. 10.7, and the load phase voltages typified in Fig. 13.7. Values are given for the cases $A = 0.5$ and $A = 1$. For all values of M the distortion factor has its maximum value when $A = 0.5$, which therefore represents the condition of minimum harmonic distortion.

13.3.4   Frequency Constraints of the Load Voltages

In the context of envelope cycloconverters the frequency ratio is related to the number of peaks in one-half of an output wave M by

$$
p = \frac{\omega_{in}}{\omega_o} = \frac{2M+1}{3}
$$

(13.15)

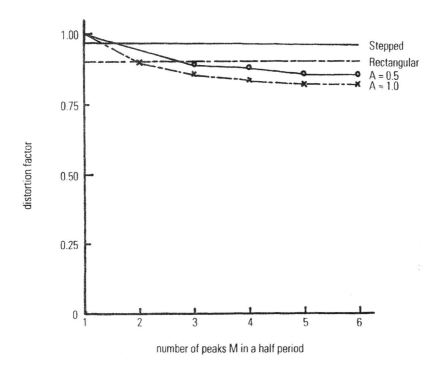

FIG. 10   Distortion factors of the load voltages [9].

With $M = 6$, for example, $p = 13/3$. Examination of the output voltage wave-forms of Fig. 13.8 shows that the frequency of (say) voltage e_u is 3/13 times the frequency of the input sine wave.

In the general form of the load voltage waveforms let the number of control waves be M_c and the number of side waves on either side of the central wave be M_s. The total number of peaks in each half wave M is then given by

$$M = M_c + 2M_s \tag{13.16}$$

irrespective of the values of wave heights A. Total M is also constrained to lie within the limits

$$3M_s - 1 \le M \le 3M_s + 1 \tag{13.17}$$

For example, Fig. 13.7b shows the case $M = 3$, where $M_c = 1$ and $M_s = 1$. Similarly, in Fig. 13.7d, $M = 5$, $M_c = 1$, $M_s = 2$. The combination of Eqs. (13.13) and (13.17) establishes limits for the frequency ratio p when certain values of M are used. When $M_s = 2$, for example, as in Fig. 13.7d and e, it is found that p must lie within the limits $11/3 < p < 9$. This is seen to be also consistent with constraints of Eq. (13.15), where $p = 13/3$. The combination of Eqs. (13.16) and (13.17) leads to a further constraint Eq. (13.19) between M_c and M_s discussed in Example 13.2.

13.3.5 Displacement Angles of the Three Load Voltages

It was pointed out in Sec. 13.3.1 that although the three load voltages e_U, e_V, and e_W of the circuit of Fig. 13.6 are equal in magnitude, they are not symmetrical in phase. This can be seen in the waveforms of Fig. 13.8. If phase voltage e_U is the datum, then e_V lags it by $110.8°$ and e_W lags it by $249.2°$.

The phase displacements θ_V and θ_W of load phase voltages e_V and e_W are given by

$$\theta_V = -\theta_W = \begin{cases} \dfrac{(M+M_c)\pi}{2M+1} & \text{for} \quad M \neq 3M_c - 1 \\[2ex] \dfrac{(M+M_c+2)\pi}{2M+1} & \text{for} \quad M = 3M_c - 1 \end{cases}$$

(13.18)

Equations (13.18) can also be expressed in terms of M and M_s or in terms of M_c and M_s. In the waveforms of Fig. 13.8, for example, $M = 6$, $M_c = 2$, and $M_s = 2$. The inequality of Eq. (13.18a) is satisfied, and the values are found to be $\theta_V = 110.8°$, $\theta_W = -110.8° = 249.2°$, which is consistent with the waveforms of Fig. 13.8. Only for the case $M = 4$, $M_c = 2$ are the phase angles found to be symmetrical at $\pm 120°$. This is the condition of sinusoidal load voltage and current, as seen in Fig. 13.9c. The asymmetry of the load voltage displacement angles causes the load voltages to be unbalanced, except when $M = 4$, $M_s = 1$. This asymmetry can be calculated in terms of the symmetrical components of the three load voltages, which is beyond the scope of the present text. It is found that the worst-case unbalance occurs when $M = 2$, is zero for $M = 3M_s + 1$, and decreases for increasing M, as shown in Fig. 13.11.

13.4 THREE-PHASE-TO-THREE-PHASE, THREE-WIRE ENVELOPE CYCLOCONVERTERS

Some three-phase loads, such as the three-phase induction motor, are usually available only as three-wire loads. The neutral point of the load is not accessible. This corresponds to operation of the circuit of Fig. 13.6 with switch SW open.

In the three-wire condition the unbalance factor of Fig. 13.11 remains unchanged, but the load voltages e_U, e_V, and e_W become unbalanced not only in

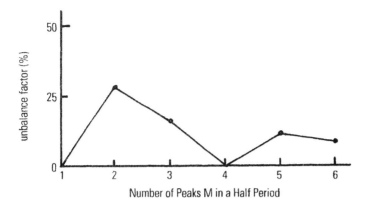

FIG. 11 Voltage unbalance factor due to asymmetry of three displacement angles [9].

phase displacement, as in four-wire operation, but also in amplitude. The further implication of this is not explored in the present text.

13.5 WORKED EXAMPLES

Example 13.1   For a range of values of parameter M calculate the corresponding number of peaks per quarter cycle m and the input–output frequency ratio p in the load voltage waveform of an envelope cycloconverter.

Using relationships (13.1) and (13.15), the following values may be calculated:

M	m	$P = \dfrac{\omega_{in}}{\omega_o}$
1	1	1
2	1	$\dfrac{5}{3}$
3	2	$\dfrac{7}{3}$
4	2	3
5	3	$\dfrac{11}{3}$
6	3	$\dfrac{13}{3}$
7	4	5
8	4	$\dfrac{17}{3}$

Example 13.2 In the load voltage waveform of an envelope cycloconverter show that the number of center peaks per half cycle M_c is related to the side peaks M_s by the relation

$$M_s - 1 \leq M_c \leq M_s + 1$$

In the text

$$M = M_c + 2M_s \tag{13.16}$$

$$3M_s - 1 \leq M_c \leq 3M_c + 1 \tag{13.17}$$

Combining the two equations,

$$3M_s - 1 \leq M_c + 2M_s \leq 3M_s + 1$$

Therefore,

$$M_s - 1 \leq M_c \leq M_s + 1 \quad \text{QED}$$

Example 13.3 For what combination of M, M_c, and M_s do the three output voltages of an envelope cycloconverter represent symmetrical three-phase operation? What are the corresponding frequency ratios?

The displacement angles between the three load voltages are represented by Eqs. (13.18). For symmetrical operation the three load voltages must be mutually displaced by 120° or $2\pi/3$ radians. The criteria of Eqs. (13.18) must satisfy the equalities

$$\frac{M + M_c}{2M + 1} = \frac{2}{3} \quad \text{for} \quad M \neq 3M_c - 1 \tag{Ex. 13.3a}$$

and

$$\frac{M + M_c + 2}{2M + 1} = \frac{2}{3} \quad \text{for} \quad M = 3M_c - 1 \tag{Ex. 13.3b}$$

Relations (Ex. 13.3a) and (Ex. 13.3b) reduce to

$$M_c = \frac{M + 2}{3} \quad \text{for} \quad M \neq 3M_c - 1 \tag{Ex. 13.3c}$$

$$M_c = \frac{M - 4}{3} \quad \text{for} \quad M = 3M_c - 1 \tag{Ex. 13.3d}$$

A table of values satisfying (Ex. 13.3c) is found to be

M	M_c	M_s	p
1	1	0	1
4	2	1	3
7	3	2	5
10	4	3	7

Values of $M_c > 4$ are found to contravene Eq. (13.17) and are not tenable. There do not appear to be any values satisfying (Ex. 13.3d). The corresponding frequency ratios p are found from Eq. (13.15) and shown in the table above.

Example 13.4   A three-phase envelope cycloconverter is supplied from a 60-Hz supply. What fundamental frequencies are available by the use of values of M up to M = 9?

From Eq. (13.9) the output frequency f_o is given by $f_o = f_{in}/p = 60/p$. But p is related to M by Eq. (13.15):

$$p = \frac{2M+1}{3}$$

Therefore,

$$\text{Therefore, } f_o = \frac{180}{2M+1}$$

The values of f_o consistent with M are found to be

M	1	2	3	4	5	6	7	8	9
f_o (Hz)	60	36	25.7	45	16.36	30	12	10.6	9.5

Example 13.5   In an envelope cycloconverter the number of peaks per half cycle of the output wave M = 9. With this constraint what options are available for the waveform pattern assuming A = 0.5 for the side waves? With the tenable patterns what will be the phase displacements θ_V and θ_W and the frequency ratios?

The frequency ratio p is given in terms of M by Eq. (13.15) and is independent of the pattern of center waves M_c and the side waves M_s

$$p = \frac{2M+1}{3} = \frac{19}{3}$$

Various options exist for the pattern of waves within a profile $M = 9$:

1. $M_c = 9$, $M_s = 0$.
2. $M_c = 7$, $M_s = 1$.
3. $M_c = 5$, $M_s = 2$.
4. $M_c = 3$, $M_s = 3$.
5. $M_c = 1$, $M_s = 4$.

Options (1), (2), (3), and (5) above fail to satisfy the criterion Eq. (13.17) and are not viable. The only option possible is option (4), which gives, from Eq. (13.18a),

$$\theta_V = -\theta_W = \frac{(M + M_c)\pi}{2M + 1} = \frac{9 + 3}{18 + 1}\pi = \frac{12}{19}\pi = 113.7°$$

PROBLEMS

13.1   In the load voltage waveform of an envelope cycloconverter what values of parameter M will result in integer frequency ratios?

13.2   Show that the load voltage waveforms of Fig. 13.7 satisfy the criteria of Eqs. (13.16) and (13.17) and Example 13.1.

13.3   Is a cycloconverter envelope consisting of $M = 11$, $M_c = 5$, $M_s = 3$ an acceptable voltage waveform?

13.4   Is a cycloconverter envelope consisting of $M = 8$, $M_c = 2$, $M_s = 3$ an acceptable voltage waveform?

13.5   A three-phase cycloconverter is supplied from a 50-Hz supply. What fundamental output frequencies are available by the use of values of M up to $M = 12$?

13.6   In an envelope cycloconverter the number of peak waves per half cycle of the output wave $M = 7$. What options are available for the output waveform patterns, and what are the values of the phase displacements θ_V and θ_W?

13.7   Repeat Problem 13.6 for the value $M = 11$.

13.8   A three-phase envelope cycloconverter is supplied from a 60-Hz supply. What fundamental output frequencies are available by use of the values (a) $M = 7$, (b) $M = 11$, and (c) $M = 13$?

13.9   Calculate the frequency ratio and the phase displacements θ_V, θ_W for the envelope waveform of Fig. 13.7e.

14

Matrix Converters

14.1 PRINCIPLE OF THE MATRIX CONVERTER

An arbitrary number of input lines can be connected to an arbitrary number of output lines directly using bidirectional semiconductor switches, as shown in Fig. 14.1. The multiple conversion stages and energy storage components of conventional inverter and cycloconverter circuits can be replaced by one switching matrix. With ideal switches the matrix is subject to power invariancy so that the instantaneous input power must always be equal to the instantaneous output power. The number of input and output phases do not have to be equal so that rectification, inversion, and frequency conversion are all realizable. The phase angles between the voltages and currents at the input can be controlled to give unity displacement factor for any loads.

In the ideal, generalized arrangement of Fig. 14.1 there do require to be significant constraints on the switching patterns, even with ideal switches. Some previous discussion of this given in Sec. 9.1. Both sides of the matrix cannot be voltage sources simultaneously since this would involve the direct connection of unequal voltages. If the input is a voltage source, then the output must be a current source, and vice versa. It is a basic requirement that the switching functions must not short-circuit the voltage sources nor open-circuit the current sources.

When the input lines are connected to an electric power utility then the source is imperfect and contains both resistance (power loss) and inductance

415

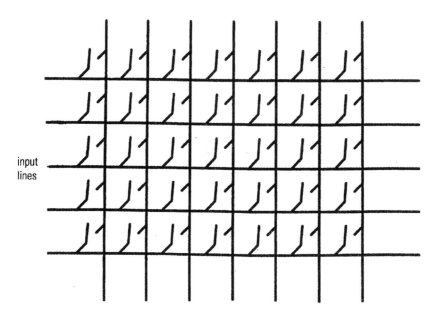

FIG. 1 General arrangement of an ideal switching matrix.

(energy storage). Many loads, especially electric motors, are essentially inductive in nature and may also contain internal emfs or/and currents. The basic premises of electric circuit theory apply also to matrix converters—it is not possible to instantaneously change the current in an inductor nor the voltage drop across a capacitor.

To achieve the operation of an ideal matrix converter, it is necessary to use ideal bidirectional switches, having controllable bidirectional current flow and also voltage blocking capability for both polarities of voltage. The detailed attributes of an ideal switch are listed in Sec. 1.2.

14.2 MATRIX CONVERTER SWITCHES

There is no such thing as an ideal switch in engineering reality. Even the fastest of semiconductor switches requires finite and different switching times for the switch-on and switch-off operations. All switching actions involve power dissipation because the switches contain on-state resistance during continuous conduction. Various options of single-phase bidirectional switches are given in Fig. 14.2.

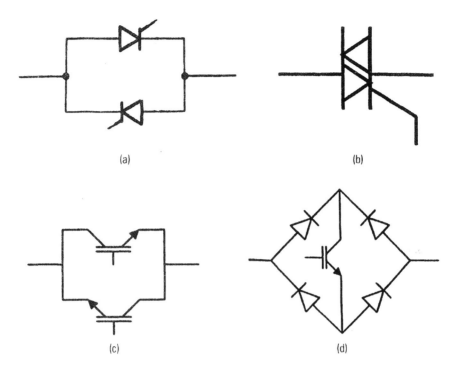

FIG. 2 Single-phase bidirectional switches: (a) two SCRs in inverse-parallel, (b) triac bidirectional switch, (c) two IGBTs in inverse parallel (probably nonviable due to limited reverse blocking), and (d) IGBT diode switch.

A fast switching pair such as Fig. 14.2c can be employed if the devices have reverse blocking capability, such as the MCT or the non-punch-through IGBT.

A fast-acting switch that has been reportedly used in matrix converter experiments is given in Fig. 14.3 [34]. Two IGBTs are connected using a common collector configuration. Since an IGBT does not have reverse blocking capability, two fast recovery diodes are connected in antiseries, each in inverse parallel across an IGBT, to sustain a voltage of either polarity when both IGBTs are switched off. Independent control of the positive and negative currents can be obtained that permits a safe commutation technique to be implemented.

The common collector configuration has the practical advantage that the four switching devices, two diodes and two IGBTs, can be mounted, without isolation, onto the same heat sink. Natural air-cooled heat sinks are used in each phase to dissipate the estimated losses without exceeding the maximum allowable junction temperatures. In the reported investigation the devices used included

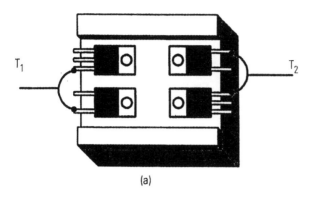

(a)

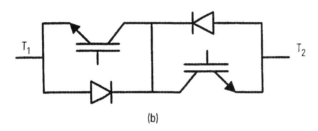

(b)

FIG. 3   Practical switch for matrix converter operation: (a) heat sink mounting and (b) equivalent circuit.

JGBT-International Rectifier (IRGBC 30 F, 600 V, 17 A)
Fast recovery diode-SSG. Thomson (STTA3006, 600 V, 18 A)

A separate gate drive circuit transmits the control signal to each IGBT. Electrical isolation between the control and the power circuits can be achieved using a high-speed opto coupler to transmit the control signal and a high-frequency transformer to deliver the power required by a driver integrated circuit.

14.3   MATRIX CONVERTER CIRCUIT

The basic circuit of a three-phase-to-three-phase matrix converter, shown in Fig. 14.4, consists of three-phase groups. Each of the nine switches can either block or conduct the current in both directions thus allowing any of the output phases

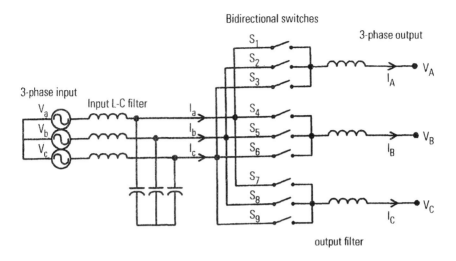

FIG. 4   Basic circuit of a three-phase matrix converter.

to be connected to any of the input phases. In a practical circuit the nine switches seen in Fig. 14.4 could each be of the common configuration of Fig. 14.3 [34]. The input side of the converter is a voltage source, and the output is a current source. Only one of the three switches connected to the same output phase can be on at any instant of time.

In general, low-pass filters are needed at both the input and output terminals to filter out the high-frequency ripple due to the PWM carrier. An overall block diagram of an experimental matrix converter system is given in Fig. 14.5.

Nine PWM signals, generated within a programmable controller, are fed to switch sequencer circuits via pairs of differential line driver receivers. In the switch sequencers the PWM signals are logically combined with current direction signals to produce 18 gating signals. Isolated gate driver circuits then convert the gating signals to appropriate drive signals capable of turning the power switches on or off. Each power circuit is protected by a voltage clamp circuit. A zero crossing (ZC) detector is used to synchronize to the input voltage controller signals. For three-phase motor loads the output filter may not be necessary.

14.4   SWITCHING CONTROL STRATEGIES FOR PWM MATRIX CONVERTERS IN THREE-PHASE MOTOR APPLICATIONS [34,35]

When a PWM matrix converter is used to control the speed of a three-phase ac motor the control system should possess the following properties:

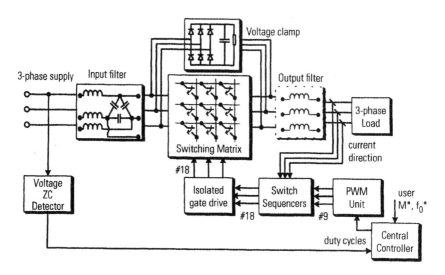

FIG. 5   Basic building blocks of the matrix converter.

Provide independent control of the magnitude and frequency of the gener-
    ated output voltages (i.e., the motor terminal voltages).
Result in sinusoidal input currents with adjustable phase shift.
Achieve the maximum possible range of output-to-input voltage ratio.
Satisfy the conflicting requirements of minimum low-order output voltage
    harmonics and minimum switching losses.
Be computationally efficient.

Many different methods have been considered as the basis of analyzing and
designing a workable matrix converter. Because of the complexity of the necessary switching the associated control logic is also complex and involves large
and complicated algorithms. General requirements for generating PWM control
signals for a matrix converter on-line in real time are

Computation of the switch duty cycles within one switching period
Accurate timing of the control pulses according to some predetermined
    pattern
Synchronization of the computational process with the input duty cycle
Versatile hardware configuration of the PWM control system, which allows
    any control algorithm to be implemented by means of the software

Microprocessor-based implementation of a PWM algorithm involves the use of
digital signal processors (DSPs).

The two principal methods that have been reported for the control of a matrix converter are discussed separately, in the following subsections.

14.4.1 Venturini Control Method [34]

A generalized high-frequency switching strategy for matrix converters was proposed by Venturini in 1980 [36,37]. The method was further modified to increase the output-to-input voltage transfer ratio from 0.5 to 0.866. In addition, it can generate sinusoidal input currents at unity power factor irrespective of the load power factor.

14.4.1.1 Principle

In the Venturini method a desired set of three-phase output voltages may be synthesized from a given set of three-phase input sinusoidal voltages by sequential piecewise sampling. The output voltage waveforms are therefore composed of segments of the input voltage waves. The lengths of each segments are determined mathematically to ensure that the average value of the actual output waveform within each sampling period tracks the required output waveform. The sampling rate is set much higher than both input and output frequencies, hence the resulting synthesized waveform displays the same low-frequency spectrum of the desired waveform.

14.4.1.2 Switching Duty Cycles

The Venturini principle can be explained initially using a single-phase output. A three-phase output is generated by three independent circuits, and the analytical expressions for all three waveforms have the same characteristics. Consider a single output phase using a three-phase input voltage as depicted in Fig. 14.6. Switching elements S_1–S_3 are bidirectional switches connecting the output phase to one of the three input phases and are operated according to a switching pattern shown in Fig. 14.6b. Only one of the three switches is turned on at any given time, and this ensures that the input of a matrix converter, which is a voltage source, is not short-circuited while a continuous current is supplied to the load.

As shown in Fig. 14.6a within one sampling period, the output phase is connected to three input phases in sequence; hence, the output voltage V_{o1} is composed of segments of three input phase voltages and may be mathematically expressed as

$$\bar{V}_{o1} = \left[\frac{t_1(k)}{T_s} \ \frac{t_2(k)}{T_s} \ \frac{t_3(k)}{T_s} \right] \begin{bmatrix} V_{i1} \\ V_{i2} \\ V_{i3} \end{bmatrix}$$

(14.1)

Symbol k represents the number of sampling intervals, $t_n(k)$ for $n = 1, 2, 3$ are the switching on times, $t_1(k) + t_2(k) + t_3(k)$ equals the sampling period T_s and

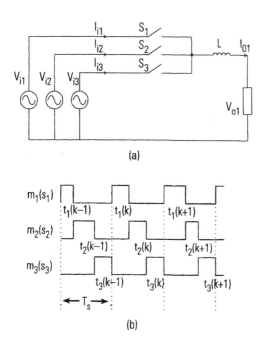

(a)

(b)

FIG. 6   One output phase of a matrix converter: (a) equivalent circuit and (b) switching pattern.

$t_n(k)/T_s$ are duty cycles. For a known set of three-phase input voltages, the waveshape of V_{o1}, within the kth sample, is determined by $t_1(k)$, $t_2(k)$, $t_3(k)$.

It should be noted that the switching control signals m_1, m_2, m_3 shown in Fig. 14.6b can be mathematically represented as functions of time

$$m_1(t) = \sum_{k=0}^{\infty} \left\{ u(kT_s) - u[kT_s + t_1(k)] \right\}$$

(14.2)

$$m_2(t) = \sum_{k=0}^{\infty} \left\{ u[kT_s + t_1(k)] - u[kT_s + t_1(k) + t_2(k)] \right\}$$

(14.3)

$$m_3(t) = \sum_{k=0}^{\infty} \left\{ u[kT_s + t_1(k) + t_2(k)] - u[(k+1)T_s] \right\}$$

(14.4)

where

$$u(t_0) = \begin{cases} 1 & t > t_0 \\ 0 & t < t_0 \end{cases}$$

is a unit step function and the switch is on when $m_n(t) = 1$ and off when $m_n(t) = 0$.

For constructing a sinusoidal output wave shape, function $m_n(t)$ defines a sequence of rectangular pulses whose widths are sinusoidally modulated. Consequently, its frequency spectrum consists of the required low-frequency components and also unwanted harmonics of higher frequencies. Since the required output voltage may be considered as the product of these functions and the sinusoidal three-phase input voltages, the Fourier spectrum of the synthesized output voltage contains the desired sinusoidal components plus harmonics of certain frequencies differing from the required output frequency.

The input current I_{i1} equals output current I_{o1} when switch S_1 is on and zero when S_1 is off. Thus the input current I_{i1} consists of segments of output current I_{o1}. Likewise, the turn-on and turn-off of switches S_2 and S_3 results in the input currents I_{i2} and I_{i3} containing segments of output current I_{o1}. The width of each segment equals the turn-on period of the switch. Corresponding to the voltage equation (14.1), the three average input currents are given by

$$\bar{I}_{i1} = \frac{t_1(k)}{T_s} I_{o1} \qquad \bar{I}_{i2} = \frac{t_2(k)}{T_s} I_{o1} \qquad \bar{I}_{i3} = \frac{t_3(k)}{T_s} I_{o1} \qquad (14.5)$$

14.4.1.3 Modulation Functions

Applying the above procedure to a three-phase matrix converter, the three output phase voltages can be expressed in the following matrix form:

$$\begin{bmatrix} V_{o1}(t) \\ V_{o2}(t) \\ V_{o3}(t) \end{bmatrix} = \begin{bmatrix} m_{11}(t) & m_{12}(t) & m_{13}(t) \\ m_{21}(t) & m_{22}(t) & m_{23}(t) \\ m_{31}(t) & m_{32}(t) & m_{33}(t) \end{bmatrix} \begin{bmatrix} V_{i1}(t) \\ V_{i2}(t) \\ V_{i3}(t) \end{bmatrix} \qquad (14.6)$$

or

$$[V_o(t)] = [M(t)][V_i(t)]$$

and the input currents as

$$\begin{bmatrix} I_{i1}(t) \\ I_{i2}(t) \\ I_{i3}(t) \end{bmatrix} = \begin{bmatrix} m_{11}(t) & m_{12}(t) & m_{13}(t) \\ m_{21}(t) & m_{22}(t) & m_{23}(t) \\ m_{31}(t) & m_{32}(t) & m_{33}(t) \end{bmatrix} \begin{bmatrix} I_{o1}(t) \\ I_{o2}(t) \\ I_{o3}(t) \end{bmatrix}, \qquad (14.7)$$

or

$$[I_r(t)] = [M^T(t)] [I_0(t)]$$

where $M(t)$ is the modulation matrix. Its elements $m_{ij}(t)$, $i,j = 1, 2, 3$, represent the duty cycle t_{ij}/T_s of a switch connecting output phase i to input phase j within one switching cycle and are called *modulation functions*. The value of each modulation function changes from one sample to the next, and their numerical range is

$$0 \le m_{ij}(t) \le 1 \qquad i,j = 1, 2, 3 \tag{14.8}$$

Bearing in mind the restriction imposed on the control of matrix switches stated above, functions $m_{ij}(t)$ for the same output phase obey the relation

$$\sum_{j=1}^{3} m_{ij} = 1 \qquad 1 \le i \le 3 \tag{14.9}$$

The design aim is to define $m_{ij}(t)$ such that the resultant three output phase voltages expressed in Eq. (14.6) match closely the desired three-phase reference voltages.

14.4.1.4 Three-Phase Reference Voltages

The desired reference-phase voltages should ensure that the maximum output-to-input voltage transfer ratio is obtained without adding low-order harmonics into the resultant output voltages. To achieve this, the reference output voltage waveform to be synthesized must, at any time, remain within an envelope formed by the three-phase input voltages, as shown in Fig. 14.7a. Thus when the input frequency ω_i is not related to the output frequency ω_o, the maximum achievable output-to-input voltage ratio is restricted to 0.5, as illustrated in Fig. 14.7a.

The area within the input voltage envelope may be enlarged by subtracting the common-mode, third harmonic of the input frequency from the input phase-to-neutral voltages. For example, when a voltage of frequency $3\omega_i$ and amplitude equal to $V_{im}/4$ is subtracted from the input phase voltages, the ratio of output to input voltage becomes 0.75 as shown in Fig. 14.7b. Note that this procedure is equivalent to adding the third harmonics of the input frequency to the target output-phase voltage. The introduction of the third-order harmonics of both the input and output frequencies into the reference output-phase voltages will have no effect on an isolated-neutral, three-phase load normally used in practice, as they will be canceled in the line-to-line output voltages.

Further improvement on the output voltage capability can be made by subtracting the third harmonics of the output frequency from the target output phase voltage. This is to decrease the peak-to-peak value of the output phase voltage, as illustrated in Fig. 14.7c. With the magnitude of the output-frequency third harmonics equivalent to $V_{om}/6$, an output-to-input voltage ratio of 0.866 can be achieved. This figure is the theoretical output voltage limit for this type of con-

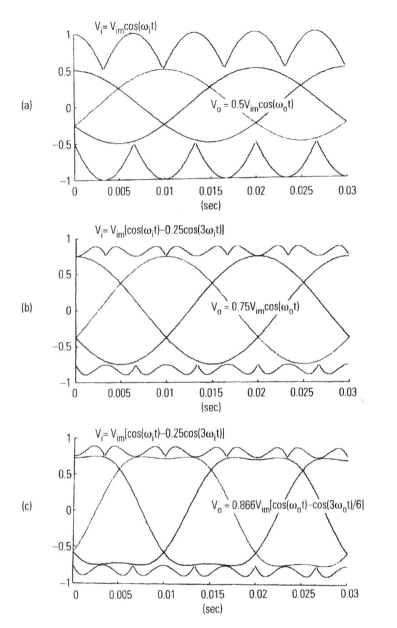

Fig. 7 Third harmonic addition to increase the maximum achievable output-voltage magnitude of matrix converter: (a) output voltages, $V_o = 0.5V_{in}$, (b) output voltages, $V_o = 0.75V_{in}$, and (c) output voltages, $V_o = 0.866V_{in}$.

verter. The two-step, third-harmonic modulation described above results in the following output phase voltages expression [38,39]

$$
\begin{bmatrix} V_{o1}(t) \\ V_{o2}(t) \\ V_{o3}(t) \end{bmatrix} = V_{om} \begin{bmatrix} \cos \omega_0 t \\ \cos\left(\omega_0 t - \dfrac{2\pi}{3}\right) \\ \cos\left(\omega_0 t - \dfrac{4\pi}{3}\right) \end{bmatrix} - \frac{V_{om}}{6} \begin{bmatrix} \cos 3\omega_0 t \\ \cos 3\omega_0 t \\ \cos 3\omega_0 t \end{bmatrix} + \frac{V_{im}}{4} \begin{bmatrix} \cos 3\omega_i t \\ \cos 3\omega_i t \\ \cos 3\omega_i t \end{bmatrix}
$$

$$(14.10)$$

where V_{om} and ω_o are the magnitude and frequency of the required fundamental output voltage and V_{im} and ω_i, are the magnitude and frequency of the input voltage, respectively. The three terms of Eq. (14.10) can be expressed as

$$
[V_o]_A + [V_o]_B + [V_o]_C \tag{14.11}
$$

14.4.1.5 Derivation of the Modulation Matrix

Having defined the three-phase output reference voltage, determination of modulation function $M[t]$ involves solving Eqs. (14.6) and (14.7) simultaneously. The three-phase input voltages with amplitude V_{im} and frequency ω_i and the three-phase output currents with amplitude I_{om} and frequency ω_o are given by

$$
[V_i(t)] = V_{im} \begin{bmatrix} \cos \omega_i t \\ \cos\left(\omega_i t - \dfrac{2\pi}{3}\right) \\ \cos\left(\omega_i t - \dfrac{4\pi}{3}\right) \end{bmatrix}
$$

$$(14.12)$$

and

$$
[I_o(t)] = I_{om} \begin{bmatrix} \cos(\omega_o t - \phi_o) \\ \cos\left(\omega_o t - \phi_o - \dfrac{2\pi}{3}\right) \\ \cos\left(\omega_o t - \phi_o - \dfrac{4\pi}{3}\right) \end{bmatrix}
$$

$$(14.13)$$

Assume the desired output voltage of Eq. (14.10) to consist only of the first term, $[V_o]_A$, then

$$
\begin{bmatrix} V_{o1}(t) \\ V_{o2}(t) \\ V_{o3}(t) \end{bmatrix} = V_{om} \begin{bmatrix} \cos \omega_o t \\ \cos\left(\omega_o t - \dfrac{2\pi}{3} \right) \\ \cos\left(\omega_o t - \dfrac{4\pi}{3} \right) \end{bmatrix}
\tag{14.14}
$$

By eliminating the other two terms of Eq. (14.10), the achievable ratio V_{om}/V_{im} becomes 0.5. Derivation of the modulation matrix for the output voltages defined in Eq. (14.10) is given in Ref. 34. Using the three-phase input voltages of Eq. (14.12) to synthesize the desired output voltage of Eq. (14.14), the modulation matrix derived must be in the form expressed as either

$$
[m(t)]_{A+} = A_1 \begin{bmatrix} \cos (\omega_o + \omega_i)t & \cos\left[(\omega_o + \omega_i)t - \dfrac{2\pi}{3}\right] & \cos\left[(\omega_o + \omega_i)t - \dfrac{4\pi}{3}\right] \\ \cos\left[(\omega_o + \omega_i)t - \dfrac{2\pi}{3}\right] & \cos\left[(\omega_o + \omega_i)t - \dfrac{4\pi}{3}\right] & \cos (\omega_o + \omega_i)t \\ \cos\left[(\omega_o + \omega_i)t - \dfrac{4\pi}{3}\right] & \cos (\omega_o + \omega_i)t & \cos\left[(\omega_o + \omega_i)t - \dfrac{2\pi}{3}\right] \end{bmatrix}
\tag{14.15}
$$

or

$$
[m(t)]_{A-} = A_2 \begin{bmatrix} \cos (\omega_o - \omega_i)t & \cos\left[(\omega_o - \omega_i)t - \dfrac{4\pi}{3}\right] & \cos\left[(\omega_o - \omega_i)t - \dfrac{2\pi}{3}\right] \\ \cos\left[(\omega_o - \omega_i)t - \dfrac{2\pi}{3}\right] & \cos (\omega_o - \omega_i)t & \cos\left[(\omega_o - \omega_i)t - \dfrac{4\pi}{3}\right] \\ \cos\left[(\omega_o - \omega_i)t - \dfrac{4\pi}{3}\right] & \cos\left[(\omega_o - \omega_i)t - \dfrac{2\pi}{3}\right] & \cos (\omega_o - \omega_i)t \end{bmatrix}
\tag{14.16}
$$

where A_1 and A_2 are constants to be determined. Substituting either Eq. (14.15) or Eq. (14.16) into Eq. (14.6) gives three output voltages which are sinusoidal and have 120° phase shift between each other.

$$
[V_o]_{A+} = [m(t)]_{A+} [V_i(t)] = \frac{3}{2} A_1 V_{im} \begin{bmatrix} \cos \omega_o t \\ \cos\left(\omega_o t - \dfrac{2\pi}{3} \right) \\ \cos\left(\omega_o t - \dfrac{4\pi}{3} \right) \end{bmatrix}
\tag{14.17}
$$

or

$$[V_o]_{A-} = [m(t)]_{A-}[V_i(t)] = \frac{3}{2}A_2 V_{im} \begin{bmatrix} \cos \omega_o t \\ \cos\left(\omega_o t - \dfrac{2\pi}{3}\right) \\ \cos\left(\omega_o t - \dfrac{4\pi}{3}\right) \end{bmatrix}.$$

(14.18)

If both $[m(t)]_{A+}$ and $[m(t)]_{A-}$ are used to produce the target output voltage, then $[V_o]_A = [V_o]_{A+} + [V_o]_{A-}$, yielding

$$A_1 + A_2 = \frac{2}{3}\frac{V_{om}}{V_{im}}$$

(14.19)

Applying the same procedure as above to synthesise the input currents, using Eq. (14.7), yields

$$[I_i]_{A+} = [m(t)]_{A+}^T [I_o(t)] = \frac{3}{2}A_1 I_{om} \begin{bmatrix} \cos(\omega_i t + \phi_o) \\ \cos\left(\omega_i t + \phi_o - \dfrac{2\pi}{3}\right) \\ \cos\left(\omega_i t + \phi_o - \dfrac{4\pi}{3}\right) \end{bmatrix}$$

(14.20)

or

$$[I_i]_{A-} = [m(t)]_{A-}^T [I_o(t)] = \frac{3}{2}A_2 I_{om} \begin{bmatrix} \cos(\omega_i t - \phi_o) \\ \cos\left(\omega_i t - \phi_o - \dfrac{2\pi}{3}\right) \\ \cos\left(\omega_i t - \phi_o - \dfrac{4\pi}{3}\right) \end{bmatrix}$$

(14.21)

This means that modulation matrix $[m(t)]_{A+}$ results in an input current having a leading phase angle ϕ_o, while $[m(t)]_{A-}$ produces an input current with a lagging phase angle ϕ_o.

Let the required sinusoidal input currents be defined as

$$[I_i(t)] = I_{im} \begin{bmatrix} \cos(\omega_i t - \phi_i) \\ \cos\left(\omega_i t - \phi_i - \dfrac{2\pi}{3}\right) \\ \cos\left(\omega_i t - \phi_i - \dfrac{4\pi}{3}\right) \end{bmatrix} \tag{14.22}$$

where ϕ_i is an input displacement angle.

Setting $[I_i(t)] = [I_i]_{A+} + [I_i]_{A-}$ and substituting $[I_i]_{A+}$ and $[I_i]_{A-}$ defined by Eqs. (14.20) and (14.21), respectively, the input phase current equation is given as

$$\frac{3}{2} I_{om} \left[A_1 \cos(\omega_i t - \gamma \frac{\pi}{3} + \phi_o) + A_2 \cos(\omega_i t - \gamma \frac{\pi}{3} - \phi_o) \right] =$$
$$I_{im} \cos\left(\omega_i t - \gamma \frac{\pi}{3} - \phi_i\right) \quad \gamma = 0,\ 2,\ 4$$

As $\cos(A + B) = \cos A \cos B - \sin A \sin B$, applying this to both sides of the equation yields

$$(A_1 + A_2)\cos\left(\omega_i t - \gamma \frac{\pi}{3}\right)\cos\phi_o = \frac{2}{3}\frac{I_{im}}{I_{om}}\cos\left(\omega_i t - \gamma \frac{\pi}{3}\right)\cos\phi_i \tag{14.23a}$$

and

$$(A_2 - A_1)\sin\left(\omega_i t - \gamma \frac{\pi}{3}\right)\sin\phi_o = \frac{2}{3}\frac{I_{im}}{I_{om}}\sin\left(\omega_i t - \gamma \frac{\pi}{3}\right)\sin\phi_i \tag{14.23b}$$

Neglecting the converter power losses, the input power of the circuit equals the output power of the circuit; hence,

$$V_{im} I_{im} \cos\phi_i = V_{om} I_{om} \cos\phi_o \tag{14.24}$$

This yields

$$\frac{I_{im}}{I_{om}} = \frac{V_{om}}{V_{im}}\frac{\cos\phi_o}{\cos\phi_i} \tag{14.25}$$

Substituting I_{im}/I_{om} from Eq. (14.23) into Eq. (14.25) gives

$$(A_1 + A_2)\cos\left(\omega_i t - \gamma \frac{\pi}{3}\right)\cos\phi_o = \frac{2}{3}\frac{V_{om}}{V_{im}}\cos\left(\omega_i t - \gamma \frac{\pi}{3}\right)\cos\phi_o$$

$$(A_2 - A_1)\sin\left(\omega_i t - \gamma \frac{\pi}{3}\right)\sin\phi_o = \frac{2}{3}\frac{V_{om}}{V_{im}}\frac{\tan\phi_i}{\tan\phi_o}\sin\left(\omega_i t - \gamma \frac{\pi}{3}\right)\sin\phi_o \tag{14.26}$$

Solving Eq. (14.26) simultaneously, coefficients A_1 and A_2 are found to be

$$A_1 = \frac{1}{3}Q\left(1 - \frac{\tan\phi_i}{\tan\phi_o}\right) \qquad A_2 = \frac{1}{3}Q\left(1 + \frac{\tan\phi_i}{\tan\phi_o}\right) \tag{14.27}$$

where $Q = V_{om}/V_{im}$ is an output-to-input voltage ratio, which also satisfies Eq. (14.21). Note that the input displacement angle and hence the input power factor of the matrix converter may be adjusted by varying these coefficients appropriately.

The overall modulation matrix may be written as the sum of $[m(t)]_{A+}$ and $[m(t)]_{A-}$. As adding three elements on the same row of the modulation matrix results in zero at any instant, the constant 1/3 must be added to each element to satisfy the constraint specified in Eq. (14.9) giving

$$[m(t)] = \frac{1}{3}\begin{bmatrix} 1 & 1 & 1 \\ 1 & 1 & 1 \\ 1 & 1 & 1 \end{bmatrix} + [m(t)]_{A+} + [m(t)]_{A-} \tag{14.28}$$

Substituting each component in Eq. (14.28) with the results derived above, the formulas for the overall modulation matrix are expressed by

$$[m(t)] = \begin{bmatrix} \dfrac{1}{3} & \dfrac{1}{3} & \dfrac{1}{3} \\[2mm] \dfrac{1}{3} & \dfrac{1}{3} & \dfrac{1}{3} \\[2mm] \dfrac{1}{3} & \dfrac{1}{3} & \dfrac{1}{3} \end{bmatrix}$$

$$\frac{Q}{3}(1-\tan\phi_i/\tan\phi_o)\begin{bmatrix} \cos[(\omega_o+\omega_i)t] & \cos[(\omega_o+\omega_i)t-2\pi/3] & \cos[(\omega_o+\omega_i)t-4\pi/3] \\ \cos[(\omega_o+\omega_i)t-2\pi/3] & \cos[(\omega_o+\omega_i)t-4\pi/3] & \cos[(\omega_o+\omega_i)t] \\ \cos[(\omega_o+\omega_i)t-4\pi/3] & \cos[(\omega_o+\omega_i)t] & \cos[(\omega_o+\omega_i)t-2\pi/3] \end{bmatrix}$$

$$+ \frac{\frac{Q}{3}(1+\tan\phi_i/\tan\phi_o)\begin{bmatrix} \cos[(\omega_o-\omega_i)t] & \cos[(\omega_o-\omega_i)t-4\pi/3] & \cos[(\omega_o-\omega_i)t-2\pi/3] \\ \cos[(\omega_o-\omega_i)t-2\pi/3] & \cos[(\omega_o+\omega_i)t] & \cos[(\omega_o+\omega_i)t-4\pi/3] \\ \cos[(\omega_o+\omega_i)t-4\pi/3] & \cos[(\omega_o+\omega_i)t-4\pi/3] & \cos[(\omega_o-\omega_i)t] \end{bmatrix}}{} \tag{14.29}$$

Alternatively, a general and simplified formula for one of the nine elements in matrix $[m(t)]$ above may be written as [39]

$$m_{ij}(t) = \frac{1}{3}\left\{1 + (1-\Theta)Q\cos\left[(\omega_o + \omega_i)t - [2(i+j)-4]\frac{\pi}{3}\right]\right.$$
$$\left. + (1+\Theta)Q\cos\left[(\omega_o - \omega_i)t - 2(i-j)\frac{\pi}{3}\right]\right\} \qquad (14.30)$$

where $i,j = 1, 2, 3$; $Q = V_{om}/V_{im}$; and $\Theta = \tan\phi_i / \tan\phi_o$ is an input-to-output phase transfer ratio. As shown in Eq. (14.29), $m_{ij}(t)$ is a function of ω_i, ω_o, Q, and Θ. Assuming that V_i and ω_i are constants, Q is a variable of V_o, while ϕ_o, the phase lag between the load current and voltage, is a nonzero value. Since unity input power factor operation is always desired, ϕ_i, is zero. This, in turn, leads to Θ being zero. Subsequently Eq. (14.30) may be simplified as

$$m_{ij}(t) = \frac{1}{3}\left[1 + Q\left(\cos\left\{(\omega_o + \omega_i)t - [2(i+j)-4]\frac{\pi}{3}\right\}\right.\right.$$
$$\left.\left. + \cos\left[(\omega_o - \omega_i)t - 2(i-j)\frac{\pi}{3}\right]\right)\right] \qquad (14.31)$$

For practical implementation, it is important to note that the calculated values of $m_{ij}(t)$ are valid when they satisfy the conditions defined by Eqs. (14.8) and (14.9). Also Eq. (14.30) gives positive results when $V_o/V_i \leq \sqrt{3/2}$

14.4.2 Space Vector Modulation (SVM) Control Method

The space vector modulation (SMV) technique adopts a different approach to the Venturini method in that it constructs the desired sinusoidal output three-phase voltage by selecting the valid switching states of a three-phase matrix converter and calculating their corresponding on-time durations. The method was initially presented by Huber [40,41].

14.4.2.1 Space Vector Representation of Three-Phase Variables

For a balanced three-phase sinusoidal system the instantaneous voltages maybe expressed as

$$\begin{bmatrix} V_{AB}(t) \\ V_{BC}(t) \\ V_{CA}(t) \end{bmatrix} = V_{ol}\begin{bmatrix} \cos\omega_o t \\ \cos(\omega_o t - 120°) \\ \cos(\omega_o t - 240°) \end{bmatrix} \qquad (14.32)$$

This can be analyzed in terms of complex space vector

$$\vec{V}_o = \frac{2}{3}\left[V_{AB}(t) + V_{BC}(t)e^{j2\pi/3} + V_{BC}(t)e^{j4\pi/3}\right] = V_{ol}e^{j\omega_o t} \tag{14.33}$$

where $e^{j\theta} = \cos\theta + j\sin\theta$ and represents a phase-shift operator and 2/3 is a scaling factor equal to the ratio between the magnitude of the output line-to-line voltage and that of output voltage vector. The angular velocity of the vector is ω_o and its magnitude is V_{ol}.

Similarly, the space vector representation of the three-phase input voltage is given by

$$\vec{V}_i = V_i e^{j(\omega_i t)} \tag{14.34}$$

where V_i is the amplitude and ω_i, is the constant input angular velocity.

If a balanced three-phase load is connected to the output terminals of the converter, the space vector forms of the three-phase output and input currents are given by

$$\vec{I}_o = I_o e^{j(\omega_o t - \phi_o)} \tag{14.35}$$

$$\vec{I}_i = I_i e^{j(\omega_i t - \phi_i)} \tag{14.36}$$

respectively, where ϕ_o is the lagging phase angle of the output current to the output voltage and ϕ_i is that of the input current to the input voltage.

In the SVM method, the valid switching states of a matrix converter are represented as voltage space vectors. Within a sufficiently small time interval a set of these vectors are chosen to approximate a reference voltage vector with the desired frequency and amplitude. At the next sample instant, when the reference voltage vector rotates to a new angular position, a new set of stationary voltage vectors are selected. Carrying this process onward by sequentially sampling the complete cycle of the desired voltage vector, the average output voltage emulates closely the reference voltage. Meanwhile, the selected vectors should also give the desired phase shift between the input voltage and current.

Implementation of the SVM method involves two main procedures: switching vector selection and vector on-time calculation. These are both discussed in the following subsections.

14.4.2.2 Definition and Classification of Matrix Converter Switching Vectors

For a three-phase matrix converter there are 27 valid on-switch combinations giving thus 27 voltage vectors, as listed in Table 14.1. These can be divided into three groups.

TABLE 14.1 Valid Switch Combinations of a Matrix Converter and the Stationary Vectors ($k = 2/\sqrt{3}$)

	On switches		Output voltages			Input currents			Voltage vector		Current vector	
			V_{AB}	V_{BC}	V_{CA}	I_a	I_b	I_c	Magnitude	Phase	Magnitude	Phase
1	S_1,	S_5,	V_{ab}	V_{bc}	V_{ca}	I_A	I_B	I_C	V_{il}	$\omega_i t$	I_o	$\omega_o t$
2	S_1,	S_6,	$-V_{ca}$	$-V_{bc}$	$-V_{ab}$	I_A	I_C	I_B	$-V_{il}$	$-\omega_i t+4\pi/3$	I_o	$-\omega_o t$
3	S_2,	S_4,	$-V_{ab}$	$-V_{ca}$	$-V_{bc}$	I_B	I_A	I_C	$-V_{il}$	$-\omega_i t$	I_o	$-\omega_o t+2\pi/3$
4	S_2,	S_6,	V_{bc}	V_{ca}	V_{ab}	I_C	I_A	I_B	V_{il}	$\omega_i t+4\pi/3$	I_o	$\omega_o t+2\pi/3$
5	S_3,	S_4,	V_{ca}	V_{ab}	V_{bc}	I_B	I_C	I_A	V_{il}	$\omega_i t+2\pi/3$	I_o	$\omega_o t+4\pi/3$
6	S_3,	S_5,	$-V_{bc}$	$-V_{ab}$	$-V_{ca}$	I_C	I_B	I_A	$-V_{il}$	$-\omega_i t+2\pi/3$	I_o	$-\omega_o t+4\pi/3$
1P	S_1,	S_5,	V_{ab}	0	V_{ab}	I_A	$-I_A$	0	kV_{ab}	$\pi/6$	kI_A	$-\pi/6$
1N	S_2,	S_4,	$-V_{ab}$	0	$-V_{ab}$	$-I_A$	I_A	0	$-kV_{ab}$	$\pi/6$	$-kI_A$	$-\pi/6$
2P	S_2,	S_6,	V_{bc}	0	V_{bc}	0	I_A	$-I_A$	kV_{bc}	$\pi/6$	kI_A	$\pi/2$
2N	S_3,	S_5,	$-V_{bc}$	0	$-V_{bc}$	0	$-I_A$	I_A	$-kV_{bc}$	$\pi/6$	$-kI_A$	$\pi/2$
3P	S_3,	S_4,	V_{ca}	0	V_{ca}	$-I_A$	0	I_A	kV_{ca}	$\pi/6$	kI_A	$7\pi/6$
3N	S_1,	S_6,	$-V_{ca}$	0	$-V_{ca}$	I_A	0	$-I_A$	$-kV_{ca}$	$\pi/6$	$-kI_A$	$7\pi/6$
4P	S_2,	S_4,	$-V_{ab}$	V_{ab}	0	I_B	$-I_B$	0	kV_{ab}	$5\pi/6$	kI_B	$-\pi/6$
4N	S_2,	S_5,	V_{ab}	$-V_{ab}$	0	$-I_B$	I_B	0	$-kV_{ab}$	$5\pi/6$	$-kI_B$	$-\pi/6$
5P	S_3,	S_5,	$-V_{bc}$	V_{bc}	0	0	I_B	$-I_B$	kV_{bc}	$5\pi/6$	kI_B	$\pi/2$
5N	S_2,	S_6,	V_{bc}	$-V_{bc}$	0	0	$-I_B$	I_B	$-kV_{bc}$	$5\pi/6$	$-kI_B$	$\pi/2$
6P	S_1,	S_6,	$-V_{ca}$	V_{ca}	0	$-I_B$	0	I_B	kV_{ca}	$5\pi/6$	kI_B	$7\pi/6$
6N	S_3,	S_4,	V_{ca}	$-V_{ca}$	0	I_B	0	$-I_B$	$-kV_{ca}$	$5\pi/6$	$-kI_B$	$7\pi/6$
7P	S_2,	S_5,	0	$-V_{ab}$	V_{ab}	I_C	$-I_C$	0	kV_{ab}	$\pi/2$	kI_C	$-\pi/6$
7N	S_1,	S_4,	0	V_{ab}	$-V_{ab}$	$-I_C$	I_C	0	$-kV_{ab}$	$\pi/2$	$-kI_C$	$-\pi/6$
8P	S_3,	S_6,	0	$-V_{bc}$	V_{bc}	0	I_C	$-I_C$	kV_{bc}	$\pi/2$	kI_C	$\pi/2$
8N	S_2,	S_5,	0	V_{bc}	$-V_{bc}$	0	$-I_C$	I_C	$-kV_{bc}$	$\pi/2$	$-kI_C$	$\pi/2$
9P	S_1,	S_4,	0	$-V_{ca}$	V_{ca}	$-I_C$	0	I_C	kV_{ca}	$\pi/2$	kI_C	$7\pi/6$
9N	S_3,	S_6,	0	V_{ca}	$-V_{ca}$	I_C	0	$-I_C$	$-kV_{ca}$	$\pi/2$	$-kI_C$	$7\pi/6$
0A	S_1,	S_4,	0	0	0	0	0	0	0	0	0	0
0B	S_2,	S_5,	0	0	0	0	0	0	0	0	0	0
0C	S_3,	S_6,	0	0	0	0	0	0	0	0	0	0

1. Group I. Synchronously rotating vectors. This group consists of six combinations (1 to 6) having each of the three output phases connected to a different input phase. Each of them generates a three-phase output voltage having magnitude and frequency equivalent to those of the input voltages (V_i, and ω_i,) but with a phase sequence altered from that of the input voltages. As the input frequency is not related to the output frequency, the SVM method does not use the above listed vectors to synthesize the reference voltage vector that rotates at a frequency ω_o.

2. Group II. Stationary vectors. The second group ($1P$ to $9N$) is classified into three sets. Each of these has six switch combinations and has a common feature of connecting two output phases to the same input phase. The corresponding space vectors of these combinations all have a constant phase angle, thus being named *stationary vector*. For example, in the first set of the group, output phases B and C are switched simultaneously on to input phases b, a, and c. This results in six switch combinations all giving a zero output line-to-line voltage V_{BC}. For the second set, the short-circuited output phases are C and A; hence, V_{CA} is zero. In the final six, output phases A and B are connected together and the zero line-to-line voltage is V_{AB}. The magnitudes of these vectors, however, vary with changes of the instantaneous input line-to-line voltages.

3. Group III. Zero vectors. The final three combinations in the table form the last group. These have three output phases switched simultaneously on to the same input phase resulting in zero line-to-line voltages and are called *zero voltage vectors*. When a three-phase load is connected to the converter output terminals, a three-phase output current is drawn from the power source. The input currents are equivalent to the instantaneous output currents; thus, all input current vectors corresponding to the 27 output voltage vectors are also listed in Table 14.1.

14.4.2.3 Voltage and Current Hexagons

A complete cycle of a three-phase sinusoidal voltage waveform can be divided into six sextants as shown in Fig. 14.8. At each transition point from one sextant to another the magnitude of one phase voltage is zero while the other two have the same amplitude but opposite polarity. The phase angles of these points are fixed. Applying this rule to the 18 stationary voltage vectors in Table 14.1, their phase angles are determined by the converter output line-to-line voltages $V_{AB}V_{BC}$, and V_{CA}. The first six, all giving zero V_{BC}, may locate either at the transition point between sextants 1 and 2 ($\omega_o t = 30°$) or that between sextants 4 and 5 ($\omega_o t = 210°$), depending upon the polarity of V_{AB} and V_{CA}. From the waveform diagram given in Fig. 14.8 the three having positive V_{AB} and negative V_{CA} are at the end of sextant 1; conversely, the other three are at the end of sextant 4.

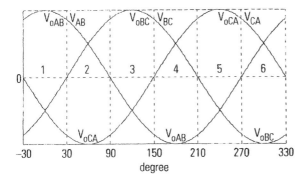

FIG. 8   Six sextants of the output line-to-line voltage waveforms.

The magnitudes of V_{AB} and V_{CA} are determined by the switch positions of the converter and can correspond to any of the input line-to-line voltages, $V_{ab}V_{bc}$, or V_{ca}. Similarly, vectors in the second set generating zero V_{CA} and nonzero V_{AB} and V_{BC} are at the end of either sextant 3 or 6. The three having positive V_{BC} and negative V_{AB} are for the former sextant, the other three are for the latter. The final set is for sextants 2 and 5.

Projecting the stationary voltage and current vectors onto the $\alpha\beta$ plane, the voltage hexagon obtained is shown in Fig 14.9a. It should be noted that this voltage vector diagram can also be obtained by considering the magnitudes and phases of the output voltage vectors associated with the switch combinations given in Table 14.1. The same principle can be applied to the corresponding 18 input current vectors, leading to the current hexagon depicted in Fig. 14.9b. Both the output voltage and input current vector diagrams are valid for a certain period of time since the actual magnitudes of these vectors depend on the instantaneous values of the input voltages and output currents.

14.4.2.4 Selection of Stationary Vectors

Having arranged the available switch combinations for matrix converter control, the SVM method is designed to choose appropriately four out of 18 switch combinations from the second group at any time instant. The selection process follows three distinct criteria, namely, that at the instant of sampling, the chosen switch combinations must simultaneously result in

1. The stationary output voltage vectors being adjacent to the reference voltage vector in order to enable the adequate output voltage synthesis
2. The input current vectors being adjacent to the reference current vector

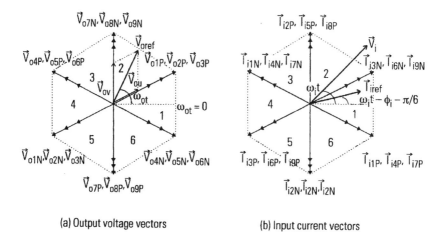

(a) Output voltage vectors   (b) Input current vectors

FIG. 9   Output-voltage and input-current vector hexagons: (a) output-voltage vectors and
(b) input-current vectors.

in order that the phase angle between the input line-to-line voltage and
phase current, and hence the input power factor, being the desired value
3. The stationary voltage vectors having the magnitudes corresponding
to the maximum available line-to-line input voltages

To satisfy the first condition above, consider the reference voltage vector that
lies in one of the six sectors at any particular time instant. One of the line-to-
line voltages in this corresponding sector is bound to be either most positive or
most negative, hence being denoted as the peak line. The vectors selected to
synthesis the reference voltage vector should be those that make the voltage of
the peak line nonzeros. For example, when the reference voltage vector $\vec{V}_{ref}$ is
in sector 2 as shown in Fig 14.9a, the peak line is V_{OCA}, and the stationary vectors
giving nonzero V_{CA} are the first and third sets in group II of Table 14.1, thus 12
in total.

Further selection takes both second and third requirements into account.
From Eq. (14.35), ϕ_i is the phase-angle between input line-to-line voltage and
phase current, which, for unity input power factor, must be kept at 30°, giving
zero phase shift angle between the input phase voltage and current. Following
the same principle for the reference voltage vector, at a particular time interval
the reference current vector locates in one of six sectors and so does the input
line-to-line voltage vector. Note that the input voltage vector leads the current
vector by 30° and transits from the same sector as that of the current to the
next adjacent one. Consequently, the maximum input voltage value is switched

between two input line-to-line voltages of these two sectors. Taking the previously selected 12 vectors and using the maximum available input voltages, there are four vectors with peak-line voltages equivalent to one of these two line-to-line voltages, and these are chosen. This can be illustrated using the input line-to-line voltage waveform in Fig. 14.10b when the reference current vector $\vec{I}_{ref}$ is in sector 1, the input line-to-line voltage vector $\vec{V}_i$ may lie in either sector 1 or 2. The maximum input line-to-line voltage in sector 1 is V_{ab}, while that in sector 2 is $-V_{ca}$. Among the 12 stationary vectors, those having V_{OCA} equivalent to either V_{ab} or $-V_{ca}$ are the suitable ones and these are vectors $1P$, $3N$, $7N$, and $9P$ in Table 14.1. Following the above stated principle, the selected sets of stationary vectors for reference voltage vector and input current vector in sextants 1 to 6 are listed in Table 14.2

14.4.2.5 Computation of Vector Time Intervals

As described above, the selected voltage vectors are obtained from two subsets of the stationary vector group. In particular, vectors $1P$ and $3N$ are from the zero V_{BC} subset, and $7N$ and $9P$ are from the zero V_{AB} subset. The sum of these, given by $|\vec{V}_{o1P}| + |\vec{V}_{o3N}|$ due to the 180° phase angle between them, defines a vector $\vec{V}_{ou}$. Similarly, the solution of $|\vec{V}_{o7N}| + |\vec{V}_{o9P}|$ gives another vector $\vec{V}_{ov}$. As shown in Fig. 14.10a both $\vec{V}_{ou}$ and $\vec{V}_{ov}$ are adjacent vectors of the output reference voltage vector $\vec{V}_{ref}$. Based on the SVM theory, the relation for these voltage vectors can be written

$$\int_{t_0}^{t_0+T_s} \vec{V}_{ref}\, dt \approx T_{ov}\vec{V}_{ov} + T_{ou}\vec{V}_{ou} \qquad (14.37)$$

where T_{ov} and T_{ou} represent the time widths for applying vectors $\vec{V}_{ov}$ and $\vec{V}_{ou}$, respectively, t_0 is the initial time, T_s is a specified sample period.

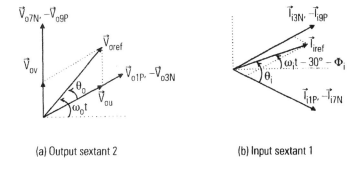

(a) Output sextant 2   (b) Input sextant 1

FIG. 10   Vector diagrams: (a) output sextant 2 and (b) input sextant 1.

TABLE 14.2 Selected Sets of Switch Combinations

	Input sextant 1	Input sextant 2	Input sextant 3	Input sextant 4	Input sextant 5	Input sextant 6
Output sextant 1	1P,4N,6P,3N	5N,2P,3N,6P	2P,5N,4P,1N	6N,3P,1N,4P	3P,6N,5P,2N	4N,1P,2N,5P
Output sextant 2	3N,9P,7N,1P	9P,3N,2P,8N	1N,7P,8N,2P	7P,1N,3P,9N	2N,8P,9N,3P	8P,2N,1P,7N
Output sextant 3	4P,7N,9P,6N	8N,5P,6N,9P	5P,8N,7P,4N	9N,6P,4N,7P	6P,9N,8P,5N	7N,4P,5N,8P
Output sextant 4	6N,3P,1N,4P	3P,6N,5P,2N	4N,1P,2N,5P	1P,4N,6P,3N	5N,2P,3N,6P	2P,5N,4P,1N
Output sextant 5	7P,1N,3P,9N	2N,8P,9N,3P	8P,2N,1P,7N	3N,9P,7N,1P	9P,3N,2P,8N	1N,7P,8N,2P
Output sextant 6	9N,6P,4N,7P	6P,9N,8P,5N	7N,4P,5N,8P	4P,7N,9P,6N	8N,5P,6N,9P	5P,8N,7P,4N

For the example given above, the space vector diagrams of output sextant 2 and input sextant 1 are shown in Fig. 14.10. Let the phase angle of the reference voltage vector, $\omega_o t$, be defined by the sextant number (1–6) and the angle within a sextant θ_o ($0° \leq \theta_o \leq 60°$). Similarly, the phase angle of the input current vector, $\omega_i t - \phi_i - 30°$, may be defined by the input sextant number and the remaining angle θ_i. The subsequent derivation is then based on these vector diagrams.

For a sufficiently small T_s, the reference voltage vector can be regarded as constant, and hence Eq. (14.37) can be expressed in two-dimensional form as

$$T_s \left| \vec{V}_{ref} \right| \begin{bmatrix} \cos(\theta_o + 30°) \\ \sin(\theta_o + 30°) \end{bmatrix} = \left(t_{1P} \left| \vec{V}_{o1P} \right| - t_{3N} \left| \vec{V}_{o3P} \right| \right) \begin{bmatrix} \cos 30° \\ \sin 30° \end{bmatrix}$$

$$+ \left(t_{7N} \left| \vec{V}_{o7N} \right| - t_{9P} \left| \vec{V}_{o9P} \right| \right) \begin{bmatrix} 0 \\ 1 \end{bmatrix} \quad (14.38)$$

where t_{1P}, t_{3N}, t_{7N}, and t_{9P} are the time widths of the associated voltage vectors. Note that in general these time widths are denoted as t_1, t_2, t_3, and t_4.

As the magnitude and frequency of the desired output voltage are specified in advance, the input voltage and phase angle are measurable and relationships for evaluating the magnitudes of the stationary voltage vectors are given in Table 14.1. Equation (14.38) can then be decomposed into the scalar equations

$$t_{1P} V_{il} \cos \omega_i t - t_{3N} V_{il} \cos(\omega_i t - 240°) = V_{ol} T_s \sin(60° - \theta_o), \quad (14.39)$$

$$t_{7N} V_{il} \cos \omega_i t - t_{9P} V_{il} \cos(\omega_i t - 240°) = V_{ol} T_s \sin\theta_o, \quad (14.40)$$

where V_{il} is the magnitude of the input line-to-line voltage and V_{ol} is that of the output line-to-line voltage.

Applying the SVM principle to control the phase angle of the reference current vector, the following equation results.

$$T_s \left| \vec{I}_{iref} \right| \begin{bmatrix} \cos(\omega_i t - \phi_i - 30°) \\ \sin(\omega_i t - \phi_i - 30°) \end{bmatrix} = \left(t_{1P} \left| \vec{I}_{i1P} \right| - t_{7N} \left| \vec{I}_{i7N} \right| \right) \begin{bmatrix} \cos 30° \\ -\sin 30° \end{bmatrix}$$

$$+ \left(t_{3N} \left| \vec{I}_{i3N} \right| - t_{9P} \left| \vec{I}_{i9P} \right| \right) \begin{bmatrix} \cos 30° \\ \sin 30° \end{bmatrix} \quad (14.41)$$

Using the magnitudes of the input current vectors given in Table 14.1, the above equation may be written as

$$(1+\alpha)T_s \left| \vec{I}_{iref} \right| \begin{bmatrix} \cos(\omega_i t - \phi_i - 30°) \\ \sin(\omega_i t - \phi_i - 30°) \end{bmatrix} = I_A \begin{bmatrix} t_{1P} + t_{3N} \\ \frac{1}{\sqrt{3}}(-t_{1P} + t_{3N}) \end{bmatrix} \quad (14.42)$$

$$-\alpha T_s \left| \vec{I}_{iref} \right| \begin{bmatrix} \cos(\omega_i t - \phi_i - 30°) \\ \sin(\omega_i t - \phi_i - 30°) \end{bmatrix} = -I_C \begin{bmatrix} t_{7N} + t_{9P} \\ \frac{1}{\sqrt{3}}(-t_{7N} + t_{9P}) \end{bmatrix},$$

$$\tag{14.43}$$

where α is an arbitrary variable that enables Eq. (14.41) to be divided into two parts. Since it is necessary to set only the phase angle of the reference input current vector, Eq. (14.42) becomes

$$\frac{\sin(\omega_i t - \phi_i - 30°)}{\cos(\omega_i t - \phi_i - 30°)} = \frac{t_{3N} - t_{1P}}{\sqrt{3}(t_{3N} + t_{1P})}$$

which can be rearranged to give

$$t_{3N} \cos(\omega_i t - \phi_i + 30°) - t_{1P} \cos(\omega_i t - \phi_i - 90°) = 0 \tag{14.44}$$

Repeating this procedure with Eq. (14.43) results in

$$t_{9P} \cos(\omega_i t - \phi_i + 30°) - t_{7N} \cos(\omega_i t - \phi_i - 90°) = 0 \tag{14.45}$$

Solving Eqs. (14.39), (14.40), (14.44), and (14.45) results in expressions for calculating the four vector time widths

$$t_{1P} = \frac{QT_s}{\cos\phi_i \cos 30°} \sin(60° - \theta_o)\sin(60° - \theta_i) \tag{14.46}$$

$$t_{3N} = \frac{QT_s}{\cos\phi_i \cos 30°} \sin(60° - \theta_o)\sin\theta_i \tag{14.47}$$

$$t_{7N} = \frac{QT_s}{\cos\phi_i \cos 30°} \sin\theta_o \sin(60° - \theta_i) \tag{14.48}$$

$$t_{9P} = \frac{QT_s}{\cos\phi_i \cos 30°} \sin\theta_o \sin\theta_i \tag{14.49}$$

where $Q = V_{ol} / V_{il}$ is the voltage transfer ratio, and θ_o and θ_i, are the phase angles of the output voltage and input current vectors, respectively, whose values are limited within 0°–60° range. The above equations are valid for when the reference output-voltage vector stays in output sextant 2 while the reference input-current vector is in input sextant 1. For different sets of vectors the same principle is applied.

In principle, each of the time widths is restricted by two rules, namely

$$0 \leq \frac{t_k}{T_s} \leq 1 \tag{14.50}$$

and

$$\sum_{k=1}^{4} \frac{t_k}{T_s} \le 1 \tag{14.51}$$

In variable speed ac drive applications, unity input power factor is desired. As a consequence, the sum of the four vector time widths is normally less than a switching period T_s when the maximum output-to-input voltage ratio is limited to 0.866. The residual time within T_s is then taken by a zero vector; thus,

$$t_0 = T_s - \sum_{k=1}^{4} t_k \tag{14.52}$$

14.5 SPECIMEN SIMULATED RESULTS [34]

14.5.1 Venturini Method

A specimen simulated result is given in Fig. 14.11 for a balanced, three-phase supply applied to a symmetrical, three-phase, series R-L load, without the use of an input filter, assuming ideal switches. With a carrier frequency of 2 kHz the output current waveform is substantially sinusoidal at 40 Hz but has a 2000/40, or 50, times ripple. It can be seen that the phase angle between the input phase voltage and the fundamental input current is zero, resulting in unity displacement factor.

14.5.2 Space Vector Modulation Method

A specimen simulated result is given in Fig. 14.12 for the case of balanced 240-V (line), 50-Hz supply with symmetrical, three-phase, series R-L load. The carrier (switching) frequency is 2 kHz, resulting in a substantially sinusoidal output current of $f_o = 40$ Hz.

Comparison of Fig. 14.12 with Fig. 14.11 shows that the two methods give very similar results. The SVM method has the advantage of simpler computation and lower switching losses. The Venturini method exhibits superior performance in terms of output voltage level and input current harmonics.

14.6 SUMMARY

The matrix converter holds the promise of being an all-silicon solution for reducing the use of expensive and bulky passive components, presently used in inverter and cycloconverter systems. Its essentially single-stage conversion may prove to be a crucial factor for improving the dynamic performance of system.

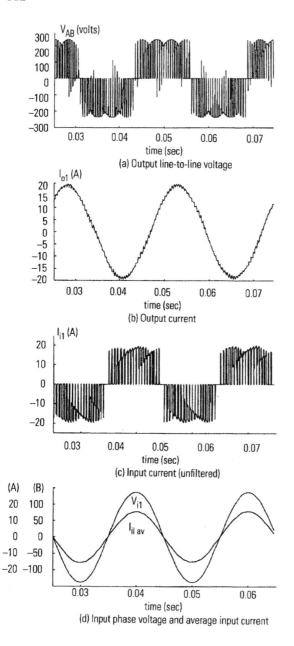

FIG. 11   Simulated matrix converter waveforms. $V_o/V_i = 0.866, f_o = 40$ Hz, $f_s = 2$ kHz. (a) Output line-to-line voltage. (b) Output current. (c) Input current (unfiltered). (d) Input phase voltage and input average current [34].

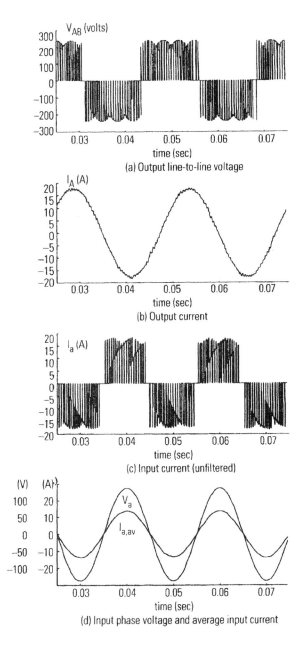

FIG. 12 Simulated matrix converter waveforms using the SVM algorithm. $V_o/V_i = 0.866$, $f_o = 40$ Hz. (a) Output line-to-line voltage. (b) Output current. (c) Input current (unfiltered). (d) Input phase voltage and input average current [34].

Matrix converters have not yet (2003) made any impact in the commercial converter market. The reasons include the difficulties in realizing high-power bidirectional switches plus the difficulties in controlling these switches to simultaneously obtain sinusoidal input currents and output voltages in real time. In addition, the high device cost and device losses make the matrix converter less attractive in commercial terms.

Recent advances in power electronic device technology and very large scale integration (VLSI) electronics have led, however, to renewed interest in direct ac-ac matrix converters. Ongoing research has resulted in a number of laboratory prototypes of new bidirectional switches. As device technology continues to improve, it is possible that the matrix converter will become a commercial competitor to the PWM DC link converter.

15

DC–DC Converter Circuits

A range of dc–dc switch-mode converters are used to convert an unregulated dc input to a regulated dc output at a required voltage level. They achieve the voltage regulation by varying the on–off or time duty ratio of the switching element. There are two main applications. One is to provide a dc power supply with adjustable output voltage, for general use. This application often requires the use of an isolating transformer. The other main application of dc–dc converters is to transfer power from a fixed dc supply, which may be rectified ac, to the armature of a dc motor in the form of adjustable direct voltage.

Three basic types of dc–dc converters are

Step-down converter
Step-up converter
Step-up-down converter

15.1 STEP-DOWN CONVERTER (BUCK CONVERTER)

15.1.1 Voltage Relationships

The basic form of a voltage step-down (buck) converter uses one controlled switch (Fig. 15.1a). This particular form of the connection does not contain an isolating transformer between the power supply and the load.

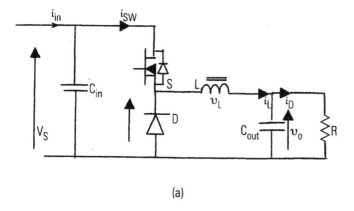

(a)

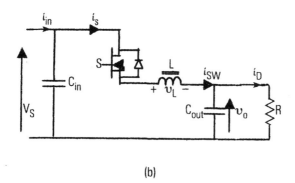

(b)

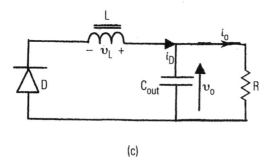

(c)

FIG. 1   Step-down converter, R load: (a) general circuit, (b) switch-on state, and (c) switch-off state.

For the purpose of analysis, the circuit resistances are assumed to be negligible and the capacitors, C_{in} and C_{out}, are very large. For steady-state operation, if the inductor current is continuous and the load current is substantially constant, the load voltage and the three circuit currents will be as given in Fig. 15.2. While the switch conducts, the ideal supply voltage, V_s, is transferred directly across the load so that $v_o(t) = V_s$. During extinction of the controlled switch the supply current, which is the switch current, is zero, but load current, $i_o(t)$, continues to flow via the diode D and the inductor L(Fig. 15.1c). The combination of inductor L and output capacitor C_{out} forms a low-pass filter. The switching frequency f_s is determined by the designer and implemented as the switching rate of the controlled switch. Typical switching frequencies are usually in the range 100 Hz $\leq$ $f_s \leq$ 20,000 Hz. The overall periodic time T_p is related to the switching frequency by

$$T_{on} + T_{off} = T_p = \frac{1}{f_s} \tag{15.1}$$

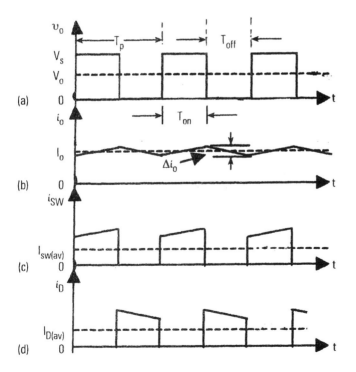

FIG. 2   Waveforms for the step-down converter. Continuous conduction, R load.

If the switch conducts for a fraction k of the total period time, then

$$T_{on} = kT_p$$
$$T_{off} = (1 - k)T_p \tag{15.2}$$

During conduction it may be seen from Fig 1.1(b) that

$$V_s - v_o = L\frac{di_{sw}}{dt} \tag{15.3}$$

The load current waveform $i_o(t)$ for repetitive steady-state operation is shown in Fig. 15.2b. During the switch *on* periods, the current is increasing and the inductor emf has the polarity shown in Fig. 15.1b.

The derivative in Eq. (15.3) can be interpreted from the waveform (Fig. 15.2b) during conduction, as

$$\frac{di}{dt} \equiv \frac{\Delta i_{sw}}{T_{on}} = \frac{\Delta i_{sw}}{kT_p} \tag{15.4}$$

Combining Eqs. (15.3) and (15.4) gives, for the conduction intervals,

$$V_s - v_o = L\frac{\Delta i_{sw}}{kT_p} \tag{15.5}$$

During the switch extinction periods, the load current $i_o(t)$ remains continuous but decreases in value. The polarity of the inductor emf assumes the form of Fig. 15.1c to support the declining current (i.e., to oppose the decrease). Correspondingly, in the switch extinction periods, the circuit voltage equation for Fig. 15.1c is

$$L\frac{di_D}{dt} = v_o \tag{15.6}$$

A typical value for inductor L would be of the order hundreds of microhenries as seen in Fig. 15.2b.

$$\frac{di_D}{dt} \equiv \frac{\Delta i_D}{T_{off}} = \frac{\Delta i_D}{(1-k)T_p} \tag{15.7}$$

Combining Eqs. (15.6) and (15.7) gives

$$L\frac{di_D}{dt} = v_o = \frac{L\,\Delta i_D}{(1-k)T_p} \tag{15.8}$$

Neglecting the very small current in the output capacitor C_{out}, it is seen in Fig 15.1 that $i_o = i_D = i_{sw}$.

The magnitude of the current change Δi_o in Fig. 15.2(b) is the same for both the *on* periods and the *off* periods. Equating Δi_o between Eqs. (15.5) and (15.8) gives

$$(V_s - v_o)\frac{kT_p}{L} = v_o(1-k)\frac{T_p}{L}$$

giving

$$kV_s = v_o = V_o + \Delta V_o \qquad (15.9)$$

The output voltage ripple ΔV_o, discussed in Sec. 15.1.3 below and shown in Fig. 15.3, has a time average value of zero so that Eq. (15.9) is true for the average output voltage V_o and

$$k = \frac{V_o}{V_s} \qquad (15.10)$$

In the absence of the filter inductor L the supply voltage is applied directly across the load during conduction. It is seen that Eq. (15.10) is then also true, because $V_o = (kT_p/T_p)V_s$. The voltage waveform $v_o(t)$ in Fig. 15.2a is defined by

$$v_o(t) = \begin{cases} V_s & 0 \le t \le kT_p \\ 0 & kT_p \le t \le T_p \end{cases} \qquad (15.11)$$

The rms value V_o (rms) can be obtained from the integral definition

$$V_o(\text{rms}) = \sqrt{\frac{1}{T_p}\int_0^{T_p} v_o^2(t)\,dt} \qquad (15.12)$$

Substituting Eq. (15.11) into Eq. (15.12) gives

$$V_o(\text{rms}) = \sqrt{\frac{1}{T_p}V_s^2 \Big|_0^{kT_p}} \qquad (15.13)$$

Therefore,

$$V_o = \sqrt{k}V_s$$

15.1.2 Current Relationships

15.1.2.1 Average Currents

Power enters the circuit of Fig. 15.1a from the ideal dc voltage source. Now power can only be obtained by combining voltage and current components of the same frequency. The component of the input current $i_{in}(t)$ that combines with

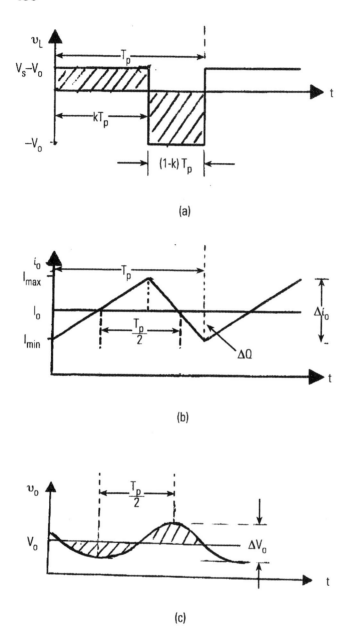

FIG. 3 Output voltage and current ripples in a step-down converter: (a) inductor voltage, (b) output current, and (c) output voltage (not sinusoidal).

ideal direct voltage V_s is therefore the zero frequency or time average value $I_{in}(av)$. This gives an input power

$$P_{in} = V_s I_{in}(av) \tag{15.14}$$

If the output voltage ripple is entirely smoothed by capacitor C_{out} in Fig. 15.1, then the output voltage is constant and $v_o(t) = V_o$. The load current (Fig. 15.2b) consists of an average value I_o plus a ripple Δi_o. The output power is then obtainable from the (presumed) constant output voltage V_o combined with the average value of the output current I_o.

$$P_{out} = V_o I_o \tag{15.15}$$

Neglecting the small switching losses in the controlled switch and the diode and also the resistive loss in the inductor, then $P_{out} = P_{in}$. Combining Eqs. (15.10), (15.14), and (15.15) gives

$$\frac{V_o}{V_s} = \frac{I_{in(ac)}}{I_o} = k \quad \text{(duty ratio)} \tag{15.16}$$

Equation (15.16) corresponds directly to the voltage and current relationships for an electrical transformer in which duty ratio k would be represented by the primary-to-secondary turns ratio. In Fig. 15.3b it may be seen that the magnitude of the current ripple Δi_o defines the maximum and minimum values of the output current in terms of the average value I_o.

$$I_{o(\max)} = I_o + \frac{\Delta i_o}{2} \qquad I_{o(\min)} = I_o - \frac{\Delta i_o}{2} \tag{15.17}$$

The peak-to-peak magnitude of the output current ripple Δi_o can be obtained from Eq. (15.17)

$$I_{o(max)} - I_{o(min)} = \Delta i_o \tag{15.18}$$

For the load current of the converter shown in Fig. 15.2b, the instantaneous value $i_o(t)$ maybe given expressed as

$$i_o(t) = \left(I_{o(\min)} + \frac{\Delta i_o}{kT_p} t \right) \Big|_0^{kT_p} \quad \text{for} \quad 0 \le t \le kT_p$$

$$+ \left[I_{o(\max)} - \frac{\Delta i_o}{(1-k)T_p}(t - kT_p) \right] \Big|_{kT_p}^{T_p} \quad \text{for} \quad kT_p \le t \le T_p \tag{15.19}$$

The average value $I_o(av)$ of the current $i_o(t)$ during any interval T_p is defined by the equation

$$I_o(av) = \frac{1}{T_p} \int_0^{T_p} i_o(t)\, dt$$

(15.20)

The average value $I_{sw}(av)$ of the switch current $i_{sw}(t)$ (Fig. 15.2c), obtained by substituting from the first bracketed term of Eq. (15.19) into Eq. (15.20), is found to be

$$I_{sw}(av) = kI_o = I_{in}(av)$$

(15.21)

which confirms Eq. (15.16). The average value $I_D(av)$ of the diode current $i_D(t)$ (Fig. 15.2d) is obtained by substituting from the second bracketed term of Eq. (15.19) into Eq. (15.20).

$$I_D(av) = (1 - k)I_o$$

(15.22)

It is seen from Eqs. (15.21) and (15.22) that the average load current $I_o(av)$ is given by

$$I_o = I_o(av) = I_{sw}(av) + I_D(av)$$

(15.23)

15.1.2.2 RMS Currents

The thermal ratings of the circuit components depend on their respective rms current ratings (not on their average current ratings). The rms value of the current during any period T_p is defined by the classical integral definition

$$I_o(\text{rms}) = \sqrt{\frac{1}{T_p} \int_0^{T_p} i_o^2(t)\, dt}$$

(15.24)

During the switch conduction intervals (Fig. 15.1b), when for example, $0 \le t \le kT_p$, then $i_o = i_{sw}$ and $i_D = 0$. During the switch extinction intervals (Fig. 15.1a), when for example, $kT_p \le t \le T_p$, then $i_o = i_D$ and $i_{sw} = 0$. Since the two components $i_{sw}(t)$ and $i_D(t)$ of $i_o(t)$ are operative for different intervals of the cycle T_p, and not simultaneously, Eq. (15.24) can be written

$$I_o^2(\text{rms}) = \frac{1}{T_p}\left[\int_0^{kT_p} i_{sw}^2(t)\, dt + \int_{kT_p}^{T_p} i_D^2(t)\, dt \right]$$

(15.25)

When the rms value $I_{sw}(\text{rms})$ of the switch current shown in Fig. 15.2c is obtained by substituting the first bracketed term of Eq. (15.19) into Eq. (15.25),

$$I_{sw}^2(\text{rms}) = k\left(I_o^2 + \frac{(\Delta i_o)^2}{12} \right)$$

(15.26)

When the ripple current Δi_o is negligibly small, it is seen from Eq. (15.26) that

$$I_{sw}(\text{rms}) = \sqrt{k}I_o$$

(15.27)

The rms value $I_D(\text{rms})$ of the diode current $i_D(t)$, shown in Fig. 15.2d, is obtained by use of the second bracketed term from Eq. (15.19) into Eq. (15.24)

$$I_D^2(\text{rms}) = (1-k)\left(I_o^2 + \frac{(\Delta i_o)^2}{12} \right)$$

(15.28)

It is seen from Eqs. (15.26) and (15.28) that

$$I_o^2(\text{rms}) = I_{sw}^2(\text{rms}) + I_D^2(\text{rms}) = I_o^2 + \frac{(\Delta io)^2}{12}$$

(15.29)

The output capacitor C_{out} in Fig. 15.1 is presumed to carry the ripple current Δi_o but is an open circuit to the average load current I_o. Instantaneous capacitor current $i_c(t)$ may be interpreted from Fig. 15.3b as

$$i_c(t) = \frac{\Delta io}{kT_P}t \bigg|_0^{kT_P} + \left(\frac{\Delta io}{(1-k)} - \frac{\Delta io}{(1-k)T_P}t \right)\bigg|_{kT_P}^{T_P}$$

(15.30)

Substituting $i_c(t)$ from Eq. (15.30) into Eq. (15.24) gives

$$I_{C_{out}}(\text{rms}) = \frac{(\Delta io)^2}{3}\frac{1-5k}{1-k}$$

(15.31)

15.1.3 Output Voltage Ripple

With a practical (noninfinite) value of capacitance C_{out} the output voltage $v_o(t)$ can be considered to consist of an average value V_o plus a small but finite ripple component with peak value ΔV_o (Fig. 15.3c). Although the two shaded areas in Fig. 15.3c are equal, the two half waves are not, in general, symmetrical about the V_o value. The waveform of the voltage ripple depends on the duty cycle k and on the nature of the load impedance, shown as resistor R in Fig. 15.1

Instantaneous variation of the inductor voltage $v_L(t)$ in Fig. 15.3a shows the variations of polarity described in Sec. 15.1.1. In this figure the two shaded areas are equal (because the average inductor voltage is zero) and represent the physical dimension of voltage multiplied by time, which is electric flux or electric charge Q.

The incremental voltage ΔV_o is associated with incremental charge ΔQ by the relation

$$\Delta V_o = \frac{\Delta Q}{C_{out}}$$

(15.32)

The area of each of the isosceles triangles representing ΔQ in Fig. 15.3b is

$$\Delta Q = \frac{1}{2}\frac{T_P}{2}\frac{\Delta i_o}{2} = \frac{T_P}{8}\Delta i_o$$

(15.33)

Combining Eqs. (15.32) and (15.33) gives

$$\Delta V_o = \frac{T_p \, \Delta i_o}{8C_{out}} \tag{15.34}$$

Eliminating Δi_o (or Δi_D) between Eqs. (15.34) and (15.8) gives

$$\Delta V_o = \frac{T_p}{8C_{out}}(1-k)\frac{V_o T_o}{L} \tag{15.35}$$

The incremental output voltage ratio $\Delta V_o / V_o$ is then

$$\frac{\Delta V_o}{V_o} = \frac{T_p^2(1-k)}{8LC_{out}} \tag{15.36}$$

Now the period T_p can be expressed in terms of the switching frequency f_s, as in Eq. (15.1). Also, the corner frequency f_c of the low-pass filter formed by L and C_{out} is

$$f_c = \frac{1}{2\pi\sqrt{LC_{out}}} \tag{15.37}$$

Expressing $\Delta V_o / V_o$ in terms of frequency is found to give

$$\frac{\Delta V_o}{V_o} = \frac{\pi^2}{2}(1-k)\left(\frac{f_c}{f_s}\right)^2 \tag{15.38}$$

Equation (15.38) shows that the voltage ripple can be minimized by selecting a frequency f_c of the low-pass filter at the output such that $f_c \ll f_s$. In switch-mode dc power supplies, the percentage ripple in the output voltage is usually specified to be less than 1%.

There is an economic case for reducing the size of the inductor (lower weight and cost), but at a given frequency this will increase (1) the current ripple and therefore the peak and rms switch currents and associated losses and (2) the ripple voltages (and currents) on the input and output capacitors implying a need for larger capacitors.

Increase of the switching frequency f_s (i.e., smaller T_p) results in lower ripple magnitude ΔV_o. This requires a smaller inductor and lower rated smoothing capacitors. However, the switching losses in the semiconductor may then become significant and electromagnetic interference problems become worse.

15.1.4 Worked Examples

Example 15.1   A semiconductor switching element in a power circuit has a minimum effective on time of 42 μs. The dc supply is rated at 1000 V. (1) If the chopper output voltage must be adjustable down to 20 V, what is the highest chopper frequency? (2) If the chopping frequency increases to 500 Hz, what will then be the minimum output voltage?

    1.   Minimum required duty cycle $= 20/1000 = 0.02 = k$.

$$T_{on} \text{ (minm)} = 42 \times 10^{-6} \text{ s}$$

But $T_{on} = kT_p$ from Eq. (15.2). Therefore,

$$T_p = \frac{T_{on}}{k} = \frac{42 \times 10^{-6}}{0.02} = 2100 \times 10^{-6} \text{ s}$$

Maximum chopper frequency is

$$f_s = \frac{1}{T_p} = \frac{10^6}{2100} = 476.2 \text{ Hz}$$

    2.   For $f_s = 500$ Hz

$$T_p = \frac{1}{f_s} = 0.002 \text{ s} = 2000 \text{ } \mu\text{s}$$

If T_{on} remains at 42 μs, the minimum duty cycle is then

$$k = \frac{T_{on}}{T_p} = \frac{42}{2000} = 0.021$$

At this value of duty cycle the converter output voltage, from Eq. (15.19), is

$$V_o = kV_s = 0.021 \times 1000 = 21 \text{ V}$$

    Example 15.2   The semiconductor switch in a dc step-down converter has a minimum effective off time of 28.5 μs. If the constant dc supply is rated at 1250 V and the chopping frequency is 2000 Hz, calculate the maximum duty cycle and the maximum output voltage that can be obtained.

$$T_p = \frac{1}{f_s} = \frac{1}{2000} = 500 \text{ } \mu\text{s}$$

But the switching off time is 28.5 μs so that the on time per cycle is

$$T_{on} = 500 - 28.5 = 471.5 \text{ } \mu\text{s}$$

The maximum duty cycle k will then be, from Eq. (15.2),

$$k = \frac{T_{on}}{T_p} = \frac{471.5}{500} = 0.943$$

From Eq. (15.10), the output voltage V_o is

$$V_o = kV_s = 0.943 \times 1250 = 1179 \text{ V}$$

Example 15.3 The dc chopper converter in Example 15.1 is connected to resistive load $R = 1.85 \; \Omega$. What is the average input current when it operates at a frequency of 500 Hz?

From Example 15.1, part (2), the average load voltage is $V_o = 21$ V. Therefore, the average load current is

$$I_o = \frac{V_o}{R} = \frac{21}{1.85} = 11.35 \text{ A}$$

As before, the duty cycle k is

$$k = \frac{T_{on}}{T_p} = \frac{42}{2000} = 0.021$$

From Eq. (15.16), the average value $I_{in}(av)$ of the input current, which is also the average value $I_{sw}(av)$ of the switch current, is

$$I_{in}(av) = kI_o = 0.021 \times 11.35 = 0.24 \text{ A}$$

Example 15.4 A step-down dc converter is used to convert a 100-V dc supply to 75-V output. The inductor is 200 μH with a resistive load 2.2 Ω. If the transistor power switch has an on time of 50 μs and conduction is continuous, calculate

1. The switching frequency and switch off time
2. The average input and output currents
3. The minimum and maximum values of the output current
4. The rms values of the diode, switch, and load currents

1. $T_{on} = 50 \times 10^{-6}$ s. Now $V_s = 100$ V, $V_o = 75$ V, so that, from Eq. (15.10),

$$k = \frac{V_o}{V_s} = \frac{75}{100} = 0.75$$

From Eq. (15.2)

$$T_p = \frac{T_{on}}{k} = \frac{50}{0.75} = 66.67 \; \mu\text{s}$$

$$T_{off} = (1-k)T_p = (0.25)T_p = 16.67 \; \mu\text{s}$$

Switching frequency f_s is given, from Eq. (15.1), by

$$f_s = \frac{1}{T_p} = \frac{10^6}{66.67} = 15 \text{ kHz}$$

2. The average output current I_o is

$$I_o = \frac{V_o}{R} = \frac{75}{2.2} = 34.1 \text{ A}$$

From Eq. (15.16), the average input current $I_{in}(av)$ is then

$$I_{in}(av) = kI_o = 0.75 \times 34.1 = 25.6 \text{ A}$$

3. From Eq. (15.15), noting that $\Delta i_{sw} = \Delta i_o$, it is seen that the peak-to-peak output current Δi_o may be expressed as

$$\Delta i_o = \frac{kT_p}{L}(V_s - V_o) = \frac{0.75 \times 66.7/10^6}{200/10^6}(100 - 75) = 6.25 \text{ A}$$

Therefore, from Eq. (15.17),

$$I_{o(max)} = 34.1 + \frac{6.25}{2} = 37.225 \text{ A}$$

$$I_{o(min)} = 34.1 - \frac{6.25}{2} = 30.975 \text{ A}$$

4. The rms output current $I_o(\text{rms})$, from Eq. (15.28),

$$I_o(\text{rms}) = \sqrt{I_o^2 + \frac{(\Delta i_o)^2}{12}} = \sqrt{(34.1)^2 + \frac{(6.25)^2}{12}} = \sqrt{1166} = 34.15 \text{ A}$$

From Eq. (15.25),

$$I_{sw}^2(\text{rms}) = kI_o^2(\text{rms})$$

so that

$$I_{sw}(\text{rms}) = \sqrt{0.75} \times 34.15 = 29.575 \text{ A}$$

Using the alternative expression (15.26), since the ripple is small,

$$I_{sw}(\text{rms}) = \sqrt{k}I_o = \sqrt{0.75} \times 34.1 = 29.53 \text{ A}$$

From Eq. (15.27),

$$I_D^2(\text{rms}) = (1 - k)I_o^2(\text{rms})$$

or

$$I_D(\text{rms}) = \sqrt{1 - k}I_o(\text{rms}) = \sqrt{0.25} \times 34.1 = 17.05 \text{ A}$$

15.2   STEP-UP CONVERTER (BOOST CONVERTER)

15.2.1   Voltage Relationship

If the controlled switch S, the diode D, and the filter inductor L are configured as in Fig. 15.4a, the circuit can operate to give an output average voltage V_o higher at all times than the supply voltage V_s. In Fig. 15.4 it is assumed that the switches are ideal and that the inductor and capacitors are lossless. Equations (15.1) and (15.2) for the basic switching conditions also apply here.

Waveforms for steady-state converter operation, with continuous conduction, are given in Fig. 15.5. The inductor voltage $v_L(t)$ and the switch voltage $v_{sw}(t)$ sum to the supply voltage V_s at all times. During conduction of the switch S the supply voltage V_s is clamped across the inductor. Because the inductor average voltage is zero the positive and negative excursions of the inductor voltage (Fig.15. 5b) are equal in area. The input current $i_{in}(t)$, in Fig 15.4a, divides between the switch (when it is conducting) current $i_{sw}(t)$ and the diode current $i_D(t)$.

During the extinction intervals, when S is on (Fig. 15.4b), inductor L charges up.

$$L\frac{di_{in}}{dt} = V_s = L\frac{\Delta i_{in}}{T_{on}} = L\frac{\Delta i_{in}}{kT_p} \tag{15.39}$$

During the extinction intervals, when S is open (Fig. 15.4c), energy from the inductor and from the supply are both transferred to the load.

$$L\frac{di_{in}}{dt} = V_s - V_o = -L\frac{\Delta i_{in}}{T_{off}} = -L\frac{\Delta i_{in}}{(1-k)T_p} \tag{15.40}$$

But the current excursions Δi_{in} are equal for both the conduction and extinction intervals (Fig. 15.5c). Therefore, eliminating Δi_{in} between Eqs. (15.39) and (15.40) gives

$$\frac{kT_pV_s}{L} = \frac{(1-k)T_p}{L}(V_s - V_o) \tag{15.41}$$

and

$$V_s = (1-k)V_o \tag{15.42}$$

While the switch S is off, load current continues to conduct via the inductor and diode in series (Fig. 15.4c). The output voltage $v_o(t)$ maps the load current $i_o(t)$.

$$v_o(t) = i_o(t)R = (I_o + \Delta i_o)R = V_o + \Delta V_o \tag{15.43}$$

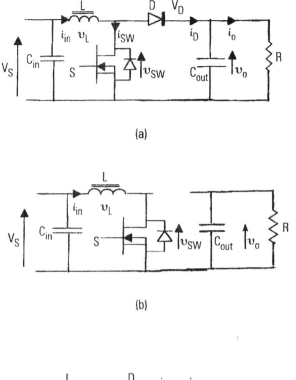

(a)

(b)

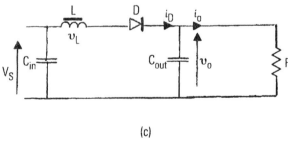

(c)

FIG. 4   Step-up converter, R load: (a) general configuration, (b) switch-on state, and (c) switch-off state.

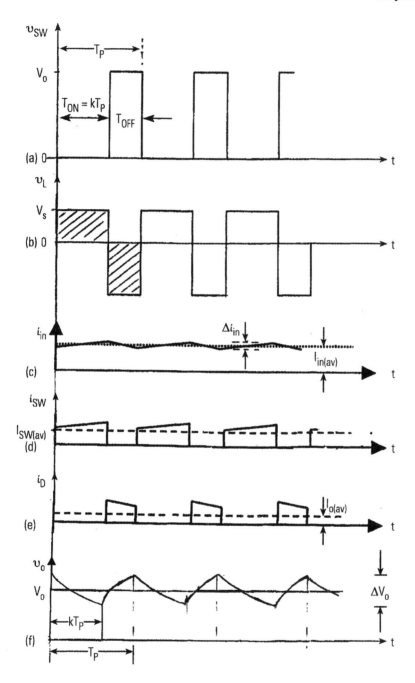

FIG. 5 Waveforms of the step-up converter: (a) switch voltage, (b) inductor voltage, (c) input current, (d) switch current, (e) diode current, and (f) output voltage.

Assume that all the ripple current component Δi_o of the load current flows through C_{out} and its average value I_o flows through the load resistor R. Then the ripple voltage ΔV_o is

$$\Delta V_o = \frac{\Delta Q}{C_{out}} = \frac{I_o k T_p}{C_{out}} = \frac{V_o}{R} \frac{k T_p}{C_{out}} \tag{15.44}$$

The output voltage ripple may be expressed in per unit as

$$\frac{\Delta V_o}{V_o} = \frac{k T_p}{R C_{out}} = \frac{k T_p}{\tau} \tag{15.45}$$

15.2.2   Current Relationships

The expressions (15.11) and (15.12) for input power P_{in} and output power P_{out} to a step-down converter remain valid here for the step-up converter. Also, for a lossless converter it is still true that $P_{in} = P_{out}$. Combining Eqs. (15.11) and (15.12) into Eq. (15.45) gives

$$\frac{\Delta V_o}{V_s} = \frac{I_{in}}{I_o(av)} = \frac{1}{1-k} \tag{15.46}$$

Equation (15.46) compares with the corresponding relation Eq. (15.16) for a step-down converter.

For any value $k < 1$ the voltage ratio V_o/V_s in Eq. (15.46) is greater than unity, corresponding to a step-up transformer.

From Eqs. (15.39) and (15.40)

$$\Delta i_{in} = \frac{k T_p V_s}{L} = \frac{k(1-k)V_o T_p}{L} = \frac{V_s}{L} t_{on} = \frac{k V_o}{L} t_{off} \tag{15.47}$$

Note that in this converter, the output current to the filter capacitor C_{out} is very unsmooth, whereas the input current is much smoother because of the inductor. This is in contrast to the buck converter, in which the opposite is true, because the inductor is in series with the output. This will strongly influence the relative ease of smoothing the input and output dc voltages (i.e., the sizes of the capacitors). From Fig. 15.5c, it can be deduced that

$$I_{in}(\max) = I_{in}(av) + \frac{\Delta i_{in}}{2} \tag{15.48}$$

$$I_{in}(\min) = I_{in}(av) - \frac{\Delta i_{in}}{2} \tag{15.49}$$

$$\Delta i_{in} = I_{in}(\text{max}) - I_{in}(\text{min}) \tag{15.50}$$

The corresponding instantaneous input current is given by an expression similar to Eq. (15.19),

$$i_{in}(t) = \left(I_{in_{min}} + \frac{\Delta i_{in}}{kT_p}t\right)\Big|_0^{kT_p} + \left[I_{in_{max}} + \frac{\Delta i_{in}}{(1-k)T_p}(t-kT_p)\right]\Big|_{kT_p}^{T_p} \tag{15.51}$$

During the switch conduction intervals, $0 \le t \le kT_p$, $i_D = 0$, and $I_m = I_{sw}$. The switch current properties, similarly to Sec. 15.1.2, are found to be

$$I_{sw}(av) = kI_{in} \tag{15.52}$$
$$I_D(av) = I_o(av) = (1 - k)I_{in} \tag{15.53}$$
$$I_{sw}(\text{rms}) = \sqrt{k}I_{in} \tag{15.54}$$

15.2.3 Worked Examples

Example 15.5   A dc–dc step-up converter requires to operate at a fixed output voltage level of 50 V. The input voltage varies between the limits of 22 V and 45 V. Assuming continuous conduction, what must be the range of duty ratios?

$$V_o = 50 \text{ V} \qquad V_s = V_{in}(av) = 22 \text{ V} \rightarrow 45 \text{ V}$$

From Eq. (15.45)

$$1 - k = \frac{V_s}{V_o}$$

If $V_s = 22$ V,

$$k = 1 - \frac{22}{50} = 1 - 0.44 = 0.56$$

If $V_s = 45$ V,

$$k = 1 - \frac{45}{50} = 1 - 0.9 = 0.1$$

Example 15.6   A step-up dc converter is used to convert a 75-V battery supply to 100-V output. As in Example 15.4, the inductor is 200 μH and the load resistor is 2.2 Ω. The power transistor switch has an on time of 50 μs, and the output is in the continuous current mode.

1. Calculate the switching frequency and switch off-time.
2. Calculate the average values of the input and output current.

3. Calculate the maximum and minimum values of the input current.
4. What would be the required capacitance of the output capacitor C_{out} in order to limit the output voltage ripple ΔV_o to 10% of V_o?

1. $V_o = 100$ V, $V_s = 75$ V. From Eq. (15.42),

$$\frac{1}{1-k} = \frac{V_o}{V_s} = \frac{100}{75} = 1.33$$

$$k = 1 - 0.75 = 0.25$$

From Eq. (15.2),

$$T_p = \frac{T_{on}}{k} = \frac{50 \times 10^{-6}}{0.25} = 200 \ \mu s$$

Switching frequency is

$$f_s = \frac{1}{T_p} = \frac{10^6}{200} = 5 \text{ kHz}$$

Switch-off period is

$$T_{off} = (1-k)T_p = 0.75 \times \frac{200}{10^6} = 150 \ \mu s$$

2. $V_o = 100$ V. The average output current I_o is given by

$$I_o = \frac{V_o}{R} = \frac{100}{2.2} = 45.45 \text{ A}$$

From Eq. (15.46), the average input current $I_{in}(av)$ is

$$I_{in}(av) = \frac{I_o}{1-k} = \frac{45.45}{0.75} = 60.6 \text{ A}$$

3. From Eq. (15.40),

$$\Delta i_{in} = \frac{(V_o - V_s)(1-k)T_p}{L}$$

$$= \frac{25 \times 0.75 \times 200 \times 10^6}{200 \times 10^6}$$

$$= 18.75 \text{ A}$$

From Eq. (15.48),

$$I_{in}(\max) = I_{in}(av) + \frac{\Delta i_{in}}{2}$$

$$= 60.6 + \frac{18.75}{2} = 69.975 \text{ A}$$

$$I_{in}(\min) = I_{in}(av) - \frac{\Delta i_{in}}{2}$$

$$= 60.6 - \frac{18.75}{2} = 51.225 \text{ A}$$

4. $\Delta V_o/V_o = 0.1$ (specified). From Eq. (15.45),

$$\frac{\Delta V_o}{V_o} = \frac{kT_p}{RC_{out}}$$

$$C_{out} = \frac{kT_p}{R} \frac{V_o}{\Delta V_o} = \frac{0.25}{2.2} \times \frac{200}{10^6} \times \frac{1}{0.1} = 227 \text{ μF}$$

15.3   STEP-DOWN–STEP-UP CONVERTER (BUCK–BOOST CONVERTER)

15.3.1   Average Voltage and Current

The circuit diagram of a buck–boost converter (Fig. 15.6a) shows that the output current $i_o(t)$ is unidirectional and that the output voltage $v_o(t)$ is negative with respect to the input. During conduction of the semiconductor switch the diode D is reverse biased and the load is electrically isolated (Fig. 15.6b), while energy is transferred from the supply to the inductor. When switch S is open, the load current flows through inductor L, and the stored energy is transferred from the inductor to the load (Fig. 15.6c). The operation of the buck–boost converter is equivalent to that of a buck (step-down) converter in cascade with a boost (step-up) converter. The following analysis presumes that the converter operates in the continuous conduction mode.

During the switch on intervals, the current increases by a value Δi_{in} in time T_{on} ($= kT_p$) through inductor L:

$$V_s = L \frac{\Delta i_{in}}{T_{on}} = L \frac{\Delta i_{in}}{kT_p} \tag{15.55}$$

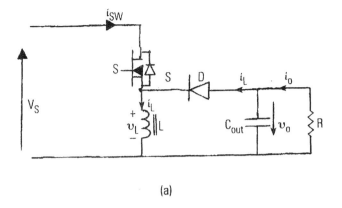

(a)

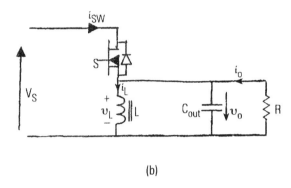

(b)

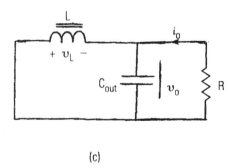

(c)

FIG. 6   Step-up-down converter, R load: (a) general circuit, (b) switch-on state, and (c) switch-off state.

But since the average inductor voltage is zero, the equal area criterion applies in Fig. 15.7 and

$$kV_sT_p = -V_o(1-k)T_p \tag{15.56}$$

It is seen from Eq. (15.56) that the voltage ratio for a buck–boost converter is

$$\frac{V_o}{V_s} = \frac{-k}{1-k} \tag{15.57}$$

For $k < 1/2$, V_o is smaller than the input voltage V_s. For $k > 1/2$, V_o is larger than the input voltage V_s. Output average voltage V_o is negative with respect to the input voltage V_s for all values of k. Once again it can be assumed that the input and output powers are equal, as in Eqs. (15.14) and (15.15), so that

$$\frac{I_o}{I_{in}(av)} = \frac{V_s}{V_o} = \frac{-(1-k)}{k} = \frac{k-1}{k} \tag{15.58}$$

Equation (15.58) demonstrates that the polarity of the output average current I_o is negative with respect to input current $I_{in}(av)$ for all values of duty ratio k. Note that the relations Eqs. (15.57) and (15.58) are valid only for fractional values of k and not at the limits $k = 0$ or $k = 1$.

15.3.2 Ripple Voltage and Current

The peak-to-peak ripple input current Δi_{in}, shown in Fig. 15.7, can be obtained from Eq. (15.55):

$$\Delta i_{in} = \frac{kT_pV_s}{L} \tag{15.59}$$

Since the periodic time is the inverse of the switching frequency f_s, the current ripple may also be expressed as

$$\Delta i_{in} = \frac{kV_s}{f_sL} \tag{15.60}$$

From Eq. (15.57) the duty ratio k can be expressed in terms of the system average voltages

$$k = \frac{V_o}{V_o - V_s} \tag{15.61}$$

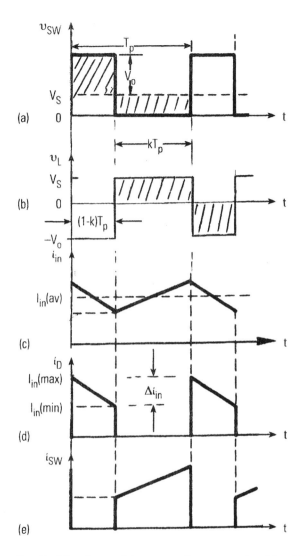

FIG. 7 Waveforms of the step-up-down converter: (a) switch voltage, (b) inductor voltage, (c) input current, (d) diode current, and (e) switch current.

Combining Eqs. (15.60) and (15.61) gives

$$\Delta i_{in} = \frac{V_o V_s}{f_s L(V_o - V_s)} \tag{15.62}$$

The load voltage ripple ΔV_o (not shown in Fig. 15.7) can be obtained by considering the charging and discharging of the output capacitor C_{out}. In the periods kT_p, while the input current $i_{in}(t)$ is increasing, the capacitor voltage or output voltage $v_o(t)$ decreases as the capacitor discharges through the load resistor R

$$v_c(t) = v_o(t) = \frac{1}{C_{out}} \int_0^{kT_p} i_c(t) \, dt \tag{15.63}$$

But if C_{out} is large, the capacitor current in the kT_p periods is almost constant at the value I_o of the load current. The capacitor voltage ripple is therefore also constant.

$$\Delta v_c(t) = \Delta V_o(t) = \frac{1}{C_{out}} \int_0^{kT_p} i_c(t) \, dt$$
$$= \frac{kT_p I_o}{C_{out}} = \frac{kI_o}{f_s C_{out}} \tag{15.64}$$

15.3.3 Worked Examples

Example 15.7   A buck–boost converter is driven from a battery $V_s = 100$ V and operates in the continuous current mode, supplying a load resistor $R = 70\ \Omega$. Calculate the load voltage and current and the input current for values of duty cycle k (1) 0.25, (2) 0.5, and (3) 0.75.
From Eqs. (15.57) and (15.58)

$$I_o = \frac{V_o}{R} = \frac{-k}{1-k} \frac{V_s}{R}$$

1. $k = 0.25$,

$$I_o = \frac{-0.25}{0.75} \times \frac{100}{70} = -0.476 \text{ A}$$
$$V_o = I_o R = -33.33 \text{ V}$$

2. $k = 0.5$,

$$I_o = \frac{-0.5}{0.5} \times \frac{100}{70} = -1.43 \text{ A}$$
$$V_o = I_o R = -100 \text{ V}$$

3. $k = 0.75$,

$$I_o = \frac{-0.75}{0.25} \times \frac{100}{70} = -4.286 \text{ A}$$

$$V_o = I_o R = -300 \text{ V}$$

The negative signs in the values above indicate reversal of polarity with respect to the supply.

From Eq. (15.58)

$$I_{in} = \frac{k}{k-1} I_o$$

Combining this with the expression for I_o above gives

$$I_{in} = \frac{k}{k-1} \frac{-k}{1-k} \frac{V_s}{R} = \left(\frac{k}{k-1}\right)^2 \frac{V_s}{R}$$

1. $k = 0.25$,

$$I_{in} = \left(\frac{0.25}{-0.75}\right)^2 \times \frac{100}{70} = +0.16 \text{ A}$$

2. $k = 0.5$,

$$I_{in} = \left(\frac{0.5}{-0.5}\right)^2 \times \frac{100}{70} = +1.43 \text{ A}$$

3. $k = 0.75$,

$$I_{in} = \left(\frac{0.75}{-0.25}\right)^2 \times \frac{100}{70} = \frac{9 \times 100}{70} = +12.86 \text{ A}$$

Example 15.8   A buck–boost converter operates in the continuous current mode at a switching frequency $f_s = 20$ kHz. $V_s = 24$ V, $C_{out} = 250$ μF, $L = 200$ μH, and the load resistance $R = 1.5$ Ω. For a duty cycle $k = 0.33$ calculate (1) the average load voltage and current, (2) the peak-to-peak input current ripple, and (3) the peak-to-peak output voltage level.

1. From (15.57)

$$V_o = \frac{-k}{1-k} V_s = \frac{-0.33}{0.66} \times 24 = -12 \text{ V}$$

$$I_o = \frac{V_o}{R} = \frac{-12}{1.5} = -8 \text{ A}$$

2. If $f_s = 20$ kHz,

$$T_p = \frac{1}{20 \times 10^3} = 50 \ \mu s$$

Using Eq. (15.62)

$$\Delta i_{in} = \frac{-12 \times 24}{20 \times 10^3 \times 200 \times 10^{-6} \times (-12 - 24)} = 2 \ A$$

3. From Eq. (15.64)

$$\Delta V_o = \frac{kI_o}{f_s C_{out}} = \frac{0.33 \times (-8)}{20 \times 10^3 \times 250 \times 10^{-6}} = -0.53 \ V$$

The negative sign has no significance in this context.

15.4   DC CHOPPER CONTROL OF A DC MOTOR

A dc converter in the voltage step-down (buck) connection is useful for controlling the armature voltage of a separately excited dc motor in order to control its speed. The armature circuit contains a small resistance R_a and a small inductance L_a. In order to ensure continuity of the armature current $i_a(t)$ (Fig. 15.8), it is often necessary to include an additional series inductance. The combination of armature inductance L_a plus the additional inductance is represented by the inductor L in Fig.15. 8a. The time constant L/R_a of the armature circuit is long compared with the switching period. It may be seen that the dc chopper circuit (Fig. 15.8a) is the same as that of the buck converter (Fig. 15.1a), differing only in the load circuit.

15.4.1   Equations of Operation for a DC Motor

During rotation of the armature an internal emf E is generated which has a polarity that opposes the flow of armature current during motoring operation. The average voltage drop on the inductor must be zero so that for steady-state motoring at constant speed,

$$V_a = I_a R_a + E \tag{15.65}$$

The torque T in newton-meters(Nm) developed by a dc motor can be represented by

$$T = K\Phi I_a \tag{15.66}$$

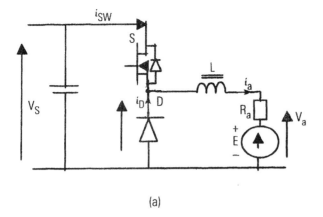

(a)

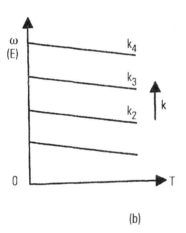

(b)

FIG. 8   Chopper control of dc motor speed: (a) equivalent circuit and (b) speed–torque characteristics in quadrant 1.

The internal emf E is related to the speed ω radians/s by

$$E = K\Phi\omega \tag{15.67}$$

The magnetic flux Φ developed in the motor depends on the current in the separate excitation winding. This winding plays no part in the following calculations. Term

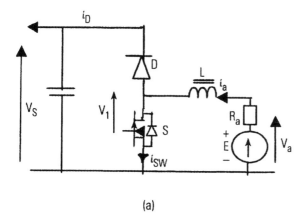

(a)

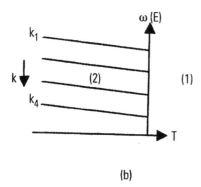

(b)

FIG. 9   Chopper control of dc motor speed in the regeneration mode: (a) equivalent circuit and (b) speed–torque characteristics in quadrant 2.

K in Eqs. (15.66) and (15.67) is a design constant; so the torque is proportional to armature current and the internal emf is proportional to the speed.

$$K\Phi = \frac{T}{I_a} = \frac{E}{\omega}$$
(15.68)

The power P_{out} delivered out of the shaft of the motor, neglecting rotational losses, is given by

$$P_{out} = T\omega \tag{15.69}$$

With the torque in newton-meters (Nm) and speed in radian per second (radian/s) the output power is in watts. It is now customary for electric motors to be rated in watts rather than in the former unit of horsepower [1 hp $\equiv$ 746 W]. But, from Eq. (15.70), it can be seen that the output power can also be represented in equivalent electrical terms.

$$P_{out} = T\omega = EI_a \tag{15.70}$$

The combination of Eqs. (15.64)–(15.66) gives an expression

$$\omega = \frac{V_a - I_a R_a}{K\Phi} = \frac{V_a}{K\Phi} - \frac{R_a}{K^2\Phi^2}T \tag{15.71}$$

Equation (15.64) represents a linear characteristic of negative slope, shown in Fig. 15.8b. The negative slope of the characteristics is constant at the value $R_a/K^2\Phi^2$, but the intercept on the speed axis is proportional to the load voltage V_a. With dc chopper control of the load (i.e., armature) voltage, it was shown in Eq. (15.10), for the buck converter, that $V_o(V_a) = kV_s$, where k is the switch duty cycle. With constant supply voltage V_s from the dc source the speed axis intercept is therefore proportional to k. The power flow in the circuit of Fig. 15.8a is unidirectional from the supply to the load. Only positive motor speed and unidirectional motor current are realizable. This particular mode of control is referred to as *one quadrant* control because the operating characteristics lie only in one quadrant of the speed–torque plane.

15.4.2 Regenerative Braking Operation

The rotating machine will behave like a generator if it is driven in the motoring direction by some external mechanical means. Energy can be transferred from the rotating machine to the electrical circuit connected to its terminals. In effect, the motor armature voltage becomes the input voltage to a boost converter and the dc system voltage V_s becomes the output voltage (Fig. 15.9a). Internal emf E becomes the source voltage. For the armature circuit the Kirchhoff voltage equation is

$$V_a = E - I_a R_a \tag{15.72}$$

During reverse current operation, the terminal voltage V_s becomes the output voltage of a generator. Compared with the motoring equation, Eq. (15.65), it is seen that Eq. (15.72) differs only in the sign of the current term. If the electrical supply is capable of receiving reverse current from the rotating machine then energy can be returned from the rotating generator into the supply. This operation is referred to as *regenerative braking*, or sometimes just *regeneration*. The energy of rotation is now reinstated into the supply, whereas for dynamic braking, it is dissipated and lost. When switch S is closed, the source voltage E drives an increasing current through R_a, L, and S. When switch S opens the continuous armature current $i_a(t)$ flows and decays through the diode D and the supply. Energy is then converted from the mechanical energy of rotation into electrical energy. The transfer of energy acts as a brake on the motion, causing speed reduction. The power generated inside the rotating dc machine during braking, is

$$P_{gen} = EI_a \tag{15.73}$$

Some part of this power is lost as heat in the armature resistance, $I_a^2(\text{rms})R_a$. The remaining power V_aI_D flows back into the dc supply With an ideal diode $V_a = V_s$ so that

$$V_sI_D(av) = EI_a(av) - I_a^2(\text{rms})R_a = P_{gen} - I_a^2(\text{rms})R_a \tag{15.74}$$

Now for boost converter operation it was shown in Eq. (15.45) that $V_s = (1 - k)V_a$ or

$$k = \frac{V_a - V_s}{V_a} \tag{15.75}$$

In most regenerative braking applications the "supply" voltage V_s remains constant. Therefore

1. For low speeds, E is low, V_a is low, so that k is required to be high.
2. For high speeds, E is high, V_a is high, so that k is required to be low.

It is a prerequisite for regenerative braking operation that the dc supply system must be capable of accepting reverse current and also be able to absorb the regenerated energy. A battery would be satisfactory but not a rectifier fed from an ac supply. If necessary, an "energy dumping" resistor must be switched across the dc system to dissipate the excess energy when V_s rises above a safe level.

15.4.3 Four-Quadrant Operation

Fully comprehensive operation of a speed-controlled motor includes both motoring and regenerative braking operation in either direction of motor speed. Such full control combines the speed-torque characteristics of Figs. 15.8b and 15.9b, which occur in quadrants (1) and (2), respectively, of the speed–torque plane, plus the corresponding reverse-speed characteristics in quadrants (3) and (4) of the speed–torque plane. The overall four-quadrant picture is demonstrated in Fig. 15.10. Four quadrant operation can be obtained by the use of the dc chopper type of converter. But reverse-speed operation requires reversal of the armature voltage

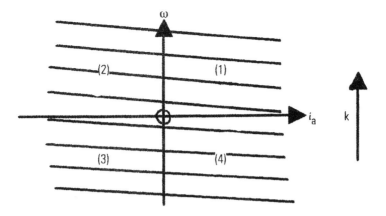

Fig. 10 Speed–torque characteristics for four-quadrant control.

polarity. It is therefore necessary either to use two choppers with independent back-to-back operation or to provide supplementary switching so that a single chopper can be applied to the motor in either of its two polarity options.

PROBLEMS

Step-Down (Buck) Converter

15.1 Sketch the basic circuit diagram for a dc–dc buck (step-down) voltage converter, incorporating a large output capacitor. Show waveforms of the load voltages for the two duty-cycle conditions (a) $k = 1/4$ and (b) $k = 3/4$. For both conditions write expressions for the average and rms values of the load voltages in terms of the supply voltage V_s.

15.2 For a buck converter with continuous load current and resistive load R, derive an expression for the output power in terms of supply voltage V_s and duty cycle k. Calculate the per-unit value of the load power for (a) $k = 1/4$ and (b) $k = 3/4$.

15.3 A buck converter with battery V_s has resistive load R and duty cycle k. Sketch waveforms of the voltages across the switch and diode and calculate their average values.

15.4 The semiconductor switch in a step-down converter has a minimum effective on time of 37.5 μs. If the output voltage range is to have a minimum value of 45 V and the dc supply is rated at 600 V, what is the highest chopper frequency?

15.5 If the chopper frequency in Problem 15.4 is reduced to 1 kHz, what will be the minimum output voltage?

15.6 A semiconductor switch has a minimum effective off time of 30.5 μs. If the ideal dc supply is rated at 1200 V and the chopping frequency is 3500 Hz, calculate the maximum duty cycle and the maximum output voltage that can be obtained.

15.7 The dc chopper convert in Problem 15.4 is connected to a resistive load $R = 2.25$ Ω. What is the average load current when it operates at a frequency of 700 Hz?

15.8 A step-down converter is used to convert a 500-V dc supply to an output voltage of 300 V. The filter inductor has a value of 200 μH, and the resistive load is 2.75 Ω. The semiconductor switch has an on time of 53.5 μs, and conduction can be assumed to be continuous. Calculate (a) the switching frequency, (b) the switch off time, and (c) the average value of the input current.

15.9 For the step-down converter of Problem 15.8, operating in the continuous conduction mode, calculate (a) the maximum and minimum values of the load current and (b) the rms values of the diode, switch, and load currents.

Step-Up (Boost) Converter

15.10 A boost converter is required to operate with a fixed output voltage V_o = 60 V. The input voltage varies in the range from 20 V to 40 V. If the conduction is continuous, what must be the range of the duty ratios?

15.11 A step-up converter supplied by a 100-V battery transfers energy to a series R-L load, where $L = 220$ μH and $R = 2.5$ Ω. The semiconductor switch has an on time of 42.7 μs. If the output voltage is required to be 130 V and the conduction is continuous, calculate (a) the switching frequency, (b) the switch off time, and (c) the average output current.

15.12 For the step-up converter of Problem 15.11 calculate the maximum, minimum, and average values of the input current.

15.13 For the step-up converter of Problem 15.11 it is required to limit the output voltage ripple $\Delta V_o/V_o$ to 10%. What value of capacitance is required of the output capacitor C_{out}?

Step-Up–Step-Down Converter

15.14 A step-up–step-down (i.e., boost–buck) converter supplied from a battery of 176 V supplies a load resistor $R = 58\ \Omega$. If the load current is continuous, calculate the average input current and the load average voltage and current for the values of duty cycle (a) $k = 0.25$, (b) $k = 0.5$, and (c) $k = 0.75$.

15.15 A buck–boost converter has the design parameters $L = 200\ \mu\text{H}$, $C_{out} = 300\ \mu\text{F}$, and $f_s = 22.4\ \text{kHz}$. It operates from a 50-V dc supply in the continuous current mode, supplying a load resistance $R = 1.85\ \Omega$. Calculate the average load voltage and current for duty cycles (a) $k = 0.25$ and (b) $k = 0.75$.

15.16 For the buck–boost converter of Problem 15.15, at the two values of the duty cycle, calculate (a) the peak-to-peak input current ripple and (b) the peak-to-peak output voltage level.

16

Switch-Mode Converter Power Supplies

A switch-mode power (SWP) converter transforms a (usually) fixed level of dc voltage to an adjustable level of dc output voltage. Voltage control is realized by variation of the duty cycle (on–off ratio) of a semiconductor switching device, as in the chopper converter described in the Chapter 15.

In a SWP system it is necessary to use an isolating transformer, as shown in the basic schematic of Fig. 16.1. The switching frequency of the dc–dc converter can be made much higher than the line frequency so that the filtering elements, including the transformer, may be made small, lightweight, efficient, and low cost. The output of the transformer is rectified and filtered to give a smoothed output voltage V_o. The output voltage may be regulated by using a voltage feedback control loop that employs a PWM switching scheme.

16.1 HIGH-FREQUENCY SWITCHING

Switch-mode power supplies use "high" switching frequencies to reduce the size and weight of the transformer and filter components. This helps to make SWP equipment portable. Consider a transformer operating at a given peak flux Φ_m, limited by saturation, and at a certain rms current, limited by winding heating. For a given number of secondary turns N_2, there is a given secondary voltage V_2 related to the primary values V_1 and N_1 by

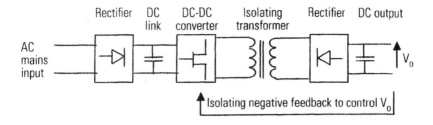

FIG. 1   Schematic of a switch-mode supply.

$$\frac{V_2}{V_1} = \frac{N_2}{N_1} \tag{16.1}$$

The sinusoidal primary (applied) voltage V_1 is related to the peak flux Φ_m and supply frequency f_1 by a relation

$$V_1 = 4.44 \ \Phi_m f_1 n \tag{16.2}$$

where n is a design constant.

If the applied frequency f_1 is increased with the same flux and the same waveforms, the voltages V_1, and V_2 will also increase proportionately. But if the transformer windings are not changed and use the same respective sizes of conductor wire and the same respective number of turns N_1, N_2, the current ratings of the two windings, will be unchanged. Increase of the voltages due to increased frequency, with unchanged current ratings, implies a proportionate increase in the winding voltampere (VA) ratings, which is equally true for both the primary and secondary windings.

16.2   HIGH-FREQUENCY ISOLATION TRANSFORMER

Transformer iron losses, due to magnetic hysteresis and to eddy currents, increase with frequency. For SWP equipment the increased losses can become considerable so that it may be desirable to use ferrite core materials rather than iron or steel laminations. The isolation transformer in Fig. 16.1 is a shell-type cored device with a form of B-H loop shown in Fig. 16.2. The peak flux density B_m is related to the flux Φ_m by

$$B_m = \frac{\Phi_m}{\text{area of flux path}} \tag{16.3}$$

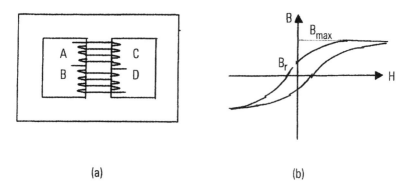

(a)   (b)

FIG. 2   Two-winding transformer: (a) shell-type core and (b) core B-H loop.

In Fig. 16.2 the intercept B_r is known as the residual flux density. The magnetizing intensity H in Fig. 16.2 is proportional to the magnetizing current in the winding. Time variations of the current and flux cause iron power losses in the transformer core, which appear as heat. When ceramic ferrite rather than iron or steel laminations is used as the magnetic core material, and when the frequency is greater than 20 kHz, the working peak flux density then falls from (say) 1.5 T to about 0.3 T. But the iron losses in the ferrite also increase with frequency so that the usable flux density then falls, and still more expensive ferrite materials may be desirable. Nevertheless, the economic benefits of using high-frequency switching are considerable, particularly in portable equipment. Present (2002) developments are moving toward operating switching frequencies up to 1 MHz, but the consequent problems of electromagnetic interference (EMI) then become significant.

In the flyback and forward converter connections, described below, unidirectional transformer core excitation is used where only the positive part of the B-H loop (Fig. 16.2b) is used.

For the push–pull, full-bridge, and half-bridge converter circuits, described below, there is bidirectional excitation of the transformer core and both the positive and negative parts of the B-H loop are used.

16.3 PUSH–PULL CONVERTER

16.3.1 Theory of Operation

The dc–dc push–pull converter (Fig. 16.3) uses a center-tapped transformer and two controlled switches, S_1 and S_2. To prevent core saturation both switches must

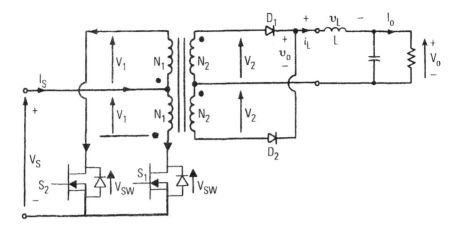

FIG. 3   Push–pull dc–dc converter.

have equal duty cycles, k. The transformer voltage ratio satisfies the turns ratio relationship Eq. (16.1) . Voltage V_1 is determined by the switch duty ratio $k = T_{on}/T_p$, as for the step-down converter in Fig. 15.2.

$$V_1 = kV_s = \frac{T_{on}}{T_p} V_s = \frac{S_{1on}}{T_p} V_s = \frac{S_{2on}}{T_p} V_s$$

(16.4)

The cycle of operation for the circuit of Fig. 16.3 is that conduction occurs sequentially through switches S_1 and S_2 with dwell periods between the sequences in which there is no conduction at all on the primary side. This action is represented in the waveforms of Fig. 16.4 in which the switch-on periods are represented by S_1 and S_4, respectively, and the dwell (off) periods by Δ.

During the intervals when switch S_1 (Fig. 16.3) conducts, then diode D_1 also conducts so that D_2 is reverse biased. Current $i_L(t)$ flows through inductor L and the output voltage $v_o(t)$ is given by

$$v_o(t)\Big|_0^{S_{1on}} = V_2 = \frac{-N_2}{N_1} V_1 = \frac{+N_2}{N_1} V_s$$

(16.5)

The signs in Eq. (16.5) refer to the voltage polarities defined by the "dots" on the transformer windings in Fig. 16.3.

During the intervals Δ when both switches are off, the inductor current $i_L(t)$ splits equally between the two diodes and the output voltage $v_o(t)$ is zero.

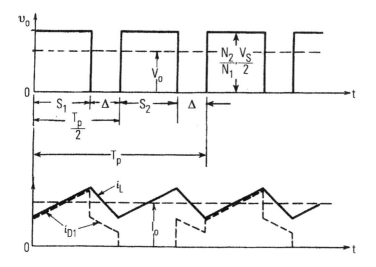

FIG. 4   Waveforms for the push–pull converter (based on Ref. 3).

$$v_0(t) \left.\begin{array}{c} \\ \end{array}\right|_{S_1 \ on}^{T_p/2} = v_0(t) \left.\begin{array}{c} \\ \end{array}\right|_{S_2 \ on}^{T_p} \tag{16.6}$$

Also

$$S_{1on} + \Delta = S_{2on} + \Delta = \frac{T_p}{2} \tag{16.7}$$

The whole period T_p in Fig. 16.4 represents a performance similar to that of the step-down converter (Fig. 15.1), but of double frequency. Combining Eqs. (16.5)–(16.7) yields, for $0 \le k \le 0.5$,

$$\frac{V_o}{V_s} = 2 \frac{N_2}{N_1} k \tag{16.8}$$

The voltage $v_L(t)$ across the inductor, which has an average value of zero, is given by

$$v_L(t) = \left[\frac{N_2}{N_1} V_s - V_o \right]_0^{\left(\frac{T_p}{2}\right) - \Delta} - V_o \left.\begin{array}{c} \\ \end{array}\right|_{\left(\frac{T_p}{2}\right) - \Delta}^{T_p/2} \tag{16.9}$$

During the conduction intervals of the switches the current $i_L(t)$ smoothly increases. When both switches are off,

$$i_L(t) = 2i_{D1}(t) = 2i_{D2}(t) \tag{16.10}$$

The system currents are given in Fig. 16.4.

16.3.2 Worked Example

Example 16.1   A resistive load $R = 8.5\ \Omega$ is supplied from a 12-V battery via a push–pull dc–dc converter in which the transformer (step-down) ratio is 1:4. Calculate the average output current for the duty ratios (1) 0.1, (2) 0.3, and (3) 0.5.

Given $N_2/N_1 = 4$. The average output voltage is, from Eq. (16.8) ,

$$V_o = 2\frac{N_2}{N_1} V_s k = 2 \times 4 \times 12 \times k = 96k$$

1. $V_o = 96 \times 0.1 = 9.6$ V, and $I_o = V_o/R = 9.6/8.5 = 1.13$ A.
2. $V_o = 96 \times 0.3 = 28.8$ V, and $I_o = 28.8/8.5 = 3.39$ A.
3. $V_o = 96 \times 0.5 = 48$ V, and $I_o = 48/8.5 = 5.65$ A.

16.4   FULL-BRIDGE CONVERTER

A particular form of single-phase voltage-fed full-bridge converter that uses four switches is shown in Fig. 16.5a. In its simplest form energy from a dc supply of voltage V_s is transferred through a transformer to a series $R\text{-}L$ load. Symmetrical switching of the four power transistors T gives an output voltage square wave (Fig. 16.5b), so that the function is an inverting operation giving an ac output voltage $v_1(t)$ across the transformer primary winding. The ideal transformer has a turns ratio 1:n, where $n = N_1/N_2$. Steady-state operation (not start-up operation) is presumed.

16.4.1   Modes of Converter Operation

1.   With switches T_1 and T_4 on the supply voltage is connected across the transformer primary and $v_1(t) = V_s$. A positive voltage $v_2(t) = V_s/n$ causes current $i_2(t)$ to increase through the load. Current $i_2(t) = ni_1(t)$ and both currents have the same wave shape.
2.   With switches T_2 and T_3 on, and T_1 and T_4 held in extinction, the rising load current (and hence primary current) $i_1(t)$ cannot suddenly reverse because of the load inductance. But this current now flows through the diode switches D_2 and D_3 and the voltage across the transformer is reversed, so that $v_1(t) = -V_s$ and $v_2(t) = -V_s/n$. Energy stored in

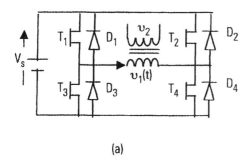

(a)

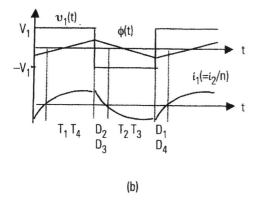

(b)

FIG. 5   Full-bridge converter: (a) circuit schematic and (b) waveforms for ideal operation.

the inductor is then fed back into the supply. The result is that the current $i_1(t)$ reduces and then reverses so that $i_1(t)$ becomes negative. The voltage relationships can be summarised as

$$v_1 = \pm V_s \tag{16.11}$$

$$v_2 = \frac{N_2}{N_1} V_1 = \frac{V_1}{n} = \pm \frac{V_s}{n} \tag{16.12}$$

The voltampere ratings of the two transformer windings must be equal so that

$$V_1 I_1 = V_2 I_2 \tag{16.13}$$

Combining Eqs. (16.12) and (16.13) gives

$$I_2 = n I_1 \tag{16.14}$$

3. If switches T_2 and T_3 are then switched off and switches T_1 and T_2 are switched on again, the negative current flows through diode D_4, the transformer primary winding, and diode D_1 back into the positive supply terminal, falling to zero and then reversing. The diagram Fig. 16.5b indicates the current paths during the actions of the transistors and diodes in each period of the cycle.

16.4.2   Operation with Full-Wave Rectifier Load

Consider the transformer of Fig. 16.5a supply a single-phase, full-wave, rectified load incorporating an L-C filter, as shown in Fig. 16.6. Assume that the filter inductor L_D is large enough to maintain the load current $i_L(t)$ effectively constant at value I_L and that the output capacitor C smooths any perturbations of the load voltage $v_L(t)$ so that $v_L(t) = V_L$. If the transformer leakage reactance is negligibly small, the primary current $i_1(t)$ can reverse instantaneously after switching. Voltage and current waveforms for the ideal condition are shown in Fig. 16.7. When switching occurs, the instantaneous transformer voltages $v_1(t)$ and $v_2(t)$ both reverse polarity, and the primary and secondary currents also reverse direction.

The practical operation of the circuit Fig. 16.5b is more complicated due to the unavoidable effects of circuit inductance. A fairly accurate picture of the circuit operation may be obtained in terms of transformer leakage inductance if the transformer magnetizing current is neglected.

When one pair of the full-bridge switches opens, the primary current $i_1(t)$ cannot suddenly reverse because of the energy stored in the transformer series leakage inductance L_1. Note that L_1 defines the total transformer series inductance, referred to as *primary turns*. When switches T_1 and T_4 turn off the current $i_1(t)$ commutates via diodes D_2 and D_3. While $i_1(t)$ remains smaller than I_L/n, the difference current $(I_L - ni_1)$ freewheels through the diode bridge and $v_1(t) =$

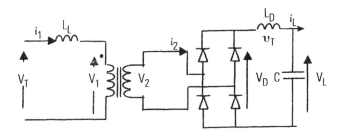

FIG. 6   Single-phase, full-wave bridge rectifier load supplied by a full-bridge converter.

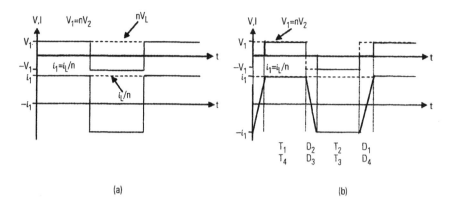

(a)   (b)

FIG. 7 Waveforms for ideal operation of a full-wave bridge rectifier load supplied by a full-bridge converter.

$v_2(t) = 0$. When $i_1(t)$ has fallen to zero, it then increases in the reverse direction through T_2 and T_3 up to the magnitude $i_1(t) = I_L/n$, when $v_1(t) = nv_2(t) = -nV_L$. During the reversal of $i_1(t)$ the freewheeling action causes the voltage $v_1(t)$ across L_1 to be $v_L(t) = \pm V_s$. It then follows that $v_1(t) = 0$ and

$$\frac{di_1}{dt} = \pm \frac{V_s}{L_1}$$

(16.15)

Although switching on takes place at zero voltage, the switching off of full current occurs at full voltage, which imposes a severe duty on the switches.

16.4.3 Worked Example

Example 16.2 A full-bridge converter with rectified load has a power supply $V_s = 300$ V and incorporates a 10:1 step-down transformer. The transformer leakage inductance is 30 μH and the switching frequency is 20 kHz. Calculate the dc output voltage when the output current is 50 A.

Neglecting transformer leakage inductance the transformer output voltage, from Eq. (16.12) , is

$$V_2 = \frac{V_1}{n} = \frac{300}{10} = 30 \text{ V}$$

The secondary current $I_2 = I_L = 50$ A. From Eq. (16.14) ,

$$I_1 = \frac{I_2}{n} = \frac{50}{10} = 5 \text{ A}$$

The peak-to-peak value of input current $i_1(t)$, from Fig. 16.5b is therefore

I_1 (peak to peak) $= 10$ A

Now $V_s = 300$ V and $L_1 = 30$ μH. During switching the peak-to-peak current is the interval described as di_1 in Eq. (16.15) . Therefore, the transient interval dt is given by

$$dt = di_1 \frac{L_1}{V_S} = 10 \times \frac{30 \times 10^{-6}}{300} = 1 \text{ μs}$$

At a switching frequency of 20 kHz each cycle period occupies a time

$$t = \frac{1}{f} = \frac{1}{20 \times 10^{-3}} = 50 \text{ μs}$$

Each switching interval dt therefore occupies 1/50 of the periodic time and there are two switchings per cycle. The switching or commutation actions take 2 μs/cycle. This means that there is zero voltage for 2 μs in each 50-μs cycle so that the net voltages $v_1(t)$ and $v_2(t)$ are lowered by 2/50, or 4%. The resultant output voltage is therefore

$V_L = 0.96V_2 = 28.8$ V

16.5 SINGLE-SWITCH FORWARD CONVERTER

16.5.1 Ideal Forward Converter

The circuit diagram of a single-switch forward converter is shown in Fig. 16.8. Assume, initially, that the transformer is ideal with zero leakage inductance and zero losses. The turns ratio is related to the voltage and current ratios by the standard relationship, using the terminology of Fig. 16.8,

$$\frac{N_1}{N_2} = n = \frac{v_1}{v_2} = \frac{i_2}{i_1}$$

$$(16.16)$$

The L-C output filter in Fig. 16.8 results in a smooth output voltage $v_o(t) = V_o$ during steady-state operation.

When the power transistor T is switched on, in the interval $0 < t < \tau_1$, the supply voltage is applied to the transformer primary winding so that $v_1(t) = V_s$ and $v_2(t) = V_s/n$. Diode D_1 is then forward biased, and D_2 is reverse biased. The inductor voltage is then positive and increasing linearly

$$v_L(t) \bigg|_0^{\tau_1} = \frac{V_s}{n} - V_o$$

$$(16.17)$$

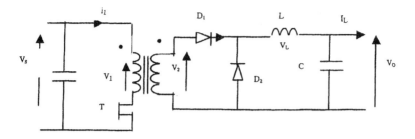

FIG. 8   Ideal single-switch forward converter.

When transistor T is switched off, the inductor current $i_L(t)$ circulates through diode D_2 so that the output voltage is negative and decreasing linearly.

$$v_L(t) \left. \right|_{\tau_1}^{T_p} = -V_o \tag{16.18}$$

Over a complete switching period T_p, the average inductor voltage is zero

$$V_L(av) = \frac{1}{T_p} \int \left[\left(\frac{V_S}{n} - V_o \right) \left. \right|_0^{\tau_1} + (-V_o) \left. \right|_{\tau_1}^{T_p} \right] = 0 \tag{16.19}$$

The solution of Eq. (16.19) shows that

$$\frac{V_o}{V_S} = \frac{1}{n} k = \frac{1}{n} \frac{\tau_1}{T_p} = \frac{N_2}{N_1} \frac{\tau_1}{T_p} \tag{16.20}$$

where k is the duty cycle.

The voltage ratio of the ideal forward converter is proportional to the duty cycle k, as in the step-down, or buck, dc–dc converter, described in Sec. 15.1.

16.5.2   Practical Forward Converter

In a practical forward converter the transformer magnetising current must be recovered and fed back to the supply to avoid the stored energy in the transformer core causing converter failure. This is realized by adding another, demagnetizing, winding with unidirectional currents, Fig. 16.9.

1. T = on. With transistor T switched on diode D_2 is forward biased and conducts current $i_3(t)$, while diodes D_1 and D_3 are reverse biased. Then

$$v_1(t) = V_s \qquad v_2(t) = V_s \qquad v_3(t) = \frac{V_s}{n} = \frac{V_s N_3}{N_1} \tag{16.21}$$

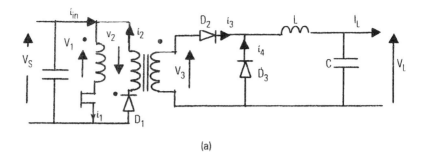

(a)

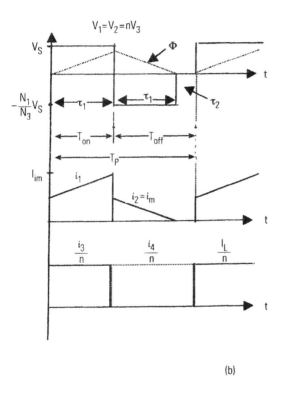

(b)

FIG. 9 Practical single-switch forward converter: (a) circuit schematic and (b) waveforms.

Also, during switch-on, the core flux Φ ramps up to satisfy the relation

$$V_s\Big|_0^{\tau_1} = N_1 \frac{d\Phi}{dt}$$

(16.22)

Even if the load current is effectively constant due to a large value of filter inductance L, the input current $i_1(t)$ will increase. Current $i_1(t)$ in Fig. 16.9b, includes the hypothetical transformer magnetizing component $i_m(t)$, flowing through the magnetizing inductance L_m and acting to increase the flux against the opposition of the magnetic reluctance. Current $i_m(t)$ increases linearly from zero to I_m. By Kirchhoff's node law

$$i_1(t) = \frac{i_3(t)}{n} + i_m(t)$$

(16.23)

2. T = off. When switch T is off, the magnetizing current must continue to flow due to the stored energy $L_m i_m^2/2$. The flux must be allowed to decay, and to support this flux, a magnetizing current must flow inward at a dotted terminal. As it can no longer flow into winding 1, it transfers to the only possible path, i.e., in at winding 2, becoming i_2. Winding 2 is then connected across V_s; thus,

$$v_1 = v_2 = nv_3 = -V_s$$

(16.24)

Current $i_L(t)$ must continue to flow in L despite being reverse biased by voltage V_s/n, so that D_3 then conducts. The core flux Φ ramps down at the same rate at which it ramps up and i_m $(= i_2)$ falls correspondingly. Flux Φ reaches zero after a second period τ_1 (Fig. 16.9b).

There is then a dwell period with the current $i_L(t)$ freewheeling through D_3, for a period τ_2, with $i_1 = i_2 = 0$, until the start of the next cycle. When T is switched on again, D_2 becomes forward biased and $i_L(t)$ commutates from D_3 to D_2.

The overall periodic frequency is

$$f = \frac{1}{T_P} = \frac{1}{T_{on} + T_{off}} = \frac{1}{2\tau_1 + \tau_2}$$

(16.25)

Any leakage flux between winding 1 and 2 will cause a voltage ringing overshoot on the device, which may be greater than $2V_s$. Therefore it is important to have very tight magnetic coupling between windings 1 and 2. Leakage between these two windings and winding 3 will cause overlap in the conduction of D_2 and D_3 during commutation, ramping the edges of the current waveforms. The presence of the leakage inductance will force a finite rate of fall of the current i_3 and growth of i_4 or a finite rate of fall of i_4 and growth of i_3 because $i_3 + i_4 = i_L$, which is constant. The coupling with winding 3 cannot be as tight as between the windings 1 and 2 because of the construction. There will usually be a different number of turns, and also because of the insulation requirements, the output must be well insulated from the incoming power supply.

Since there can be no average dc voltage across the ideal filter inductor L, when T is off, the input voltage to L is zero, and when T is on, it is V_s/n. Therefore,

$$V_L = \frac{f\tau_1 V_s}{n} \tag{16.26}$$

For a lossless system the input power is equal to the output power, $P_{out} = P_{in}$.

$$P_{out} = V_L I_L$$
$$P_{in} = V_s I_{in(av)} = V_s[I_{1(av)} - I_{2(av)}]$$
$$V_L I_L = V_s[I_{1(av)} - I_{2(av)}] \tag{16.27}$$

The energy stored in the transformer core is incidental and basically undesirable for the operation of this converter. The only inductive stored energy necessary for operation is that in L, the output filter. Since only one quadrant of the magnetic B-H loop is employed this form of converter is generally restricted to lower power applications.

16.5.3 Maximum Duty Ratio

If $N_1 \neq N_2$, then during the time interval when $i_2(t)$ is flowing through D_1 into the power supply in Fig. 16.9a,

$$i_2(t) = \frac{N_1}{N_2} i_m(t) \tag{16.28}$$

The voltage $v_1(t)$ across the transformer primary is also the voltage $v_m(t)$ across the transformer magnetising reactance L_m, when $i_2(t)$ is flowing,

$$v_1(t) = \frac{-N_1}{N_2} V_s \tag{16.29}$$

Once the transformer is demagnetized, the hypothetical current $i_m(t) = 0$ and also $v_1(t) = 0$. This occurs during the interval τ_2 in Fig. 16.9b. The time interval T_m during demagnetization can be obtained by recognizing that the time interval of voltage $v_1(t)$ across L_m must be zero over a complete switching period:

$$kV_s - \frac{N_1}{N_2} \frac{V_s T_m}{T_p} = 0 \tag{16.30}$$

Time interval T_m is therefore given by

$$\frac{T_m}{T_p} = k \frac{N_2}{N_1}$$

(16.31)

For a transformer to be totally demagnetised before the next cycle begins, the maximum value that T_m/T_p can have is $1 - k$. Thus the maximum duty ratio with a given turns ratio is

$$(1 - k_{max}) = \frac{N_2}{N_1} k_m$$

(16.32)

or

$$k_{max} = \frac{1}{1 + N_2 / N_1} = \frac{n}{n+1}$$

With an equal number of turns for the primary and the demagnetizing windings $n = 1$ and the maximum duty ratio in such a converter is limited to 0.5.

16.5.4   Worked Example

Example 16.3   A forward converter is operating at 35 kHz with an on time of 12 μs. The supply voltage of 300 V drops by 10%, but the output voltage must remain constant. Calculate a new on time. If the magnetizing inductance of the primary is 45 mH, calculate the peak magnetizing component of the primary current when the supply voltage is 300 V.

The output average voltage V_L is given by Eq. (16.26)

$$V_L = \frac{f \tau_1 V_S}{n} = \frac{35 \times 10^3 \times 12 \times 10^{-6} \times 300}{n} = \frac{126}{n}$$

The turns ratio n is fixed, but its value is not given. The problem specification is that V_L must remain constant.

In the above relation, if V_s drops by 10%, then the switch on time τ_1 must increase by 10% to $1.1 \times 12 = 13.2$ μs. The transformer magnetizing reactance L_m is related to V_s by

$$V_s = L_m \frac{di}{dt}$$

Since the on time $dt = 13.2$ μs, the change of current di in that interval is

$$di = \frac{V_S}{L_m} dt = \frac{300 \times 13.2 \times 10^{-6}}{45 \times 10^{-3}} = 0.088 \text{ A} = 88 \text{mA}$$

If the actual regulated supply voltage is used in the calculation the value V_s becomes

$$V_s = 0.9 \times 300 = 270 \text{ V}$$

Then

$$di_m = \frac{270}{300} \times 0.088 = 0.0792 \text{ A}$$

16.6 SINGLE-SWITCH FLYBACK CONVERTER

Another converter system for dc–dc applications requiring only one semicondutor switch is the flyback converter, which is derived from the buck–boost converter described in Sec. 15.3. The transformer is required to store energy and so includes an air gap in its magnetic circuit, that is made of high-permeability material. In addition the transformer must be of high-quality production with good magnetic coupling between the windings.

16.6.1 Ideal Flyback Converter with Continuous Current

The circuit schematic diagram of a flyback converter is given in Fig. 16.10a. If the transformer is ideal having zero leakage flux, and operates in the steady state, the waveforms are shown in Fig. 16.10b.

1. T = on. When the transistor switch T is on the supply voltage V_s is applied across the transformer primary winding N_1. For $0 \le t \le \tau_1$

$$v_1(t) = V_s \qquad v_2(t) = \frac{V_S}{n} = V_S \frac{N_2}{N_1} \tag{16.33}$$

While T is switched on diode D_1 becomes reverse biased by a potential difference $V_s/n + V_L$ and the transformer secondary current $i_2(t) = 0$. The core flux $\Phi(t)$ ramps up from $\Phi(0)$ to $\Phi(\tau_1)$, satisfying the relation

$$N_1 \frac{d\Phi}{dt} = V_s \tag{16.34}$$

The primary current $i_1(t)$ also ramps to provide the mmf needed to drive the flux (largely) across the airgap, which contains a stored magnetic energy $\frac{1}{2}L_1 i_1^2$.

In the switch-on intervals the time variation of the transformer core flux is obtained from Fig. 16.10b and Eq. (16.34)

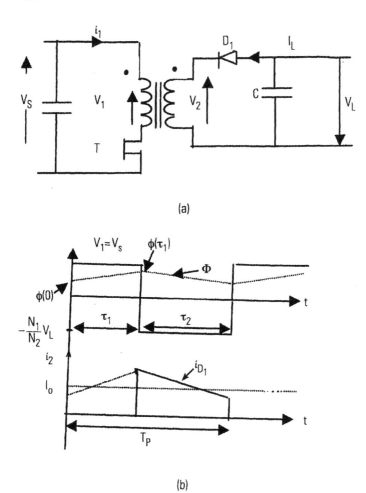

FIG. 10   Ideal flyback converter in continuous current mode: (a) circuit schematic and (b) ideal waveforms.

$$\Phi(t) = \Phi(0) + \frac{V_S}{N_1} t \tag{16.35}$$

The flux reaches its maximum value $\Phi(t)$ when $t = \tau_1$

$$\hat{\Phi}(t) = \Phi(\tau_1) = \Phi(0) + \frac{V_s}{N_1} \tau_1 \tag{16.36}$$

2. $T = $ off. When $i_1(t)$ falls to zero there can be no instantaneous change of the core flux and no instantaneous change of the stored energy. The unidirectional secondary current $i_2(t)$ continues to flow through diode D_1 so that for $\tau_1 \le t \le T_p$,

$$v_2(t) = -V_L \qquad v_1(t) = -nV_L = N_1 \frac{d\Phi}{dt} \qquad (16.37)$$

The voltage across the open transistor switch T is

$$v_T(t) = V_s + nV_L \qquad (16.38)$$

In the turn-off intervals the flux decrease is defined by the equation

$$\Phi(t) \bigg|_{\tau_1}^{T_p} = \Phi(\tau_1) - \frac{V_L}{N_2}(t - \tau_1)$$

$$\Phi(T_p) = \Phi(\tau_1) - \frac{V_L}{N_2}(T_p - \tau_1)$$

$$= \Phi(\tau_1) - \frac{V_L}{N_2}(\tau_2) \qquad (16.39)$$

But the net change of flux in the transformer core over one complete switching period must be zero in the steady state. Equating the flux transitions between Eqs. (16.36) and (16.39) gives

$$\frac{V_S}{N_1}\tau_1 = \frac{V_L}{N_2}\tau_2$$

or

$$\frac{V_L}{V_S} = \frac{N_2}{N_1}\frac{\tau_1}{\tau_2} = \frac{N_2}{N_1}\frac{k}{1-k} \qquad (16.40)$$

where k is the duty ratio $\tau_1/T_P = \tau_1/(\tau_1 + \tau_2)$. The voltage ratio realised in a flyback converter depends on k such that

$$V_L < \frac{V_S}{n} \qquad \text{for } k < 0.5$$

$$V_L > \frac{V_S}{n} \qquad \text{for } k > 0.5$$

$$V_L = \frac{V_S}{n} \qquad \text{for } k = 0.5$$

The effect of duty ratio k on the voltage transfer relationship corresponds exactly to the performance of the buck–boost dc–dc chopper circuit, described in Chapter 15.

16.6.2 Ideal Flyback Converter with Discontinuous Current

When the transformer flux $\Phi(t)$ in the circuit of Fig. 16.10a falls to zero during the switch-off period τ_2, the voltage waveform $v_1(t)$ assumes the form shown in Fig. 16.11. There is a nonconduction or dwell period τ_3 of zero current. Evaluating the voltage relationship from the flux variation is found to give

$$\frac{V_L}{V_S} = \frac{N_2}{N_1} \frac{k}{1 - k - \tau_3/T_p} \tag{16.41}$$

This differs from the equation for continuous current, Eq. (16.40) , only in the denominator τ_3 term. If nV_L is smaller than V_s, the rate of fall of flux is slower than the rate of rise, and it will take longer, as in Fig. 16.11 ($\tau_2 > \tau_1$). Once the flux has "reset" to zero, transistor T can be switched on again. The rise of flux equals the fall of flux; so the positive voltseconds applied equals the negative voltseconds applied. In Fig. 16.11,

$$V_s\tau_1 = nV_L\tau_2 \tag{16.42}$$

Also

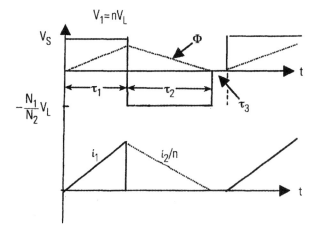

FIG. 11 Waveforms for ideal flyback converter with discontinuous current.

$$f = \frac{1}{\tau_1 + \tau_2 + \tau_3} = \frac{1}{T_P} \tag{16.43}$$

Now the power flow is equal to the energy stored per cycle multiplied by the frequency. From Eq. (16.43) an increase of τ_1 would decrease the frequency. In circuit design it would seem logical to choose a turns ratio so that nV_L is greater than V_S, but this choice is not always made. If the design choice is $n = V_S/V_L$, then $\tau_1 = \tau_2$, and the power is a maximum when $\tau_3 = 0$ and $f = 1/(\tau_1 + \tau_2)$.

16.6.3 Comparison of Continuous and Discontinuous Conduction Performance

It is reasonable to consider whether continuous current operation is superior to discontinuous current operation. The two modes are shown in Fig. 16.12 necessarily carrying the same average input current for the same load power. The particular design chosen for comparison has $\tau_1 = \tau_2$ and a 10% ripple for the continuous current mode.

16.6.3.1 Peak Currents and Current Ripples

The peak-to-peak switch current ripple $i_m(t)$ is twice the average value I_o in the discontinuous case but specified as 10% of the average value for the continuous case

$$
\begin{aligned}
i_{ma} &= 2I_o && \text{for discontinuous operation} \\
i_{mb} &= 0.1I_o && \text{for continuous operation}
\end{aligned}
\tag{16.44}
$$

The peak values I_L of the load current, illustrated in Fig. 16.12, are seen to be

$$\hat{I}_{La} = I_o + \frac{i_{ma}}{2} = 2I_o \qquad \text{for discontinuous operation}$$

$$\hat{I}_{Lb} = I_o + \frac{i_{mb}}{2} = 1.05I_o \qquad \text{for continuous operation} \tag{16.45}$$

16.6.3.2 Transformer Inductances

For discontinuous operation

$$L_a = \frac{V_s \tau_1}{i_{ma}} = \frac{V_s \tau_1}{2I_o} \tag{16.46}$$

For continuous operation

$$L_b = \frac{V_s \tau_1}{i_{mb}} = \frac{V_s \tau_1}{0.1I_o} \tag{16.47}$$

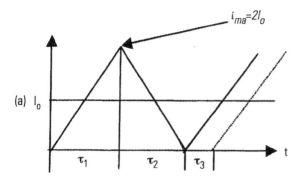

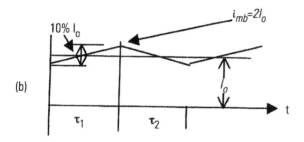

Fig. 12 Comparison of flyback converter operation ($\tau_1 = \tau_2$): (a) discontinuous operation and (b) continuous operation.

Comparing the necessary inductance values for the two cases, from Eqs. (16.46) and (16.47) gives

$$\frac{L_b}{L_a} = 20$$

(16.48)

The result of Eq. (16.48) does not mean that the inductor must be physically 20 times bigger for the continuous conduction case. Comparison of the maximum stored energy is a better indicator of the necessary physical size.

16.6.3.3 Peak Stored Energy in the Transformer

The peak energy stored can be expressed in terms of the transformer inductance L and the peak current of the load. For the discontinuous conduction condition of Sec. 16.6.3.1,

$$W_a = \frac{1}{2} L_a \hat{I}_{La}^2 \qquad (16.49)$$

Substituting from Eqs. (16.45) and (16.46) into Eq. (16.49) gives

$$W_a = \frac{1}{2} \times \frac{V_s \tau_1}{2I_o} \times 4I_o^2 = V_s \tau_1 I_o \qquad (16.50)$$

For the continuous conduction condition of Sec. 16.6.3.2,

$$W_b = \frac{1}{2} L_b \hat{I}_{Lb}^2 \qquad (16.51)$$

Substituting from Eqs. (16.45) and (16.47) into Eq. (16.51) gives

$$W_b = \frac{1}{2} \times \frac{V_s \tau_1}{0.1 I_o} \times (1.05)^2 \times I_o^2$$

$$= V_s \tau_1 I_o \times \frac{1.1}{0.2} = 5.51 V_s \tau_1 I_o \qquad (16.52)$$

The ratio of the peak stored energies for the two cases is

$$\frac{W_b}{W_a} = 5.51 \qquad (16.53)$$

Equation (16.53) shows that in the continuous conduction condition about 5.5 times the stored energy capability is required, compared with discontinuous operation, although only about one half of the peak current is needed.

16.6.3.4 Energy Storage in the Transformer

The type of transformer used in the circuits of Figs. 16.8 and 16.9 and 16.10 is the shell type of structure shown in Fig. 16.2, with the addition of an air gap of length g in the center leg. An ideal transformer with an infinitely permeable core stores no energy. It is a poor material for storing energy but is good for guiding flux. In a transformer or inductor with a ferromagnetic core, an air gap is required in which to store the magnetic energy. The design requirement is tight magnetic coupling between primary and secondary windings but also a small air-gap length. As discussed in Sec. 16.2 above, the ferrite core then has a peak flux density B_m = 300 mT at frequencies higher than 20 kHz.

The current density in the windings around the central limb of the core is limited by the $i^2 R$ heating. The volumes of copper in the primary and secondary are about the same in a well-designed transformer, even though the numbers of turns and wire sizes may be very different. Let the flux path in the core center leg have the contour dimensions width W times length L. Then the maximum energy stored in an air gap of volume $W \times L \times g$ is found to be

$$\text{Maximum energy} = \frac{B_m H_m WLg}{2} = \frac{B_m^2 WLg}{2\mu_o} \quad\quad (16.54)$$

The magnetic permeability of free space, $\mu_o = 4\,\pi \times 10^{-7}$ SI units, is applicable here because all of the stored energy is in the air gap and not in the core. Typical practical dimensions for a transformer core are $W = L = 10$ mm, $g = 1$ mm and $B_m = 0.3$ T. The maximum energy stored in the airgap can be found from Eq. (16.54)

$$\text{Maximum energy} = \frac{0.3^2 \times 0.1 \times 0.1 \times 0.01}{2 \times 4\pi \times 10^{-7}} = 3.58 \text{ mJ}$$

The periodic time T_p at 20 kHz is

$$T_p = \frac{1}{20 \times 10^3} = 0.05 \text{ ms}$$

For discontinuous operation the rate of energy conversion or power is therefore

$$P_b = \frac{W_b}{T_p} = \frac{3.58}{0.05} = 71.6 \text{ W}$$

But, from Eqs. (16.50) and (16.52), the energy conversion rate with continuous operation is therefore

$$P_a = \frac{W_a}{T_p} = \frac{P_a}{5.51} = \frac{71.6}{5.51} = 13 \text{ W}$$

16.6.3.5 Comparison Summary

In this comparison of designs for continuous and discontinuous operation, the continuous current case would require to have an air gap 5.5 times larger. In fact, the whole transformer would need to be bigger by this factor and so would be the copper losses. With continuous operation the range of excursion of the flux density is only about one-tenth that in the discontinuous case, and so the iron losses are lower, even though the core is five times bigger and works to the same peak flux density.

16.6.3.6 Addition of Tertiary Winding

If the power being taken from the output is less than that being fed in, the output voltage will rise. The maximum output voltage can be limited by using a third winding, closely coupled to winding 1 with the same number of turns (Fig. 16.13). If V_L tries to rise to a voltage greater than V_s/n, then a negative voltage greater than V_s is induced in the windings 1 and 3. The effect of winding 3 and D_2 is to clamp the maximum voltage of windings 1 and 3 at $-V_s$ and winding 2 at $-V_s/n$, i.e., it will prevent V_L becoming greater than V_s/n.

When T is switched off, the current commutates from winding 1 to winding 3 (instead of 2) and the stored energy is fed back to the supply instead of into the output circuit. This prevents the rise of V_L from becoming greater than V_s/n and also limits v_{DS} to $2V_s$.

Note that it is easier to wind windings 1 and 3 closely coupled because they have the same numbers of turns and can be wound together fully interlaced, so that the leakage flux between them and the corresponding leakage reactance, is very small. This is called a *bifilar winding*.

Windings 1 and 3, which are "live" on the incoming mains supply, have to be well insulated from winding 2, which is connected to the output circuit of the power supply. This insulation is governed by international standards. It means that the leakage between windings 1, 3, and 2 cannot be as small as between 1 and 3 because of the amount of insulation and the differing number of turns and wire sizes.

16.6.4 Worked Examples

Example 16.4   A flyback converter operates in the discontinuous mode with an input voltage of 300 V and a turns ratio of 10:1 at a frequency of 50 kHz. The output voltage is 30 V and the maximum output current is 1 A. Calculate (1) the peak input current, (2) the transformer stored energy, and (3) the inductance.

1. $V_s = 300$ V, $V_L = 30$ V, and $n = 10$. From Eq. (16.40) ,

$$\frac{V_L}{V_s} = \frac{30}{300} = \frac{1}{10}\frac{\tau_1}{\tau_2}$$

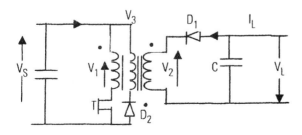

FIG. 13   Single-switch flyback converter with added tertiary winding.

Therefore, $\tau_1 = \tau_2$. Correspondingly,

$$\frac{I_{L(av)}}{I_{1(av)}} = \frac{n\tau_2}{\tau_1} = 10$$

Therefore,

$$I_{1(av)} = \frac{1}{10} = 0.1 \text{ A}$$

For a switching frequency $f = 50$ kHz,

$$T_p = \frac{1}{f} = \frac{1}{50 \times 10^3} = 0.02 \text{ ms}$$

From Fig. 16.11 it may be deduced that

$$I_{1(av)} \times T_p = \hat{I}_1 \times \tau_1 \times \frac{1}{2} = \hat{I}_1 \times \frac{T_p}{2} \times \frac{1}{2}$$

$$\hat{I}_1 = 4I_{1(av)} = 0.4 \text{ A}$$

2. For a lossless converter the input power must be equal to the output power

$$P_{in} = V_L I_L = 30 \times 1 = 30 \text{ W}$$

The transformer stored energy is obtained from the input power and the periodic switching time

$$W = P_{in} \times T_p = 30 \times \frac{0.02}{1000} = 0.6 \text{ mJ/cycle}$$

3. The stored energy can also be expressed in term of the peak input current and the transformer inductance

$$W = \frac{1}{2}L\hat{I}_1^2 = \frac{1}{2}L(0.4)^2 = 0.6 \text{ mJ}$$

The inductance L is required to be

$$L = \frac{0.6 \times 2}{0.4^2} = 7.5 \text{ mH}$$

which is a typical practical value.

Example 16.5   A flyback converter operates from a 300-V dc supply in discontinuous mode at a frequency of 35 kHz. The transformer turns ratio is 10:

1. When the duty cycle is 0.4, the output voltage is 25 V. Calculate the three time periods shown in Fig. 16.11.

$$V_s = 300 \text{ V} \qquad V_L = 25 \text{ V} \qquad n = 10 \qquad k = 0.4$$

From Eq. (16.42),

$$\frac{\tau_1}{\tau_2} = \frac{nV_L}{V_s} = \frac{100 \times 25}{300} = 0.833 \qquad\qquad \text{(Ex. 16.5a)}$$

From Eq. (16.44),

$$\frac{nV_L}{V_s} = \frac{k}{1 - k - \tau_3 / T_p}$$

Therefore,

$$\frac{\tau_3}{T_p} = 1 - k - \frac{kV_s}{nV_L}$$

$$= 1 - 0.4 - \frac{0.4 \times 300}{10 \times 25}$$

$$= 1 - 0.4 - 0.48 = 0.12$$

$$T_p = \frac{1}{f} = \frac{1}{35 \times 10^3} = 28.6 \text{ }\mu\text{s}$$

Therefore,

$$\tau_3 = 0.12 \times 28.6 = 3.43 \text{ }\mu s$$

From Eq. (16.43),

$$T_p = \tau_1 + \tau_2 + \tau_3$$

Therefore,

$$\tau_1 + \tau_2 = Tp - \tau_3 = 28.6 - 3.43 = 25.17 \text{ }\mu\text{s} \qquad\qquad \text{(Ex. 16.5b)}$$

Combining Eqs. (Ex. 16.5a) and (Ex. 16.5b),

$$0.833\tau_2 + \tau_2 = 25.17$$

$$\tau_2 = \frac{25.17}{1.833} = 13.73 \text{ }\mu\text{s}$$

$$\tau_1 = 0.833\tau_2 = 0.833 \times 13.73 = 11.44 \text{ }\mu\text{s}$$

The calculated values of τ_1, τ_2, τ_3 can be checked by the use of Eq. (16.43)

$$\tau_1 + \tau_2 + \tau_3 = 11.44 + 13.73 + 3.43 = 28.6 \ \mu s = T_p$$

Example 16.6 For the flyback converter of Example 16.5 the output voltage is adjusted to be 30 V. What is the required value of duty cycle and what are the three time periods τ_1, τ_2, and τ_3?

$$V_s = 300 \ V \qquad V_L = 30 \ V \qquad n = 10$$

Now $nV_L = V_s$. From Eq. (16.42) or (16.40) it is seen that $\tau_1 = \tau_2$, and when $V_s = nV_Lk = 0.5$. From Eq. (16.41) ,

$$\frac{nV_L}{V_s} = \frac{k}{1-k-\tau_3/T_p} = 1$$

Therefore,

$$k = 1 - k - \frac{\tau_3}{T_p}$$

$$\frac{\tau_3}{T_p} = 1 - 2k = 1 - 1 = 0$$

Therefore,

$$\tau_3 = 0$$

From Eq. (16.43) ,

$$T_p = \tau_1 + \tau_2 = 2\tau_1$$

Therefore,

$$\tau_1 = \tau_2 = \frac{T_p}{2} = \frac{28.6}{2} = 14.3 \ \mu s$$

Example 16.7 In the flyback converter of Example 16.4 the load current falls by 50% while the frequency and output voltage remain constant. Calculate the time duration τ_3 of the zero flux period in the transformer

When the load current I_L falls by 50%, the load power P_L also falls by 50%, and so does the load energy W_{out}:

$$W_{out} = 0.5 \times \frac{0.6}{10^3} = 0.3 \ mJ$$

If the input and the load powers are equal, then

$$W_{in} = W_{out} = \frac{1}{2}LI_1^2$$

$$I_1^2 = \frac{2W_{out}}{L} = 2 \times \frac{0.3}{10^3} \times \frac{10^3}{7.5} = \frac{0.6}{7.5}$$

Therefore,

$$I_1 = 0.283 \text{ A}$$

In the Fig. 16.12, with $\tau_1 = \tau_2$, the time periods τ_1 and τ_2 that had a value 0.01 ms in Example 16.4 reduce by a factor $1/\sqrt{2}$. Therefore,

$$\tau_1 = \tau_2 = 0.01 \times 0.707 = 0.00707 \text{ ms}$$

But the overall period T_p remains at 0.02 ms because the frequency is constant. From Fig. 16.12,

$$\tau_3 = T_p - 2\tau_1 = 0.02 - 0.014 = 0.006 \text{ ms}$$

PROBLEMS

Push–Pull Converter

16.1   A push–pull dc-to-dc converter has a transformer step-down ratio 1:3.5 and is supplied by a 60 V battery. The industrial load has a dc resistance $R_L = 23.4 \ \Omega$. Calculate the average load current with the duty ratio values (a) 0.05, (b) 0.25, and (c) 0.45.

16.2   For the push–pull converter of Problem 16.1, what is the maximum inductor voltage when $k = 0.45$?

16.3   For the push–pull converter of Problem 16.1, what is the average value of the diode currents when both controlled switches are off, if $k = 0.45$?

Full-Bridge Converter

16.4   A full-bridge converter with rectified load operates from a dc power supply of 120 V. The transformer has a step-down ratio of 8.33:1 and a leakage reactance of 20.4 μH. At a switching frequency of 20 kHz the converter delivers a load current of average value 40 A. What is the average output voltage?

16.5   A full-wave bridge converter with a rectified load operates from a dc power supply $V_s = 120$ V. The transformer has a step-down turns ratio of 8.33:1. At a switching frequency of 20 kHz the average load current is required to be 50 A. What value of transformer primary inductance L_1 will limit the output voltage regulation to 5%?

Single-Switch Forward Converter

16.6   A forward converter incorporating a demagnetizing winding operates from a 300-V dc supply. If the primary and secondary numbers of turns, what is the maximum permitted duty cycle?

16.7   A forward converter operates at a frequency 40 kHz from a 300-V dc supply. The controlled switch is set to provide an on time of 12 μs. If the transformer turns ratio N_1/N_2 is 3:1, calculate the average load voltage.

16.8   For the forward converter of Problem 16.7 what is the dwell period τ_2 (in Fig. 16.9b) during which $i_1 = i_2 = 0$?

16.9   A forward converter incorporating a demagnetizing winding operates from a dc supply, nominally of $V_s = 300$ V but subject to 7% regulation. The output voltage must remain constant at its nominal value. The converter operates at a frequency 35 kHz with a switch on time of 13.2 μs. If the magnetizing inductance of the transformer is 37.5 mH, calculate the peak magnetizing component of the primary current.

Single-Switch Flyback Converter

16.10   An ideal flyback converter has a transformer turns ratio of 8:1 and operates from a 600-V dc supply. Calculate the output voltage for values of duty cycle k equal to (a) 0.25, (b) 0.5, and (c) 0.75.

16.11   For an ideal flyback converter, with a unity turns ratio, what value of duty cycle k will cause the output voltage to be equal to the supply voltage?

16.12   A flyback converter with a transformer turns ratio of 10:1 operates in the continuous current mode from a 600-V dc supply. If the operating frequency is 35 kHz and the transistor switch on time is 19.5 μs, calculate the necessary duty cycle and the output voltage.

16.13   A flyback converter with a transformer turns ratio of 6.5:1 operates in the continuous current mode at a frequency of 35 kHz from a 400-V dc supply. What value of transistor switch on time is required to cause an output voltage of 75 V.

16.14   A flyback converter with a transformer turns ratio of 8:1 and inductance 7.5 mH operates in the continuous current mode at a frequency of 50 kHz from a 300-V dc supply. The output voltage is 30 V and the maximum output current is 1 A. Calculate (a) the switch on time, (b) the peak primary current, and (c) the maximum stored energy.

16.15 A flyback converter operates from a 330-V dc supply in discontinuous mode at a frequency of 42 kHz. The transformer turns ratio is 10:1. When the duty cycle is 0.60, the output voltage is 52.3 V. Calculate the three time periods in the switching cycle.

16.16 At a certain value of duty cycle the output voltage from the flyback converter of Problem 16.15 is 33 V. Calculate the duty cycle and the time periods of the switching cycle.

Appendix: General Expressions for Fourier Series

If a periodic function $e(\omega t)$, of any wave shape, is repetitive every 2π radians, it may be expressed as a summation of harmonic terms

$$e(\omega t) = \frac{a_o}{2} + \sum_{n=1}^{\infty} (a_n \cos n\omega t + b_n \sin n\omega t)$$

$$= \frac{a_o}{2} + \sum_{n=1}^{\infty} c_n \sin(n\omega t + \psi_n) \tag{A.1}$$

It follows from Eq. (A.1) that

$$c_n = \sqrt{a_n^2 + b_n^2} = \text{peak value of } n\text{th harmonic} \tag{A.2}$$

$$\psi_n = \tan^{-1} \frac{a_n}{b_n} = \text{phase displacement of the } n\text{th harmonic} \tag{A.3}$$

Also,

$$a_n = n\sin\psi_n \tag{A.4}$$

$$b_n = n\cos\psi_n \tag{A.5}$$

508

The various coefficients in Eq. (A.l) are defined by the expressions

$$\frac{a_0}{2} = \frac{1}{2\pi} \int_0^{2\pi} e(\omega t)\, d\omega t$$
$$= \text{time value average} = \text{dc term value} \tag{A.6}$$

$$a_n = \frac{1}{\pi} \int_0^{2\pi} e(\omega t) \cos n\omega t\, d\omega t \tag{A.7}$$

$$b_n = \frac{1}{\pi} \int_0^{2\pi} e(\omega t) \sin n\omega t\, d\omega t \tag{A.8}$$

For the fundamental component, $n = 1$, so that

$$a_1 = \frac{1}{\pi} \int_0^{2\pi} e(\omega t) \cos \omega t\, d\omega t \tag{A.9}$$

$$b_1 = \frac{1}{\pi} \int_0^{2\pi} e(\omega t) \sin \omega t\, d\omega t \tag{A.10}$$

$$c_1 = \sqrt{a_1^2 + b_1^2} \tag{A.11}$$

$$\psi_1 = \tan^{-1} \frac{a_1}{b_1} \tag{A.12}$$

Note that the definition Eqs. (A.7) and (A.8) represent a sign convention. Some authors use a reverse definition, whereby the right-hand side of Eq. (A.7) is defined as b_n. The values of c_n and ψ_n are not affected by the sign convention used.

Answers to Problems

CHAPTER 2

2.1
  (g) Asymmetrical line current $\dfrac{V}{R_1}\sin\omega t\Big|_{0}^{90°}$ and $\dfrac{V}{R_1}\sin\omega t\Big|_{90°}^{180°}$

  (h) Half-wave rectified $i_L = \dfrac{E_m}{2R}\Big|_{0}^{90°} + 0\Big|_{90°}^{180°}$

2.2 Equations (2.6) and (2.8)

2.3 Equations (2.54) and (2.55)

2.4 $1/\sqrt{2} = 0.707$

2.5 Textbook proof

2.6 $a_1 = 0, b_1 = c_1 = \dfrac{E_m}{2R}$

  $\psi_1 = 0, \cos\psi_1 = 1.0$

2.7 $E_{av} = \dfrac{2E_m}{\pi} = 210.1 \text{ V}$

$E_L = \dfrac{E_m}{\sqrt{2}} = 233.3 \text{ V}$

$RF = 0.48$

2.8 $i_L(\omega t) = \dfrac{E_m}{R}\left(\dfrac{2}{\pi} - \dfrac{4}{3\pi}\cos 2\omega t - \dfrac{4}{15\pi}\cos 4\omega t\right)$

2.9 $\dfrac{I_{av_{V=E_m}}}{I_{av_{V=0}}} = \dfrac{I_m}{I_m/\pi} = \pi$

2.10 $I_{av} = \dfrac{E_m}{2\pi R}(0.685) = 0.11 \text{ A}$

2.11 $I_L = 0.208 \text{ A}$
$P = 9.83 \text{ W}$
$RF = 0.668$

2.12 I_{av} (supply) $= 0$
I_{av} (load) $= 0.872 \text{ A}$
(a), (b) Half-wave operation, $I_{av} = 0.436 \text{ A}$

2.13 See Fig. 2.5.

2.14 v_L is same as i_R in Fig. 2.5.

$v_D = 0 \left.\begin{vmatrix} x \\ \\ x-\theta_c \end{vmatrix}\right. + E_m \sin \omega t \left.\begin{vmatrix} 2\pi \\ \\ x \end{vmatrix}\right.$

2.15 2.15 See Sec. 2.1.2.

2.16 Solution not given here—refer to library texts on applied electronics

2.17 $i_L(\omega t) = \dfrac{E_m}{R}\sin \omega t \left.\begin{vmatrix} \pi \\ \\ 0 \end{vmatrix}\right. + \dfrac{E_m}{X_c}\sin(\omega t + 90°) \left.\begin{vmatrix} 2\pi \\ \\ 0 \end{vmatrix}\right.$

$|I_L| > |I_R|$ for some P; therefore, PF reduces.

2.18 See Sec. 2.3.1.

2.19 $\theta_c = 225.8°$

2.20   (a) Equation (2.72)

       (b) Figure 2.11

       (c) Differentiate Eq. (2.72) wrt time and equate to zero

2.21   Section 2.3.1 and Fig. 2.11

2.22 $\theta_c \cong 212°$

2.23 $\theta_c = 210.05°$ (by iteration)

       $I_{av} = 4.83$ A

2.24   Section 2.31

       $\theta_c = 180° + 45° + 4° = 229°$

2.25   9.06 A

2.26   When $de_L/dt = 0$, $e_L(\omega t) \cong 0.5E_m$.

2.27 $X = 210.05°$

       $\theta_c = 225.8°$

       $I_{av} = 0.764$ A

       Taking the dc, fundamental, plus second harmonic components of the voltage

$$I_L = \sqrt{(0.9)^2 + \frac{1}{2}(2+0.36)} = 1.41 \text{ A}$$

       Distortion factor $= \dfrac{1}{1.41} = 0.71$

2.29   With positive supply voltage there are positive load and supply currents. With negative supply voltage the secondary current is zero and the primary current is the magnetizing current.

2.30   Section 2.3.1 and Fig. 2.10

2.31   QED question

2.32   Section 2.4 and Fig. 2.13

2.33   See Sec. 2.4. Current waveform can be deducted from Fig. 2.13.

CHAPTER 3

3.1   See Sec. 3.1.1.

       $I_s(av) = 2.387$ A

3.3 At $\alpha = 0$, $P = 50$ W.

At $\alpha = 180°$, $P = 25$ W.

$P_{est} = 37.5$ W.

DC, even and odd order harmonics.

3.4 $I_{av} = -0.08$ A

3.5 See Sec. 3.1.1.

3.6 $$a_n = \frac{E_m}{2\pi R} \left(\frac{\cos(n+1)\alpha - (-1)^{n+1}}{n+1} + \frac{\cos(1-n)\alpha - (-1)^{n-1}}{1-n} \right)$$

$$b_n = \frac{E_m}{2\pi R} \left(\frac{\sin(n+1)\alpha}{n+1} - \frac{\sin(1-n)\alpha}{1-n} \right)$$

At $\alpha = \pi/2$,

$$c_1 = 0.295 \frac{E_m}{R} \qquad c_2 = 0.237 \frac{E_m}{R} \qquad c_3 = \frac{E_m}{2\pi R} = 0.16 \frac{E_m}{R}$$

3.7 $I_1 = 0.296 \dfrac{E_m}{R}$ peak value

$I = \dfrac{E_m}{2\sqrt{2}R} = 0.354 \dfrac{E_m}{R}$ rms value

$$\left(\frac{I_1/\sqrt{2}}{I} \right)^2 = \left(\frac{0.29}{0.5} \right)^2 = 0.35$$

3.8 Equation (3.7)

3.9 No. The supply current waveform will remain nonsinusoidal.

3.10 105.2°

3.11 Section 3.1.1

3.12 Section 3.1.1

3.13 Judging *PF* from the waveforms could be misleading. Better to undertake the necessary calculations.

3.14 Use Eqs. (3.4) and (3.10) for the uncompensated distortion factor. Displacement factor $\cos\psi_1 = b_1/c_1$; so use Eqs. (3.9) and (3.10).

3.15 The diode has no effect at all.

3.16 During thyristor conduction the transformer has little effect on the load and supply current waveforms. During thyristor extinction the secondary current is zero and the primary winding draws its magnetizing current (i.e., 2–5% of full-load current at 90° lagging).

3.17   0.48, 0.613, 0.874, 1.21, 1.69, indeterminate at 180°

3.18   Use Eqs. (3.4) and (3.6).

3.19   Distortion factor = 0.707. Displacement factor = 1.0, PF = 0.707. Since ψ_1 = 0°, no power factor improvement is possible, at α = 90°, by the use of capacitance.

3.20   Equations (3.32)–(3.34)

3.21   (b) 1.592 A

3.22   Connection (b) contains only one controlled switch and is likely to be the cheapest. Connection (d) contains four controlled switches and is expensive both for the switches and the firing circuits.

3.23   See Figs. 3.5 and 3.6.

3.24 I_{av} = 2.173 A I_L = 3.643 A RF = 1.34 P = 1327 W

3.25 $a_1 = -\dfrac{E_m}{2\pi R}$ $b_1 = 5.712\dfrac{E_m}{2\pi R}$

3.26   Equation (3.38)

3.27   Figure 3.13

3.28   Figure 3.16

3.29   (a) 209.85°
       (b) 209.8°
       (c) 210°

3.30   239.5°

3.31 Φ = 57.5° X = 237.2° θ_c = 177.2° Z = 18.62 Ω
       I_{av} = 5.626 A I_L = 6.18 A RF = 0.455 P = 381.9 W
       PF = 0.26

3.32 $i_L(\alpha)$ = 0.83 A $i_L(\pi)$ = 11.65 A I_{av} = 7.16 A

3.33   The diode cannot conduct while the controlled switch is on—it acts as a freewheel diode to provide continuity of the load current when the controlled switch is off.

3.34   The comments for Problem 3.33 still apply.

3.35   Yes. The supply current waveform is that part of the load current wave-
       form (Fig. 3.20) in the intervals $\alpha < \omega t < \pi$. The fundamental component
       of this lags the supply voltage. Hence there is a lagging displacement
       angle that can be compensated by a capacitor of appropriate value.

3.36   (a) Figure 3.18, Eq. (3.60)

       (b) Figure 3.19

3.37   Equation (3.64)

3.38 $a_1 = -\dfrac{2}{\pi} I_d \sin\alpha$ $b_1 = -\dfrac{2}{\pi} I_d (1+\cos\alpha)$ $\psi_1 = \dfrac{\pi}{2}$

3.39   The waveforms are not a reliable indicator of power factor correction. It
       is always better to undertake a calculation.

3.40   Equation (3.65), $\psi_1 = \alpha/2$.

3.41 $Q = \dfrac{E_m I_d}{\pi} \sin\alpha$

3.42   This is a QED question.

3.43 $a_1 = -\dfrac{4}{\pi} I_d \sin\alpha$ $b_1 = \dfrac{4}{\pi} I_d \cos\alpha$ $\psi_1 = \alpha$

3.44   No solution available.

3.45 $Q = \dfrac{2}{\pi} E_m I_d \sin\alpha$

3.46   QED question

3.47   QED question

3.48 $c_n = \dfrac{\sqrt{2} E_m}{\pi}\left(\dfrac{1+\cos(1+n)\alpha}{(1+n)^2} + \dfrac{1+\cos(1-n)\alpha}{(1-n)^2} + \dfrac{1+\cos 2\alpha + 2\cos\alpha\cos n\alpha}{(1-n)^2} \right)^{1/2}$

       $c = 166.6$ V $c_4 = 41.62$ V $c_6 = 18.54$ V

       $I_{av} = 10.08$ A $I_L = 10.24$ A $P = 2097$ W $RF = 0.334$

3.49   graphical solution

3.50   graphical solution

CHAPTER 4

4.1 See Eq. (4.5).

4.2 See Eqs. (4.7) and (4.9).

4.3 See Eq. (4.11).

4.4 $\quad a_1 = 0 \qquad b_1 = c_1 = \dfrac{0.471\, E_m}{2R} \qquad \psi_1 = 0$

Displacement factor $\quad = \cos \psi_1 = 1.0$

4.5 $\quad I = 0.485\, \dfrac{E_m}{R} \qquad I_1 = 0.471\, \dfrac{E_m}{\sqrt{2}R}$

Distortion factor $\quad = \dfrac{I_1}{I} = 0.687 = PF$

4.6 $\quad a_n = \dfrac{E_m}{\pi R}\left(\dfrac{\left[\sin(1+n)\pi/2\right]\times\sin(1+n)\pi/3}{1+n} - \dfrac{\left[\sin(1-n)\pi/2\right]\times\sin(1-n)\pi/3}{1-n}\right)$

$\qquad = 0 \qquad \text{(for } n \text{ odd)}$

$\quad b_n = \dfrac{E_m}{\pi R}\left(\dfrac{\left[\sin(1-n)\pi/3\right]\times\cos(1-n)\pi/2}{1-n} - \dfrac{\left[\sin(1+n)\pi/3\right]\times\cos(1+n)\pi/2}{1+n}\right)$

$\qquad = 0 \qquad \text{(for } n \text{ even)}$

4.7 1.886 kW

4.8 $i_a = 0$ The load current has the waveshape of wave e_{D_a} in Fig. 4.6, but is positive.

4.9 $\quad I_s(av) = 0.276\dfrac{E_m}{R} \qquad RF = 1.446$

4.10 Figure 4.5b to e

4.11 See Eq. (4.22).

4.12 See Eqs. (4.25) and (4.11).

4.13 $\quad I_{av}(\text{supply}) = \dfrac{I_{av}(\text{load})}{3} = \dfrac{\sqrt{3}}{2\pi}\dfrac{E_m}{R}$

$\quad I_a = 0.477\dfrac{E_m}{R} \qquad RF = 1.41$ compared with 1.446 for R load (Problem 4.9)

4.14 2628 W compared with 2716 W.

4.15 $a_1 = 0$ $b_1 = c_1 = \dfrac{\sqrt{3}I_{av}}{\pi} = 0.456\dfrac{E_m}{R}$

Displacement factor = 1.0

4.16 $i_a(\omega t) = 0.456 \dfrac{E_m}{R} \sin\omega t$

[$i_a(\omega t)$ is in time phase with $e_a(\omega t)$.]

4.17 $I_a = 0.477\dfrac{E_m}{R}$ $I_{a_1} = 0.322\dfrac{E_m}{R}$

Distortion factor = 0.676, $PF = 0.676 \times 1.0 = 0.676$

4.18 $a_n = \dfrac{2I_{av}}{\pi} \times \sin\dfrac{n\pi}{3} \times \cos\dfrac{n\pi}{2} = 0$ (for n odd)

$b_n = \dfrac{2I_{av}}{\pi} \times \sin\dfrac{n\pi}{3} \times \sin\dfrac{n\pi}{2} = 0$ (for n even)

4.19 Substitute $e_L(\omega t)$ into the defining equations in the Appendix. Note that it is a very long and tedious calculation.

4.20 The load current consists of rectangular pulses with conduction period 120°.

4.21 9.35 A, 339.5 V

4.22 Figure 4.8, Eq. (4.32)

4.23 Equations (4.26) and (4.29) or (4.48)
E_{aN} (rms) varies from $E_m/\sqrt{2}$ at $\mu = 0$ to $0.599E_m$ at $\mu = 90°$.

4.24 $\psi(\mu)$ varies almost linearly for $0° < \mu < 60°$.

4.25 graphical solution

4.26 133.74 V (cf. 196 V), 8.88 A (cf. 9.36 A)

4.27 2627.7 W (unchanged), $PF = 0.74$ (cf 0.676)

4.28 Note in Eq. (4.33) that cos μ is negative for $\mu > 90°$.

4.29 Waveforms similar to Fig. 4.10. Waveform of $i_a(\omega t)$ during overlap is given in Example 4.8.

CHAPTER 5

5.1 $E_{av} = 203.85$ V   E = 238.9 V   RF = 0.611

5.2 (a) 3.24 A
(b) 2.16 A

(c) 1.08 A

(d) 0.289 A

5.3   574 W

5.4   1.838 A at 45°

5.5

α	0	30	60	90	120	150
I_{av}	3.74	3.24	2.16	1.08	0.29	0
I_{rms}	3.8	3.72	2.765	1.73	0.662	0
RF	0.186	0.564	0.8	1.25	2.061	No meaning

5.6   QED question

5.7 $P = VI_1 \cos\psi_1$ W/phase $\psi_1 = -32.5°$ 574 W

5.8

α	0	30	60	90	120	150
PF	0.686	0.634	0.5	0.443	0.121	0

The fundamental (supply frequency) component of the supply current lags the applied phase voltage in time-phase. But the lagging current is not associated, in any way, with energy storage in a magnetic field.

5.9   The appearance of a current waveform can be deceptive. If the waveform contains severe discontinuities, it may have a low distortion factor and therefore a low power factor.

5.10   QED question

5.11 $c < \dfrac{1-\cos(2\alpha + 3\pi)}{4\pi^2 fR}$

5.12

α	60	90	120
I_{min}/I	5.2%	7.2%	5.66%

5.13   No energy storage occurs in the thyristors or in the load resistor. Nevertheless, if $a > 0$, there is a component of reactive voltamperes $Q = 3VI_1 \sin \psi_1$ entering the load. All of the voltamperes entering the capacitors, $Q_c = 3VI_c \sin 90°$, are reactive voltamperes.

5.14 The capacitors have no effect on the average value of the supply current.

5.15

α	0	30	60	90
I_{av}	3.736	3.235	1.868	0

5.16

α	0	30	60	90
I_s	2.157	1.868	1.078	0
P	1047	785	262	0
PF	0.675	0.585	0.338	0

5.17 Expressions for a_1 and b_1 are given in Table 5.2

α	0	30°	60°	90°
ψ_1	0	$-30°$	60°	$-90°$
$\cos \psi_1$	1	-0.866	.0.5	0

5.18 QED question

5.19 QED question

5.20

α	0	30	60	90
PF_L	0.675	0.585	0.338	0
PF_R	0.686	0.634	0.5	0.443

5.21 QED question

5.22 QED question

5.23 The degree of power factor improvement is the ratio of PF_c [Eq. (5.52)] to PF [Eq. (5.44)].

α	0	30	60	90
PF_c/PF	1.355	1.063	1.063	0.83

Note: For $\alpha \geq 90°$ the ratio $PF_c/PF < 1.0$, implying a reduction of power factor due to the presence of capacitance.

5.24 $Q = \dfrac{3}{2} E_m^2 \left(\dfrac{1}{X_c} - \dfrac{9}{2\pi^2} \dfrac{1}{R} \sin^2 \alpha \right)$

For uncompensated operation, $X_c = \infty$.

Q is zero if $C = \pi R/9 f \sin^2 \alpha$ where $f =$ frequency.

5.25   The presence of compensating capacitors makes no difference to the average value of the supply current

5.26 $I_1 = \dfrac{9}{2\sqrt{2}\pi^2} \dfrac{E_m}{R} \cos \alpha$ distortion factor $\dfrac{3}{\sqrt{2}} \pi$

5.27   QED question

5.28   QED question

5.29

α	30°	45°	60°	75°
$\dfrac{E_{av_R}}{E_{av_L}}$	1.00	1.028	1.155	1.654

5.30   The reduction of area due to μ in Fig. 5.13 is given by

$$\text{Area} = \int_{\alpha+30°}^{\alpha+\mu+30°} \left[e_{aN} - \frac{(e_{AN} + e_{CN})}{2} \right]$$

5.31   Evaluation of the integral gives

$$\text{Area} = \frac{\sqrt{3}E_m}{2} \left[\cos\alpha - \cos(\alpha + \mu) \right]$$

5.32

α	0°	30°	60°	90°
E_{av}	274.23	241.7	134.11	−6
R	6.856	6.043	3.353	meaningless

5.33   Similar in style to Fig. 5.13

5.34   QED question

CHAPTER 6

6.1   The load current waveform (Fig. 6.2 e contains six pulsations in each supply voltage period. The lowest order alternating current term in the Fourier series is therefore of sixth harmonic frequency.

6.2   See Figs. 6.1 and 6.2.

$$I_{av} = 1.654 \, \frac{E_m}{R} \text{ (load)} \qquad I_{av} = 0 \text{ (supply)}$$

6.3   Equations (6.8)–(6.11)

6.4 $$I_{av} = 1.827 \, \frac{E_m}{R} \text{ (load)}, \, I_{av} = 0$$

6.5   6292 W, 11.25 A

6.6   9.156 A, 6.47 A.

6.7   QED question

6.8   Section 6.1, Eq. (6.11)

6.9   The fundamental component of the load current is in time phase with its respective phase voltage. Any parallel connected reactor at the supply point would draw a component of supply current in time quadrature with the respective phase voltage, without affecting the load current and power. The net supply current would increase causing reduction of the power factor.

6.10   See Example 6.6.

6.11   The peak amplitude, relative to the maximum dc value, of any voltage harmonic of order n, for firing angle α, in a controlled bridge, is

$$\frac{\sqrt{2}E_{L_n}}{E_{av_o}} = \frac{1}{(n+1)^2} + \frac{1}{(n-1)^2} - \frac{2\cos\alpha}{(n+1)(n-1)}$$

At $\alpha = 0$,

$$\frac{\sqrt{2}E_{L_n}}{E_{av_o}} = \frac{2}{(n^2-1)}$$

6.12   From the relationship of Problem 6.11, with $n = 6$, it is seen that $E_{L_n}/E_{av_o} = 5.71\%$.

6.13   See Fig. 6.14.

6.14   See Eqs. (6.16)–(6.19).

6.15   See Eqs. (6.16)–(6.18).

6.16   See the answer to Problem 6.9. Because $a_1 = 0$, the fundamental compo-
       nent of the load current is in time phase with its respective phase voltage.

6.17 $I_D = 57.74$ A PRV $= 339.4$ V

6.18   Since $a_1 = 0$, the displacement factor $\cos\psi_1 = 1$.

$$PF = \text{distortion factor} = \frac{I_1}{I} = \frac{3}{\pi} \text{ (using the expressions in Table 6.1)}$$

6.19 $$i(\omega t) = \frac{18}{\pi^2}\frac{E_m}{R}\sin\omega t \qquad I_1 = \frac{18}{\pi^2\sqrt{2}}\frac{E_m}{R}(rms) \qquad \psi_1 = 0$$

6.20   See Eqs. (6.19) and (6.22).

6.21 $$i_a(\omega t) = \frac{18}{\pi^2}\frac{E_m}{R}\left(\sin\omega t - \frac{1}{5}\sin 5\omega t - \frac{1}{7}\sin 7\omega t + \frac{1}{11}\sin 11\omega t + \frac{1}{13}\sin 13\omega t + \cdots\right)$$

$$I_{a_1} = \frac{18}{\pi^2\sqrt{2}}\frac{E_m}{R} = 1.29\frac{E_m}{R}$$

$$I_{a_5} = \frac{I_{a_1}}{5} = 0.258\frac{E_m}{R}$$

$$I_{a_7} = \frac{I_{a_1}}{7} = 0.184\frac{E_m}{R}$$

6.22   315.11 V, 19.2°

6.23   (a) $\mu = 19.2°$ $\dfrac{E_{av}}{E_{av_o}} = 0.972$

       (b) $\mu = 52°$ $\dfrac{E_{av}}{E_{av_o}} = 0.81$

6.24 $\dfrac{E_{av}}{E_{av_o}} = 0.933$

6.25   Use Eq. (6.29).

6.26   Section 6.3

CHAPTER 7

7.1   See Fig. 7.8.

7.2   (a) 2.81 A, 814.5 W.

      (b) 1.621 A, 338 W.

      (c) 0.434 A, 49.84 W

7.3 Use Fig 7.3b.

7.4 Section 7.1.2

7.5 Section 7.1.1 and 7.1.2

7.6 7.7

α	a_1	b_1	c_1	I_{a_1}	ψ_1	$\cos\psi_1$	P
30°	−140.4	277.1	310.64	2.197	−26.87	0.892	814.8
60°	−140.4	115	181.5	1.283	−50.7	0.633	337.7
90°	−46.8	16.96	49.78	0.352	−70	0.342	50

The values of P are seen to agree with those of Problem 7.2.

7.8 Use Eq. (7.20) in the defining equations for a_1 and b_1 in Sec. 7.1.3.

7.9 See the relevant expressions in Table 7.2.

7.10 Use Eqs. (7.25) and (7.26).

7.11

α	ψ_1	$\sin\psi_1$	Q	P	$\sqrt{P^2 + Q^2}$	I_a	S
30	−26.9	0.45	411	814.8	912.6	2.33	968.8
60	−50.7	0.774	412.9	337.7	533.4	1.5	623.7
90	−70	0.94	137.6	50	146.4	0.576	239.5

It is seen that $\sqrt{P^2 + Q^2}$ is less than $S\,(= 3EI_{a1})$ for all α. The knowledge of P and Q does not account for all the apparent voltamperes. See Refs. 10 and 20.

7.12 The use of waveforms alone is not reliable indicator of power factor.

7.13 (a) 45.6 μF

(b) 45.6 μF

(c) 15.1 μF

7.14 (a) 22.8 μF

(b) 22.8 μF

(c) 7.5 μF

7.15 (a) $PF = 0.84$, $PF_c = 0.93$

(b) $PF = 0.542$, $PF_c = 0.722$

(c) $PF = 0.21$, $PF_c = 0.254$

7.16 See Fig. 7.8.

7.17 See Eqs. (7.54)–(7.58)

7.18 (a) 235.8 W, 2436.5 W (b) 786 W, 1011 W

7.19 See Eq. (7.59).

7.20 QED question

7.21 See Eqs. (7.60) and (7.61).

7.22 Use Eq. (7.49), which has the value unity for $\alpha = 0$.

7.23 (a) 3.96A at 30°
      (b) 2.29A at 60°

7.24 587 V, 3.24 A

7.25 QED question

7.26 Use Eqs. (7.54) and (7.55).

7.27 Without compensation, $PF_{30} = 0.827$, $PF_{60} = 0.477$.
      When $X_c = R$, $PF_{30} = 0.932$, $PF_{60} = 0.792$, $\alpha = 33.25°$.

7.28 See Eq. (7.68).

7.29 25.11 μF $PF_{30} = 0.941\ (0.827)$ $PF_{60} = 0.848\ (0.477)$

7.30 Waveforms questions-subjective

7.31
$$\text{Current distortion factor} = \frac{I_1}{I} = \frac{\left(R/X_c\right)^2 - \left(18/\pi^2\right)\left(R/X\right)\sin 2\alpha + \left(18/\pi^2\right)\cos^2\alpha}{\left(R/X_c\right)^2 - \left(18/\pi^2\right)\left(R/X\right)\sin 2\alpha + \left(36/\pi^2\right)\cos^2\alpha}$$

When the limiting condition of Eq. (7.66) is satisfied, then $R/X_c = 18 \sin 2\alpha/\pi^2$. This gives $I_1/I = 3/\pi$, which is the condition for resistive load. If $R/X_c > 18\sin 2\alpha/\pi^2$, then $I_1/I \rightarrow 3/\pi$.

7.32 Section 7.3.1

7.33 QED question

7.34 25.14 mH, 59.5°

7.35 Eqs. (7.89)–(7.91)

7.36 3000 W, compared with 7358 W

7.37 $\dfrac{I_a\ (m\ =\ 15^{\circ})}{I_a\ (m\ =\ 0^{\circ})} = 0.975$

7.38 (a) 0.927

   (b) 0.802

   (c) 0.47

7.39 Section 7.3.1, Eq. (7.77)

7.40 Section 7.3.1, part (b), Eq. (7.97)

CHAPTER 8

8.1 Use the basic integral definition Eq. (2.7).

8.2 (a) 1.03 V

   (b) 0.9 V

8.3 $V_n = \dfrac{4V}{n\pi}\left(1 - \cos\dfrac{n\pi}{6} + \cos\dfrac{n\pi}{3}\right)$

   $\delta = 120^{\circ} \qquad V_1 = 0.81\ \text{V(cf. 1.1 V)}$

8.4 $\alpha_1 = 17.8^{\circ} \qquad \alpha_2 = 40^{\circ}$

8.5 QED question

8.6 $\alpha_1 = 23.6^{\circ} \qquad \alpha_2 = 33.3^{\circ}$

8.7 Graphical question

8.8 $b_1 = 0.99$ V, $b_3 = 0.004$ V, $b_5 = -0.001$ V, $b_7 = -0.03$ V, $b_9 = -0.21$ V, $b_{11} = -0.184$ V, $b_{13} = 0.11$ V, $b_{15} = 0.14$ V, $b_{17} = -0.02$ V, $b_{19} = -0.12$ V, $b_{21} = -0.01$ V, $b_{\text{rms}} = 0.743$ V.

CHAPTER 9

9.1 0, 12051 W, 11840 W, 10133 W

9.2 622 V, 37.24 A

9.3 154°, $P_{165} = 0$, 52.6 kVA

9.4 933 V, 32.5 A

9.5 146.4°

9.6   0.675

9.7 $I_{av} = 228\ 5A\ E_{av} = 218.81$ kV

9.8 $\alpha = 155°$

9.9 $\mu = 13.24°$

CHAPTER 10

10.1   See Figs. 10.3 and 10.4.

10.2   See Figs. 10.5 and 10.6.

   V_{ab} leads V_{an} by 30°

10.3   See Figs. 10.5–10.7

10.4 $V_{NO} = \left(V_{dc} - V_{AN}\right)\Big|_0^{\pi} = \left(V_{dc} - V_{BN}\right)\Big|_{\pi}^{2\pi}$

   (a) V_{NO} is square wave $\pm V_{dc}/3$ with three times supply frequency.
   (b) I_{NO} is square wave $\pm V_{dc}/R$ with three times supply frequency.

10.5 $a_1 = 0$ $b_1 = \dfrac{3E_m}{\pi}$ $V_1 = \dfrac{3E_m}{\pi\sqrt{2}}$ $V_{av} = \dfrac{2E_m}{3}$ $V_{av_1} = \dfrac{6E_m}{\pi^2}$

10.6 $V_{AB}(\omega t) = 2.21V \sin(\omega t + 30°) = \dfrac{4\sqrt{3}}{\pi}V \sin(\omega t + 30°)$

   $V_{AB}(\omega t) = \dfrac{1}{2}$ (wave average) $= \dfrac{4V}{3}$

   $V_{AB}(\omega t) = \dfrac{1}{2}$ (wave average) $= \dfrac{2.21V}{\pi} = 0.703$ V

   Current wave form is shown in Fig. 10.7.

10.7   See Example 10.1

10.8 $a_1 = 0$ $b_1 = \dfrac{4E_m}{\pi} = c_1$ $\cos \psi_1 = 1$ $E_{rms} = E$

   Distortion factor $= 0.9$

10.9 $\dfrac{\pi}{3} \dfrac{6V}{\pi} \int_{\pi/6}^{\pi/3} \sin \omega t \, d\omega t = 0.732$

10.10 $V_{rms} = 0.732V$ $b_1 = 1.023$ V

   Distortion factor $= 0.99$

10.11   4.35 A, $I_{dc} = 5.67$ A, 1135 W

10.12   8.165 A, $I_{dc} = 20$ A ,400 W

10.13   8.723 A, 0.38 A

CHAPTER 11

11.1 $b_1 = 0.99$ V, $b_3 = 0.004$ V, $b_5 = -0.001$ V, $b_7 = -0.03$ V, $b_9 = -0.21$ V, $b_{11} = -0.184$ V, $b_{13} = 0.11$ V, $b_{15} = 0.14$ V, $b_{17} = -0.02$V, $b_{19} = -0.12$ V, $b_{21} = -0.01$ V, $b_{rms} = 0.743$ V

11.2 $I_{rms} = 5.013$ A, CDF = 0.999, VDF = 0.942

11.3   628 W, 0.42

11.4   See Example 11.3 and Fig. 11.13.

11.5 $I_{rms} = 32.72$ A, $I_1 = 32.65$ A, $I_{11} = 1.72$A, $I_{13} = 0.87$ A, $I_{19} = 0.65$ A

11.6   There are no triplen harmonic currents.
       $P_{in} = 15421$W, $V_{rms} = 171\ 5$V, $I_{rms} = 32.72$A, $PF = 0.915$ lagging

11.7   (a) 89.6%
       (b) 81.1%

CHAPTER 12

12.1   (a) 114 V
       (b) 103.2 V
       (c) 81.5 V
       (d) 51.02 V
       (e) 19.9 V

12.2   Lowest harmonic is the second-order harmonic.

12.3   (a) 109.8 V
       (b) 95.1 V
       (c) 54.9 V
       (d) 0

12.4 $E_{rms} = 302.3$ V $P_{in} = 5.25$ kW/phase
       $I_{out} = 70$ A rms $PF = 0.43$ lagging

12.5   (a) $\alpha = 36°$

      (b) $I_o = 53.67$ A (rms)

      (c) $P_1 = 2880$ W

      (d) $I_{in} = 31$ A(rms)

      (e) $Irh = 21.92$ A (rms)

      (f) $PF = 0.244$ lagging

12.6   See section 12.3.1

12.7 $3f_{in} \pm 2f_o = 193.33 - 166.67$ Hz

      $3f_{in} \pm 4f_o = 206.66 - 153.34$ Hz

      $3f_{in} \pm 6f_o = 220 - 140$ Hz

      $3f_{in} \pm 8f_o = 233.33 - 126.67$ Hz

12.8 $6f_{in} \pm f_o = 380 - 340$ Hz

      $6f_{in} \pm 3f_o = 420 - 300$ Hz

      $6f_{in} \pm 5f_o = 460 - 260$ Hz

12.9 $P_1 = P_{in} = 151.4$ kW

      $E_o = 420$ V

      $\cos\phi_{in} = 0.77$ lagging

      $I_o = 120$ A

CHAPTER 13

13.1 $M = 1, 4, 7, 10, 13$

13.2   The criteria are satisfied in all cases.

13.3   The combination $M = 11, M_c = 5, M_s = 3$ does not satisfy Eq. (13.17) and is not viable.

13.4   The combination $M = 8, M_c = 2, M_s = 3$ satisfies both Eqs. (13.17) and (13.19) and is an acceptable waveform.

13.5

M	1	2	3	4	5	6	7	8	9	10
f_o (Hz)	50	30	21.4	16.67	13.6	11.54	10	8.82	7.9	7.14

13.6 $M_c = 3, M_s = 2, M = 7$

      $\theta_v = -\theta_w = 120°$

13.7 $M_c = 3, M_s = 4, M = 11$

$\theta_v = -\theta_w = 109.6°$

13.8 (a) 12 Hz

(b) 7.83 Hz

(c) 6.67 Hz

13.9 $M = 6, M_c = 2, M_s = 2$

$\theta_{..} = -\theta_w = 110.8°$

$p = \dfrac{13}{3}$

CHAPTER 14

There are no end-of-chapter problems for this section.

CHAPTER 15

15.1 $V_{av} = kV_s$ $V_{rms} = \sqrt{k}V_s$

15.2 $P = kV_s^2 / R$

(a) $\dfrac{V_s^2}{4R}$

(b) $\dfrac{3V_s^2}{4R}$

15.3 $V_{switch} = (1 - k)V_s$ $V_{diode} = -kV_s$

15.4 $T_p = 500 \; \mu s$ $f = 2000$ Hz

15.5 22.5 V

15.6 $k = 0.893$ $V_o = 1072$ V

15.7 $I = 7$ A

15.8 (a) 4486 Hz

(b) 169.42 μs

(c) 26.18 A

15.9 (a) 135.85 A, 82.35 A

(b) $I_o = 110.2$ A, $I_{sw} = 53.98$ A, $I_d = 96.1$ A

15.10   0.667–0.333

15.11   (a) 185 μs
         (b) 142.15 μs
         (c) 52 A

15.12 $I_{in}(\text{max}) = 77.32$ A, $I_{in}(\text{min}) = 57.92$ A, $I_{in}(av) = 67.62$ A

15.13   171 μs

15.14   (a) 0.337 A, 58.67 V
         (b) 3.034 A, 176 V
         (c) 27.31 A, 528 V

15.15   (a) 16.67 V, 9 A
         (b) 150 V, 81.1 A

15.16   (a) 2.79 A, 0.335 V at $k = 0.25$
         (b) 8.37 A, 9.05 V at $k = 0.75$

CHAPTER 16

16.1   (a) 0.9 A
         (b) 4.49 A
         (c) 8.08 A

16.2   − 18 V

16.3   4.04 A

16.4 $I_2 = 40$ A, $I_1 = 4.8$ A
         $t = 50$ μs, $V_L = 13.46$ V

16.5   15.6 μH

16.6   1/2

16.7   48 V

16.8 $\tau_1/2 = 1$ μs

16.9 $di_m = 105$ mA

16.10   (a) 25 V
         (b) 75 V
         (c) 225 V

16.11 $k = 1/2$

16.12 $V_L = 128.6$ V, $k = 0.682$

16.13 $k = 0.549$, $\tau_1 = 15.71$ μs

16.14 (a) 8.89 μs
(b) 0.125 A
(c) 3.16 mJ

16.15 14 μs, 9.04 μs, 0.8 μs

16.16 $\tau_1 = 11.905$ μs, 0.5 μs, 11.9 μs, 0 μs

References

1. Rissik H. Mercury Arc Current Converters. London: Sir Isaac Pitman & Sons, 1963.
2. Davis RM. Power Diode and Thyristor Circuits. Cambridge, England: Cambridge University Press, 1971.
3. Pelly BR. Thyristor Phase Controlled Converters and Cycloconverters. New York: Wiley-Interscience, 1971.
4. McMurray W. The Theory and Design of Cycloconverters. Cambridge, MA: MIT Press, 1972.
5. Mazda FF. Thyristor Control. London, England: Newnes-Butterworth, 1973.
6. Csaki F, Gansky K, Ipsits I, Marti S. Power Electronics. Budapest, Hungary: Academic Press, 1975.
7. Dewan SB, Straughen A. Power Semiconductor Circuits. New York: Wiley-Interscience, 1975.
8. Ramamoorty M. Introduction to Thyristors and their Applications. London, England: The MacMillan Press, 1978.
9. Hirane Y, Shepherd W. Theoretical Assessment of a Variable Frequency Envelope Cycloconverter, Trans. IEEE. 1978; IECI–25 (No. 3):238–246.
10. Shepherd W, Zand P. Energy Flow and Power Factor in Nonsinusoidal Circuits. Cambridge, England: Cambridge University Press, 1979.
11. Lander CW. Power Electronics. 1st ed. 3rd ed. London, England: McGraw-Hill, 1993.
12. Sugandhi RK, Sugandhi KK. Thyristors—Theory and Applications. New York: J. Wiley & Sons, 1981.

13. De G. Principles of Thyristorised Converters. India: Oxford and JBH Publishing Co, 1982.

14. Dewan SB, Straughen AR, Siemon GR. Power Semiconductor Drives. New York: Wiley-Interscience, 1984.

15. Bird BM, King KG. An Introduction to Power Electronics. 1st ed. 2nd ed. London, England: Wiley Interscience, 1984 and 1993 (with D G Pedder).

16. Dubey GK, Doradla SR, Joslu A, Sinha RMK. Thyristorised Power Controllers. New Dehli. India: J Wiley and Sons, 1986.

17. Seguier G. Power Electronic Converters, Vol. 1-AC/DC Converters. England: North Oxford Academic Press, 1986.

18. Hoft RG. Semiconductor Power Electronics. New York: Van Nostrand Reinhold, 1986.

19. Bose BK. Power Electronics and AC Drives. NJ, Englewood Cliffs: Prentice-Hall, 1986.

20. Shepherd W, Hulley LN. Power Electronics and Motor Control. 1st ed. 2nd ed. (with D. T. W. Liang). Cambridge. England: Cambridge University Press, 1995.

21. Murphy JMD, Tumbull FG. Power Electronic Control of A. C. Motors. Oxford, England: Pergamon Press, 1988.

22. Thorborg K. Power Electronics. London, England: Prentice-Hall (UK) Ltd, 1988.

23. Mohan N, Undeland TM, Robbins WP. Power Electronics: Converters, Applications and Design. New York: J. Wiley and Sons, 1989.

24. Kassakian JG, Schlect MF, Verghese GC. Principles of Power Electronics. Reading, MA: Addison-Wesley, 1989.

25. Dubey GK. Power Semiconductor Controlled Drives Prentice-Hall. Englewood Cliffs, NJ, 1989.

26. Gray CB. Electrical Machines and Drive Systems. Longman, England, 1989.

27. Williams BW. Power Electronics. 2nd ed. London, England: The Macmillan Press, 1993.

28. Rashid MH. Power Electronics: Circuits, Devices and Applications. 2nd ed. Englewood Cliffs, NJ: Prentice-Hall, 1993.

29. Ramshaw RS. Power Electronic Semiconductor Switches. 2nd ed. London, UK: Chapman and Hall, 1993.

30. Barton TH. Rectifiers, Cycloconverters and AC Controllers. Oxford, England: Oxford Science Publications, 1994.

31. Vithayathil J. Power Electronics: Principles and Applications. New York: McGraw-Hill, 1995.

32. Novotny DW, Lipo TA. Vector Control and Dynamics of AC Drives. New York: Oxford Science Publications, 1995.

33. Krein PT. Elements of Power Electronics. New York: Oxford University Press, 1998.

34. Watthanasarn C. Optimal Control and Application of AC–AC Matrix Converters, Ph.D. Thesis, University of Bradford, UK. 1997.

35. Zhang L, Watthanasarn C, Shepherd W. Analysis and Comparison of Control Techniques for AC–AC Matrix Converters. Proc. IEE 1998; 145(No. 4):284–294.

36. Venturini MGB, Alesina A. A New Sine Wave In–Sine Wave Out Conversion Technique Eliminates Reactive Elements. Proc. Powercon 7 1980:E.3.1–3.15.

37. Alesina A, Venturini MGB. Solid State Power Conversion: A Fourier Analysis Approach to Generalized Transformer Synthesis. IEEE Trans 1981; CAS-28(No 4): 319–330.
38. Maytum MJ, Colman D. The Implementation and Future Potential of the Venturini Converter, Conference Proceedings on Drives/Motors/Controls 83. York, England, Oct. 1983:108–117.
39. Venturini MGB, Alesina A. Analysis and Design of Optimum-Amplitude, Nine-Switch, Direct AC–AC Converters. IEEE Trans Jan. 1989; PE-4(No. 1):101–112.
40. Huber L, Borojevic D. Space Vector Modulator for Forced Commutated Cycloconverters IEEE 1AS Conf Record, Part I, 1989:871–876.
41. Huber L, Borojevic D. Space Vector Modulation With Unity Input Power Factor for Forced Commutated Cycloconverter, IEEE IAS Conf. Record, 1991:1032–1041.

Index